AF581786

Microalgae Cultures: Environmental Tool and Bioenergy Source

Microalgae Cultures: Environmental Tool and Bioenergy Source

Editors

José Carlos Magalhães Pires
Ana Luísa Gonçalves

MDPI • Basel • Beijing • Wuhan • Barcelona • Belgrade • Manchester • Tokyo • Cluj • Tianjin

Editors
José Carlos Magalhães Pires
University of Porto
Portugal

Ana Luísa Gonçalves
University of Porto
Portugal

Editorial Office
MDPI
St. Alban-Anlage 66
4052 Basel, Switzerland

This is a reprint of articles from the Special Issue published online in the open access journal *Energies* (ISSN 1996-1073) (available at: https://www.mdpi.com/journal/energies/special_issues/microalgae_cultures).

For citation purposes, cite each article independently as indicated on the article page online and as indicated below:

LastName, A.A.; LastName, B.B.; LastName, C.C. Article Title. *Journal Name* **Year**, *Volume Number*, Page Range.

ISBN 978-3-0365-2908-0 (Hbk)
ISBN 978-3-0365-2909-7 (PDF)

Cover image courtesy of José Carlos Magalhães Pires

Contents

About the Editors

José Carlos Magalhães Pires (Assistant Researcher) graduated in Chemical Engineering from the Faculty of Engineering of the University of Porto (FEUP) in 2004. He worked in two chemical companies and then started his Ph.D. in Environmental Engineering in 2006 at FEUP. In 2010, he began studying the environmental applications of microalgal cultures, including CO_2 capture and bioenergy production. Performed studies showed that microalgal cultures are a promising solution to mitigate the atmospheric CO_2 concentration. JCM Pires is currently Assistant Researcher at LEPABE-FEUP. Since 2017, JCM Pires has also tested innovative PBR designs to maximize microalgal biomass productivities. JCM Pires published 64 papers in international peer-reviewed journals, 5 books, and 25 book chapters, and his research work was discussed in 45 international meetings and 3 national meetings.

Ana Luísa Gonçalves (Junior Researcher) graduated in Bioengineering (Biological Engineering Branch) from the Faculty of Engineering of the University of Porto in 2012. In 2013, she began her Ph.D. in Chemical and Biological Engineering at FEUP. The Ph.D., entitled "Microalgal Cultivation for Biomass Production, CO_2 Capture and Nutrients Uptake", aimed to optimize the culturing conditions of microalgae and cyanobacteria on a laboratory and pilot-scale to improve both CO_2 capture and nutrients uptake from the culture medium. Before finishing their Ph.D., she worked for 9 months as the Wet Biomass Production Manager in a microalgal production company. From 2017 to 2019, AL Gonçalves worked as a postdoctoral researcher in the TexBoost project at CITEVE, and currently, AL Gonçalves is working as a Junior Researcher in the PIV4Algae project at LEPABE-FEUP. AL Gonçalves published 28 papers in international peer-reviewed journals, 9 book chapters, and her research work was discussed in 15 international meetings.

Preface to "Microalgae Cultures: Environmental Tool and Bioenergy Source"

Microalgae have been intensively studied for CO_2 capture, nutrient removal from wastewater, and biofuel production. Microalgae are prokaryotic or eukaryotic photosynthetic microorganisms presenting a unicellular or simple multicellular structure, which allows them to grow in a wide variety of environmental conditions, even harsh ones. Photosynthesis is a key process for microalgal growth, whereby they convert solar radiation and CO_2 (absorbed from the atmosphere or large emission sources) into organic matter and energy. Therefore, microalgal cultivation can play an important role in climate change mitigation, reducing the major greenhouse gas concentration in the atmosphere. Besides CO_2 sequestration, microalgae can also be used in wastewaters bioremediation. They can grow in these low-quality waters and use their contaminants (mainly nitrogen and phosphorus species, which are macronutrients for microalgal growth) as nutrients' sources. Moreover, the produced biomass, which is very rich in a wide variety of interest compounds, can be used as raw material for diverse applications, such as the production of nutraceuticals, human food and animal feed, pharmaceuticals, cosmetics, fine chemicals, bioenergy, and biofertilizers. Regarding bioenergy production, the fatty acids produced by microalgae can be extracted and used for biodiesel production, and products, such as proteins and residual biomass, can be fermented to produce ethanol or methane. The wide range of applications described for microalgae, together with the high growth rates described for these microorganisms, have contributed to the growing interest in microalgal biomass production in the last decades.

This book reflects some research studies that have been dedicated to the environmental applications and bioenergy production potential of microalgae.

Regarding the environmental applications of microalgae, the first study evaluates the nutrient removal ability of the green alga *Coelastrum morus* under different nitrogen and phosphorus concentrations and different N:P ratios. In the second study, waste glycerol from biodiesel production was used as an organic carbon source to optimize biomass and docosahexaenoic acid production by *Schizochytrium* sp.

In terms of bioenergy production, this book project covers several process aspects, from the optimization of culturing conditions to biomass processing. The third to sixth studies focus on the optimization of microalgal culturing conditions to improve lipid accumulation and, hence, biofuels production. In the third study, an accessible microfluidic platform was developed to facilitate the screening process and optimize the biofuel production of microalgae. In the fourth study, different operation regimes (batch, continuous, and semi-continuous) were compared during the Spring/Summer seasons (in 2.6 m^3 tubular photobioreactors) to select the most suitable one for the production of the oleaginous microalga *Nannochloropsis oceanica*. In the fifth study, the authors investigated *Dunaliella tertiolecta* growth, pigment production, and bioenergetic parameters in different culture media in order to select the best cultivation conditions to overproduce such added-value compounds. In the sixth study, the mutualistic interactions between the oleaginous yeast *Lipomyces starkeyi* and the green microalga *Chloroidium saccharophilum* in mixed cultures were investigated to exploit possible synergistic effects in terms of lipid content and productivity. The seventh, eighth, and ninth studies focus on the development of effective harvesting, cell disruption and lipid recovery procedures to improve biofuels production from microalgae. In the seventh study, the potential of consortium biomass formation between *Mucor circinelloides*, an oleaginous filamentous fungal species, and *Chlorella vulgaris* was evaluated to improve the harvest efficiency of the microalga and study the lipid productivity of the consortium system. In the eighth study, the harvesting and hydrocarbon recovery from the microalga Botryococcus braunii were evaluated: harvesting was evaluated based on filtrate experiments using an inclined solid–liquid separator and hydrocarbon recovery was evaluated based on solvent extraction experiments on a concentrated algal slurry. In the ninth study, an effective and easily controlled cell wall disruption method based on electro-Fenton reactions was used to enhance lipid extraction from the wet biomass of *Nannochloropsis oceanica* IMET1.

In the tenth study, the energetic efficiency of biomass and lipids production by *Chlorella pyrenoidosa* was evaluated in multi-tubular, helical-tubular, and flat-panel airlift pilot-scale photobioreactors to ascertain the sustainability of microalgae-based biofuels.

Finally, a review study (the eleventh study) is presented on the opportunities of microalgae and cyanobacteria immobilization to intensify microalgal biotechnology and bioprocessing.

We hope that the content of this book project presents valuable information regarding the environmental and bioenergy production application of microalgal cultures.

José Carlos Magalhães Pires, Ana Luísa Gonçalves
Editors

 *energies*

Article

Effects of Nutrient Content and Nitrogen to Phosphorous Ratio on the Growth, Nutrient Removal and Desalination Properties of the Green Alga *Coelastrum morus* on a Laboratory Scale

Aida Figler [1,2], Kamilla Márton [1], Viktória B-Béres [3] and István Bácsi [1,*]

[1] Department of Hydrobiology, University of Debrecen, Egyetem Sqr. 1, H-4032 Debrecen, Hungary; figler.aida@science.unideb.hu (A.F.); kamillamarton@yahoo.com (K.M.)
[2] Pál Juhász-Nagy Doctoral School of Biology and Environmental Sciences, University of Debrecen, Egyetem Sqr. 1, H-4032 Debrecen, Hungary
[3] Department of Tisza Research, Institute of Aquatic Ecology, Centre for Ecological Research, Bem Sqr. 1, H-4026 Debrecen, Hungary; beres.viktoria@gmail.com
* Correspondence: istvan.bacsi@gmail.com; Tel.: +36-52512-900 (ext. 22634)

Abstract: In wastewater, nutrient concentrations and salinity vary substantially, however, the optimal N:P ratio for the treatment using microalgae is not well described. In this study, the effects of higher and lower nitrate and phosphate contents and N:P ratios on growth, nutrient removal ability and halotolerance of the common green alga *Coelastrum morus* were investigated in model solutions. The results suggest that high nitrate content (above 100 mg L^{-1}) with a similarly high phosphate concentration (resulting low N:P ratio) is not favorable for growth. The studied isolate can be considered as a halotolerant species, showing remarkable growth up to 1000 mg L^{-1} NaCl and it seems that despite the negative effects on growth, higher nutrient content contributes to higher halotolerance. A significant amount of nitrate removal was observed in media with different nutrient contents and N:P ratios with different salt concentrations. High N:P ratios favor phosphate removal, which is more inhibited by increasing NaCl concentration than nitrate uptake. Overall, with a relatively higher nutrient content and a favorable (5 or higher) N:P ratio, a common green algal species such as *C. morus* could be a promising candidate next to species from the Chlorellaceae and Scenedesmaceae families.

Keywords: green alga; nutrient content; N:P ratio; salt tolerance; nutrient removal; salt content reduction

Citation: Figler, A.; Márton, K.; B-Béres, V.; Bácsi, I. Effects of Nutrient Content and Nitrogen to Phosphorous Ratio on the Growth, Nutrient Removal and Desalination Properties of the Green Alga *Coelastrum morus* on a Laboratory Scale. *Energies* **2021**, *14*, 2112. https://doi.org/10.3390/en14082112

Academic Editor: José Carlos Magalhães Pires

Received: 26 February 2021
Accepted: 6 April 2021
Published: 9 April 2021

1. Introduction

Untreated wastewater released to water bodies can result in changes in chemical and physical parameters which finally can affect the vital functions of living organisms living there. Large amounts of nutrients, especially nitrogen and phosphorous are responsible for eutrophication can enter to water resources, leading to changes in trophic status and nutrient supply. These shifts are followed by changes in dominance relationships and succession processes, which ultimately result in changes in the aquatic community [1].

The most common and most frequently used wastewater treatment system is the traditional activated sludge treatment system. Mechanical treatment of inflow wastewater is followed by a biological one, during which organic and inorganic contaminants are removed from the wastewater with the help of microorganisms [2]. However, due to the variety of contaminants, it is not possible to remove all of them with the traditional method, so systems are constantly being developed to effectively remove contaminants other than nutrients. Examples include algae-based purification systems, which have been under development for decades [3,4].

The autotrophic, heterotrophic, and mixotrophic metabolic properties of several algal species can be used in the wastewater biological treatment process [5,6]. Due to their photoautotrophic metabolism, algae are able to produce O_2 in the presence of light using CO_2, thus providing the amount of oxygen (aerobic conditions) required for the functioning of microorganisms in activated sludge and reducing the amount of CO_2 in the atmosphere [7,8]. At the same time, algae have a pivotal role in reducing the amount of inorganic substances in wastewater that serve as nutrients for them (nitrogen, phosphorus), as well as removing other contaminants [3,4,9,10].

Algae-based wastewater treatment systems primarily use eukaryotic green algae and prokaryotic cyanobacteria. The efficiency of *Chlorella* species is reported in several studies, as nitrogen is removed by 45–97%, phosphorus by 28–96%, chemical oxygen demand (COD) is decreased by 61–86% from different types of wastewater (agricultural, communal, industrial—e.g., textile industry), and they are able to accumulate large amounts of lipids, especially under mixotrophic conditions [11]. Among cyanobacteria, filamentous *Arthrospira* species are the most studied, removing large amounts of nitrogen and phosphorus under autotrophic and mixotrophic conditions, while producing large amounts of biomass [12].

Due to the diversity of wastewater parameters, it is important to select appropriate microalgae species for algal-based wastewater treatment, as the efficiency of traditionally used algal species may decrease under changing conditions. In this case, the use of extremophilic, i.e., specialist microalgae species, which are viable even under changing or extreme conditions, should be preferred. Metabolic processes in some specialist microalgae species have been shown to be more effective at high temperatures (thermophilic) or lower temperatures (psychophilic) and under strongly acidic (acidophilic) or alkaline conditions (alkalophilic), other microalgae species adapted well to high salt (halophilic) or sugar (osmophilic) concentrations [13]. Among algae, halotolerant species are those ones, which are able to ploriferate in the presence or in the absence of a certain amount of salt, while halophilic algae require the presence of a certain, usually extreme amount of salt for optimal growth. Depending on the required salt content, halophilic species can be divided into three groups: weak halophiles: optimal growth is achieved with 1–3%, i.e., 10,000–30,000 mg L^{-1}—NaCl; moderate halophiles: optimal growth is achieved with 3–15%, i.e., 30,000–150,000 mg L^{-1}—NaCl and extreme halophiles (optimal growth is achieved above 15%, i.e., 150,000 mg L^{-1}—NaCl [13,14].

Several studies report the importance of the N:P ratio from both ecological and biotechnological perspectives. The well-known Redfield's Ratio (atomic or molar ratios of carbon, nitrogen, and phosphorus in phytoplankton [15]) can be considered a global average with significant variance for different phytoplankton species [16]. It is globally accepted that this ratio could be the optimal nutrient ratio required for growth, obviously with taxon specific differences: the average optimum N:P ratios for eukaryotic algae range from 16 to 23 N to 1 P, while the average optimum ratio for cyanobacteria found to be 10-16N:1P [17]. Some experimental studies proved that the optimal N:P ratio for certain algal species is indeed around the ratio widely accepted in the literature [18,19].

In wastewater, nutrient concentrations vary substantially. However, the optimal N:P ratio for the treatment of municipal wastewater using microalgae is not well described. High nutrient concentrations with unfavorable N:P ratios could lead inadequate growth or nutrient removal: e.g., in our recent study all the tested nine green algae isolates showed poor phosphate removal from Bold's Basal Medium [20].

Coelastrum species are increasingly studied recently, thanks to their lipid and pigment accumulation ability and applicable capability in wastewater treatment. It was found that N-limitation promoted higher lipid contents in a *Coelastrum* strain (CORE-1), resulting in the highest lipid content (48% *w*/*w*) among five studied green algal strains [21]. Similarly, increasing lipid content occurred as a function of decreasing nitrogen concentration also in the case of another *Coelastrum* strain (HA-1) [22]. Úbeda et al. [23] reported remarkable growth of *Coelastrum* cf. *pseudomicroporum* in wastewater, furthermore they observed carotenoid accumulation in the used isolate under salt stress. A *Coelastrum* sp. isolated from cattle manure leachate showed high carbon fixation and nutrient removal abilities cultured in wastewater [24]. *Coelastrum* sp. TISTR 9501RE (closely related to *Coelastrum morus* strain SAG217-5) seemed to be a potent accumulator of astaxanthin, canthaxanthin, and lutein as major carotenoids under nutrient limited conditions; in addition, the strain was able to tolerate a wide range of salinity (up to 500 mM–29.25 g L^{-1}) [25]. Tharek et al. [26] reported induction of astaxanthin accumulation by 3 g L^{-1} salinity (set with NaCl) and moderate nitrogen supply (nitrate) in the case of a Malaysian *Coelastrum* sp. isolate. Certain members of the genera are able to accumulate lipids, i.e., serve as a suitable feedstock for biodiesel production in heterotrophic cultivation using synthetic wastewater with molasses as carbon source [27].

Because of the incontestable importance of the topic and the above mentioned uncertainties about the relations of nutrient removal abilities, salinity tolerance and N:P ratios, we aimed to study the effects of higher and lower nitrate and phosphate content and N:P ratios (ranging within the "nutrient rich" category) on growth, nutrient removal ability and salt tolerance of the common green alga *Coelastrum morus*. The experiments were carried out in model solutions with slightly different nitrate and phosphate concentrations and N:P ratios to represent how usual, not harsh fluctuations in wastewater composition could effect on biomass production and removal characteristics. On the basis of literature data and our previous findings [20] we hypothesized that:

- Higher nutrient content with higher N:P ratio is better for growth and salt tolerance than higher nutrient content with lower ratio, or lower nutrient contents.
- Higher N:P ratios favor more efficient nutrient removal, independently from the initial nutrient contents.
- More favorable growth conditions favor conductivity reduction (salt removal ability).

2. Materials and Methods

The coenobial green alga *Coelastrum morus* was isolated from a small aquatic habitat in northeastern Hungary, maintained as an axenic isolate in the Algal Culture Collection of the Department of Hydrobiology, University of Debrecen (ACCDH-UD) as standing culture under 14 h light (40 µmol photons m^{-2} s^{-1})—10 h dark photoperiod at 24 °C.

The experiments were carried out within the same circumstances in shaken cultures (SOH-D2 circular shaker, 90 rpm), in 100 mL Erlenmeyer flasks with 50 mL final volume. Duration of the experiments was 14 days. As model solutions, two generally used culturing media were chosen: Bold's Basal Medium (BBM; Table S1; CCAP media recipes a [28]) and Jaworski's Medium (JM; Table S1; CCAP media recipes b [29]). Both chosen media can be considered as media with low N:P ratio in relation to Redfield's Ratio, and both of them can be considered as nutrient rich media, as culturing media are in general. But relative to each other, they represent different types from the point of view of nutrient content and N:P ratio. BBM can be considered as a medium with high nutrient content (182.4 mg L^{-1} nitrate and 163.2 mg L^{-1} phosphate) and low N:P ratio (2.93 mM N:1.7 mM P; i.e., N:P = 1.7). It was designated as higher content—lower ratio medium (HC-LR medium; Table 1). The modified version of this medium (with decreased phosphate content; Table S1) was used as a medium with high nutrient content (184.4 mg L^{-1} nitrate and 47.9 mg L^{-1} phosphate) and high N:P ratio (2.93 mM N:0.505 mM P; i.e., N:P = 5.8). It was designated as higher content—higher ratio medium (HC-HR medium; Table 1). JM was considered as a medium with lower nutrient content (68.86 mg L^{-1} nitrate and 18.2 mg L^{-1} phosphate) and higher N:P ratio (1.1 mM N:0.19 mM P; i.e., N:P = 5.8). It was designated as lower content—higher ratio medium (LC-HR medium; Table 1). The modified version of this medium (with increased phosphate content; Table S1) was used as a medium with lower nutrient content (68.86 mg L^{-1} nitrate and 60.35 mg L^{-1} phosphate) and lower N:P ratio (1.1 mM N:0.655 mM P; i.e., N:P = 1.7). It was designated as lower content—lower ratio medium (LC-LR medium; Table 1).

Table 1. Nitrogen and phosphorous contents and N:P ratios of the applied culturing media.

Medium	Nitrogen		Phosphorous		N:P Ratio (Molar)
	mg L^{-1}	mmol	mg L^{-1}	mmol	
HC-LR (BBM)	41.06	2.93	53	1.7	1.7
HC-HR (modified BBM)	41.06	2.93	15.5	0.5	5.8
LC-HR (JM)	15.64	1.1	5.94	0.19	5.8
LC-LR (modified JM)	15.64	1.1	19.7	0.635	1.7

HC: higher content; LC: lower content; HR: higher ratio; LR: lower ratio; BBM: Bold's basal medium (CCAP Media Recipes a); JM: Jaworski's Medium (CCAP Media Recipes b).

To study the salinity tolerance and desalination abilities of the *C. morus* strain, cultures supplemented with different amounts of NaCl were assembled for all four media. Control cultures were prepared without additional NaCl. To reach 500, 1000, 5000 and 10,000 mg L^{-1} (0.05–1%) salt concentrations, NaCl stock solution of 300 g L^{-1} was used. NaCl supplemented cultures were called as treated cultures. The composition of the different cultures is shown in Table 2. Accurate conductivity values and chloride contents at the beginning of the experiments were measured from so called "negative control" compositions (culturing media + salt, without algae). The results obtained from the algae cultures were corrected with changes in these compositions without algae.

Table 2. Salt treatment of *Coelastrum morus* cultured in media with different nutrient content and N:P ratios.

NaCl Treatment	Medium (mL)	Alga Inoculum (mL)	dH_2O (μL)	300 g L^{-1} NaCl Stock Solution (μL)
Control	41.7	5	3300	0
500 mg L^{-1}	41.7	5	3215	85
1000 mg L^{-1}	41.7	5	3135	165
5000 mg L^{-1}	41.7	5	2475	825
10,000 mg L^{-1}	41.7	5	1650	1650

2.1. Measurement of the Growth of the Cultures

For growth measurements, individual numbers (i.e., coenobia numbers in this case) were counted (according to the European Standard EN 15204 [30]. Samples of 1 mL were taken on every 2nd days of the experiments, 143 μL conc. formaldehyde was added to each sample for preservation (5% final concentration). Individual numbers were counted from 10 μL of the preserved samples in Bürker chamber at 400× magnification (BX50F-3 microscope, Olympus Optical Co. Ltd., Tokyo, Japan). To give To present the NaCl concentrations causing 50% growth inhibition (EC_{50} values), firstly the extent of growth inhibitions were calculated in percentage compared to control. Than the extent of growth inhibition was plotted as functions of NaCl concentrations obtaining growth inhibition—salt concentration curves. Trend lines were fitted showing a second order relationship between growth inhibition and salt concentration. The concentrations causing 50% inhibition were calculated from the quadratic equations of the trend lines.

The cultures were filtered (vacuum filter, Pall Corporation, New York, NY, USA GF/C™ 693 filter paper) at the end of the experiments (on the 14th day), the cell free media were used for the measurements of nitrate and phosphate concentrations, conductivity, and chloride contents.

2.2. Measurement of Nitrate, Phosphate, Conductivity and Chloride

Nutrient (nitrate and phosphate) concentrations, conductivity and chloride content of the cultures were measured at the beginning and at the end of the experiments. Data from negative control compositions on day 0 were used as baseline values, data measured at day 14 were used for the corrections of values measured in the cultures. Dataset of negative compositions proved that no precipitation occurred among the applied circumstances.

Nutrient concentrations were measured from 1 mL of cell-free samples collected at the beginning and at the 14th day of the experiments. Both nitrate and phosphate were measured by spectrophotometric methods: the salicylic acid colorimetric method [31] was applied for nitrate, and the acidic phosphorous molybdate method [32] was applied for phosphate content measurements. Both methods were reduced in volume to be possible to carry out in Eppendorf tubes in order to minimize sample requirements.

Conductivity was measured from the cell free media by a HQ30d portable multimeter (Hach Lange GmbH, Düsseldorf, Germany) equipped with an Intellical^{TM} CDC401 conductivity measuring electrode.

Chloride concentrations of the cell free media were measured by precipitation titration [33]. Aliquots of 12.5 mL cell free culturing media were used in the case of control, 500 and 1000 mg L^{-1} NaCl treated cultures and 1 mL was used for 5000 and 10,000 mg L^{-1} NaCl treated cultures. The titrations were done with silver nitrate measuring solution, the end point of the titration was indicated with potassium chromate indicator. After precipitation of the silver chloride precipitate equivalent to the amount of chloride ion, the excess of the measuring solution gives a reddish-brown silver chromate precipitate with the chromate ion, the yellow to reddish-brown color change indicate the end point of the titration. The amount of chloride was calculated with the formula given in the method and was expressed as mg L^{-1} chloride.

2.3. Statistical Analysis

Average values and standard deviations of three independent replicates of all experiments are presented. One-way analysis of covariance (ANCOVA [34,35]) was used to analyze differences among growth curve tendencies based on individual (coenobia) numbers in control and in treated cultures. One-way analysis of variance (ANOVA) was used for the comparison of conductivity reduction values (% and μS cm^{-1}) and extents of chloride and nutrient removals (% and mg L^{-1}). Paired T-tests were applied to analyze that nutrient concentrations, conductivity and chloride contents changed significantly from day 0 to day 14, or not. The Past software was used for all the statistical analyses [35].

3. Results

3.1. Growth and Salt Tolerance of the Cultures

In HC-LR medium, only cultures containing 10,000 mg L^{-1} NaCl showed significantly lower growth compared to control cultures ($p < 0.05$; Figure 1a). The amount of NaCl required for 50% growth inhibition in the treated cultures decreased from day 4 to day 14 (Table 3), so over time, the cultures showed an increasing sensitivity to the presence of salt.

In HC-HR medium, *C. morus* showed significantly lower growth compared to control in cultures treated with 1000, 5000 and 10,000 mg L^{-1} NaCl. Treatments with higher NaCl concentrations resulted in significantly lower individual numbers also compared to each other ($p < 0.05$; Figure 1b). The amount of NaCl required for 50% growth inhibition decreased from day 4 to day 7 and then increased again to day 14 (Table 3), suggesting growth regeneration after day 7.

In LC-HR medium, the growth of *C. morus* cultures treated with NaCl was significantly lower compared to control cultures in the cases of all treatments ($p < 0.01$; Figure 1c). Treatments with higher NaCl concentrations resulted in significantly lower individual numbers also compared to each other, except the 5000 and 10,000 mg L^{-1} treatment (Figure 1c). During the experiments, the amount of salt concentration required to achieve 50% growth inhibition decreased, so it can be concluded that the salt tolerance of *C. morus* cultures decreased in time (Table 3).

In LC-LR medium all treated cultures showed a significantly lower growth compared to the control, and in parallel with the increase of NaCl content, the growth of the cultures decreased significantly ($p < 0.01$; Figure 1d). The amount of NaCl required to achieve 50% growth inhibition decreased from day 4 to day 7 and then increased to day 14 (Table 3), so a slight regeneration of growth was observed also in this case.

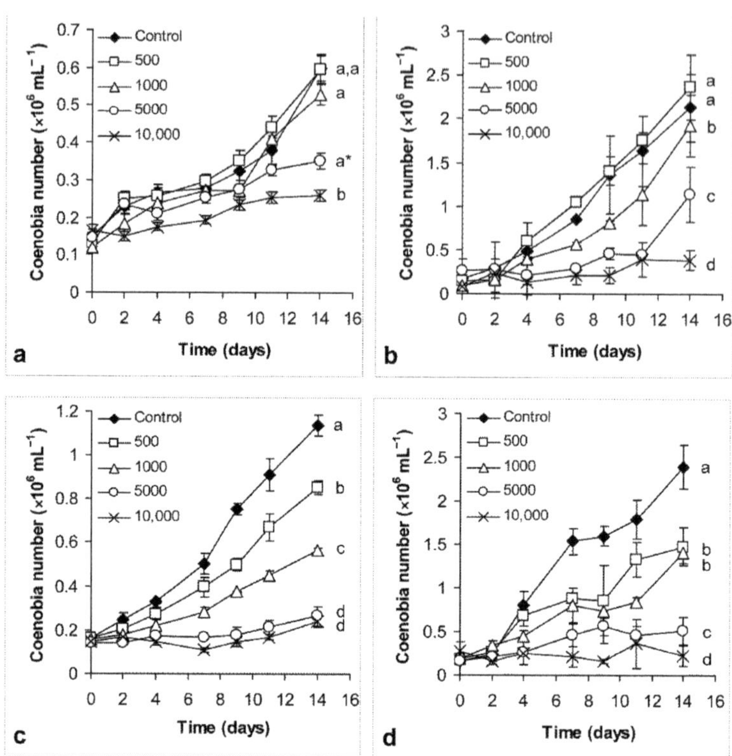

Figure 1. Growth of control and NaCl-treated *Coelastrum morus* cultures in: (**a**) HC-LR; (**b**) HC-HR; (**c**) LC-HR and (**d**) LC-LR media based on coenobia numbers. HC: higher content; LC: lower content; HR: higher ratio; LR: lower ratio. Mean values (n = 3) and standard deviations are plotted. 500; 1000; 5000 and 10,000: NaCl concentrations in mg L^{-1}. Significant differences ($p < 0.05$) among growth curve tendencies are indicated by different lowercase letters. Lowercase letters with asterisks show partial significant difference when it cannot be marked by a different letter.

Table 3. NaCl concentrations causing 50% growth inhibition in media with different nutrient concentrations and N:P ratios.

Medium	**EC_{50} (mg L^{-1} NaCl)**		
	Day 4	**Day 7**	**Day 14**
HC-LR	n.c.	n.c.	6720
HC-HR	3870	2790	5430
LC-HR	5300	2970	1000
LC-LR	2010	1620	1860

HC: higher content; LC: lower content; HR: higher ratio; LR: lower ratio; n.c.: not calculable.

Comparing the growth in different media under different treatments, control cultures showed the following order: LC-LR > HC-HR > LC-HR > HC-LR. There were significant differences among all media ($p < 0.05$). Cultures in 500 and 1000 mg L^{-1} treatments could be ranked as follows: HC-HR > LC-LR > LC-HR > HC-LR. Growth in all media differed significantly from each other in the presence of 500 mg L^{-1} NaCl ($p < 0.01$). It was significantly higher in HC-HR and LC-LR media, than in LC-HR and HC-LR ($p < 0.05$) in the presence of 1000 mg L^{-1} NaCl. In 5000 and 10,000 mg L^{-1} salt treated cultures the growth order was only slightly modified: HC-HR > LC-LR > HC-LR > LC-HR; only the growth in LC-HR was significantly lower than in other media.

The results suggest, that in general, higher N:P ratio is favorable for growth, while higher nutrient content is favorable for higher salt tolerance.

3.2. Nutrient (Nitrate and Phosphate) Removal

3.2.1. Nitrate Removal

Nitrate content decreased significantly ($p < 0.05$) in all media in control and in NaCl treated cultures (Table S2).

Nitrate content decreases in HC-LR medium ranged from 68.2 (10,000 mg L^{-1} NaCl treatment) to 97.9% (control). There was a significantly lower decrease of nitrate content compared to control in the 10,000 mg L^{-1} NaCl treatment ($p < 0.05$; Figure 2a; Table S3). There were lower decreases of nitrate concentrations with the increasing salt concentration, but the differences were not significant (Figure S1a).

The extent of nitrate content reduction in HC-HR medium ranged from 85.7 (10,000 mg L^{-1} NaCl treatment) to 93.1% (500 mg L^{-1} NaCl treatment). There was no significant difference between control and NaCl treated cultures, but compared to the 500 mg L^{-1} NaCl treated cultures, the 5000 and 10,000 mg L^{-1} treatments showed significantly lower nitrate removal ($p < 0.05$; Figure 2b; Table S3). On nitrate concentration basis, the decrease in 10,000 mg L^{-1} NaCl treated culture was significantly lower than in control and in 500 mg L^{-1} treatment (Figure S1b).

Nitrate removal was complete in control and treated cultures in LC-HR medium, there were no measurable amounts of nitrate in the culturing media on the 14th day of the exposition (Figure 2c; Table S3).

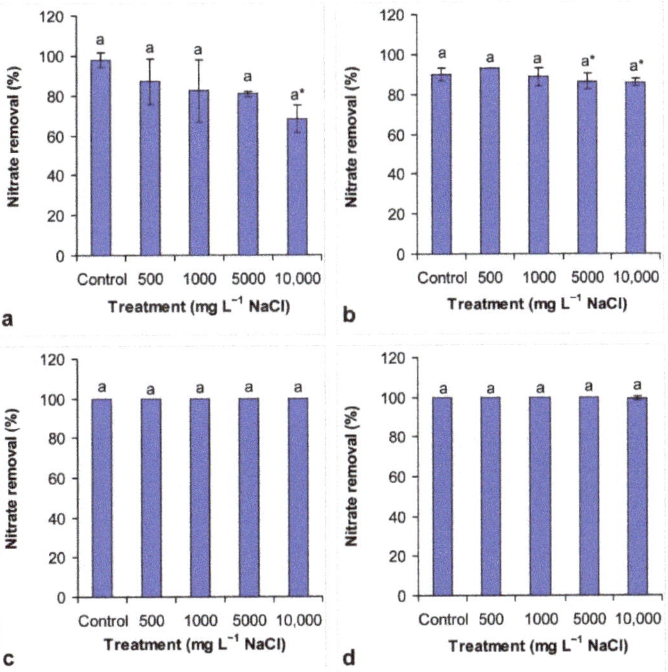

Figure 2. Nitrate removal (%) of control and NaCl-treated *Coelastrum morus* cultures in: (**a**) HC-LR; (**b**) HC-HR; (**c**) LC-HR and (**d**) LC-LR media. HC: higher content; LC: lower content; HR: higher ratio; LR: lower ratio. Mean values ($n = 3$) and standard deviations are plotted. Significant differences ($p < 0.05$) are indicated by different lowercase letters. Lowercase letters with asterisks show partial significant difference when it cannot be marked by a different letter.

Nitrate removal in LC-LR medium ranged from 99.2 (10,000 mg L^{-1} NaCl treatment) to 100% (control, 500 and 1000 mg L^{-1} NaCl treatment; Figure 2d; Table S3).

Comparing the nitrate removal efficiency in different media, the results show that lower nitrate content favored complete nitrate removal regardless of the N:P ratio. However, significantly less nitrate was removed only from 5000 and 10,000 mg L^{-1} NaCl treatments cultured in media with high nitrate content (HC-LR and HC-HR). On nitrate concentration basis, nitrate removal obviously was higher from media with higher initial nitrate concentration, suggesting the high nitrate removal efficiency of the studied green algal strain. Overall, nitrate removal was significant in all media both in control and in NaCl treated cultures (Table S3), showing that N:P ratio or salt concentration has no direct effect on nitrate uptake of the applied *C. morus* strain.

3.2.2. Phosphate Removal

The phosphate content of all media was significantly reduced in all cultures to day 14 compared to the values measured at the beginning of the experiments ($p < 0.05$), except 10,000 mg L^{-1} NaCl treatments in LC-HR and LC-LR media (Table S4).

The extent of phosphate removal in HC-LR medium ranged from 25.8 (10,000 mg L^{-1} NaCl treatment) to 43% (control). As the salt concentration increased, the extent of phosphate removal decreased, but this decrease was not significant (Figure 3a; Table S5). Analysing phosphate removal on phosphate concentration basis, similar results were obtained (Figure S2a).

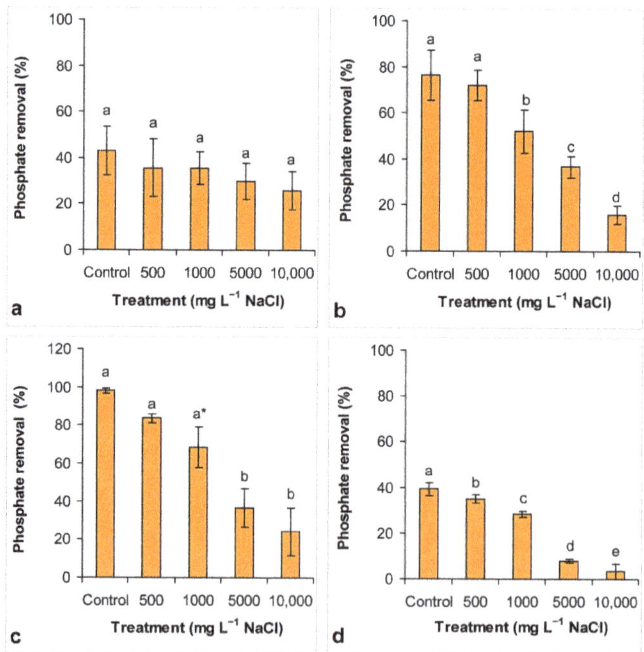

Figure 3. Phosphate removal (%) of control and NaCl-treated *Coelastrum morus* cultures in: (**a**) HC-LR; (**b**) HC-HR; (**c**) LC-HR and (**d**) LC-LR media. HC: higher content; LC: lower content; HR: higher ratio; LR: lower ratio. Mean values ($n = 3$) and standard deviations are plotted. Significant differences ($p < 0.05$) are indicated by different lowercase letters. Lowercase letters with asterisks show partial significant difference when it cannot be marked by a different letter.

The decrease of phosphate content in HC-HR medium ranged from 15.8 (10,000 mg L^{-1} NaCl treatment) to 76.3% (control). In 1000, 5000 and 10,000 mg L^{-1} NaCl treated cultures phosphate removal was significantly lower compared to the control and to each other ($p < 0.01$; Figure 3b; Table S5). Analysing phosphate removal on phosphate concentration basis, similar results were obtained (Figure S2b).

The extent of phosphate removal in LC-HR medium ranged from 24.2 (10,000 mg L^{-1} NaCl treatment) to 98.2% (control). Phosphate removal was significantly lower in the 1000, 5000 and 10,000 mg L^{-1} treatments compared to the control, but in the presence of 1000 mg L^{-1} NaCl it was significantly higher than in 5000 and 10,000 mg L^{-1} treatments (Figure 3c; Table S5). On phosphate concentration basis, phosphate removal was significantly lower only in 5000 and 10,000 mg L^{-1} NaCl treated cultures compared to control and other treatments (Figure S2c).

The decrease of phosphate content in LC-LR medium ranged from 3.4 (10,000 mg L^{-1} NaCl treatment) to 39.8% (control). Significantly lower phosphate removal was observed in treated cultures compared to control and compared to each other along the increasing salt concentration ($p < 0.05$; Figure 3d; Table S5). On phosphate concentration basis, phosphate removal was significantly lower than in control only from 500 mg L^{-1} NaCl treatment (Figure S2d).

Comparing the different media, it can be said that significantly higher ($p < 0.05$) phosphate removal occurred in media with high N:P ratio in control, 500 and 1000 mg L^{-1} NaCl treated cultures, although in parallel with the increase of NaCl concentration, the extent of phosphate removal decreased. Lower phosphate removal was observable in media with low N:P ratio, significantly less ($p < 0.05$) phosphate was removed from LC-LR medium compared to the others in the case of all treatments (Table S5). On phosphate concentration basis, phosphate removal obviously was higher from media with higher initial phosphate concentration. Overall, these results suggest that higher N:P ratio is favorable to achieve more effective phosphate removal, and increasing salinity affects significantly the process.

3.3. Conductivity Reduction

Conductivity was significantly reduced to day 14 in all cultures compared to the initial values ($p < 0.05$; Table S6).

The decreases of conductivity in HC-LR medium ranged from 47.4 (control) to 69.7% (5000 mg L^{-1} treatment). The conductivity reduction increased with increasing salt concentration (Figure 4a; Table S7). Although percentage values did not differed significantly among different treatments, taking into account the initial concentrations, the amount of removed ionic compounds was significantly higher in 5000 and 10,000 mg L^{-1} NaCl treated cultures (Figure S3a).

The decrease of conductivity in HC-HR medium ranged from 10.3 (control) to 29.6% (10,000 mg L^{-1} treatment). Compared to the control, the extent of conductivity decrease was significantly higher in the treated cultures ($p < 0.05$; Figure 4b; Table S7), this was more remarkable on µS cm^{-1} basis (Figure S3b).

The decreases of conductivity in LC-HR medium ranged from 18 (10,000 mg L^{-1} treatment) to 24.7% (control; Figure 4c; Table S7). The decrease in conductivity in the treated cultures decreased with increasing salt concentration, the differences among percentage values were not significant. Taking into account the initial concentrations, the amount of removed ionic compounds was significantly higher in 5000 and 10,000 mg L^{-1} NaCl treated cultures (Figure S3c).

Conductivity reduction in LC-LR medium ranged from 32.7 (10,000 mg L^{-1} treatment) to 35.5% (5000 mg/L treatment). There was no significant difference in the percentage values of the treated cultures compared to the control (Figure 4d; Table S7), but on µS cm^{-1} basis conductivity reduction was significantly higher in 5000 and 10,000 mg L^{-1} NaCl treated cultures (Figure S3d).

Comparing the extent of the conductivity reduction in the different media, it can be concluded that significantly higher decrease occurred in HC-LR medium in all treatments ($p < 0.05$) than in the same treatment in other media. The lowest conductivity reduction occurred in LC-HR medium (except in controls, where lowest reduction was observed in HC-HR medium), with significant differences in the case of 5000 and 10,000 mg L^{-1} treated cultures in other media ($p < 0.05$; Table S7). Overall, conductivity reduction was higher in media with higher ionic contents and lower N:P ratio favored a more effective conductivity reduction.

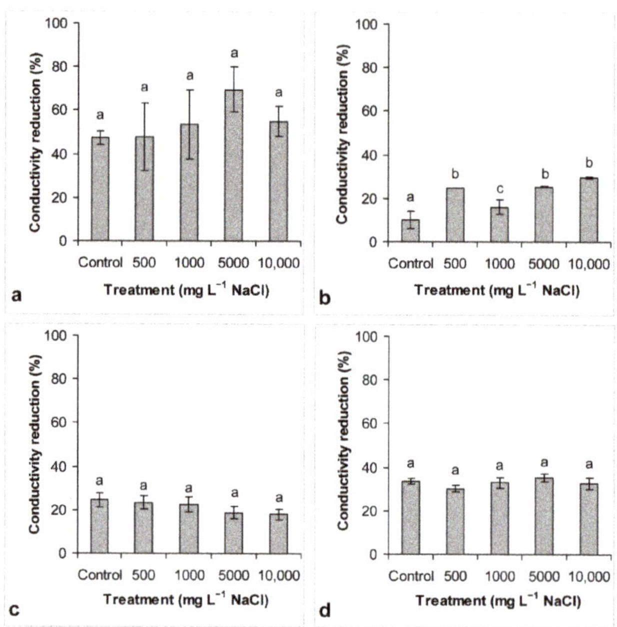

Figure 4. Conductivity reduction (%) of control and NaCl-treated *Coelastrum morus* cultures in: (**a**) HC-LR; (**b**) HC-HR; (**c**) LC-HR and (**d**) LC-LR media. HC: higher content; LC: lower content; HR: higher ratio; LR: lower ratio. Mean values ($n = 3$) and standard deviations are plotted. Significant differences ($p < 0.05$) are indicated by different lowercase letters.

3.4. Chloride Removal

Chloride content decreased significantly ($p < 0.05$) from the beginning of the experiments to day 14 in every cases (except in 500 mg L^{-1} NaCl treatment in LC-HR; Table S8).

In HC-LR medium, chloride removal ranged from 23.5 (control) to 55.7% (5000 mg L^{-1} treatment). Not surprisingly, treated cultures removed significantly higher amount of chloride compared to the control cultures ($p < 0.01$; Figure 5a; Table S9). The proportion of removed chloride increased with increasing salt content to 5000 mg L^{-1} treatment, but there were no significant differences between the individual treatments (Figure 5a; Table S9). On concentration basis, chloride removal was significantly higher in the case of cultures treated with 5000 and 10,000 mg L^{-1} NaCl concentrations (Figure S4a).

Chloride removal ranged in HC-HR medium from 24.9 (control) to 66.3% (10,000 mg L^{-1} treatment). The extent of chloride removal increased with the increasing NaCl concentration; there was significantly higher chloride removal in the 1000, 5000 and 10,000 mg L^{-1} treatments compared to the control and to each other in several cases ($p < 0.001$; Figure 5b; Table S9). Similar results were obtained by analyzing the results on chloride concentration basis (Figure S4b).

Chloride removal in LC-HR medium ranged from 13.5 (1000 mg L^{-1} treatment) to 50% (control). Compared to the control, the extent of chloride removal was significantly lower in the treated cultures ($p < 0.01$; Figure 5c; Table S9). There was no clear trend in chloride removal along NaCl concentrations and there were no significant differences between treatments (Figure 5c; Table S9). Analyzing the results on chloride concentration basis, a different picture can be seen: chloride removal was significantly higher in the case of cultures treated with 5000 and 10,000 mg L^{-1} NaCl concentrations (Figure S4c).

Chloride removal in LC-LR ranged from 17.5 (500 mg L^{-1} treatment) to 72.4% (control); characteristic of chloride content changes were very similar to that of observed in LC-HR medium. Compared to the control, chloride removal was significantly lower in all treated cultures ($p < 0.01$; Figure 5d; Table S9). Similarly to LC-HR medium, analyzing the results on chloride concentration basis, a different picture can be seen: chloride removal was significantly higher in the case of cultures treated with 5000 and 10,000 mg L^{-1} NaCl concentrations (Figure S4d).

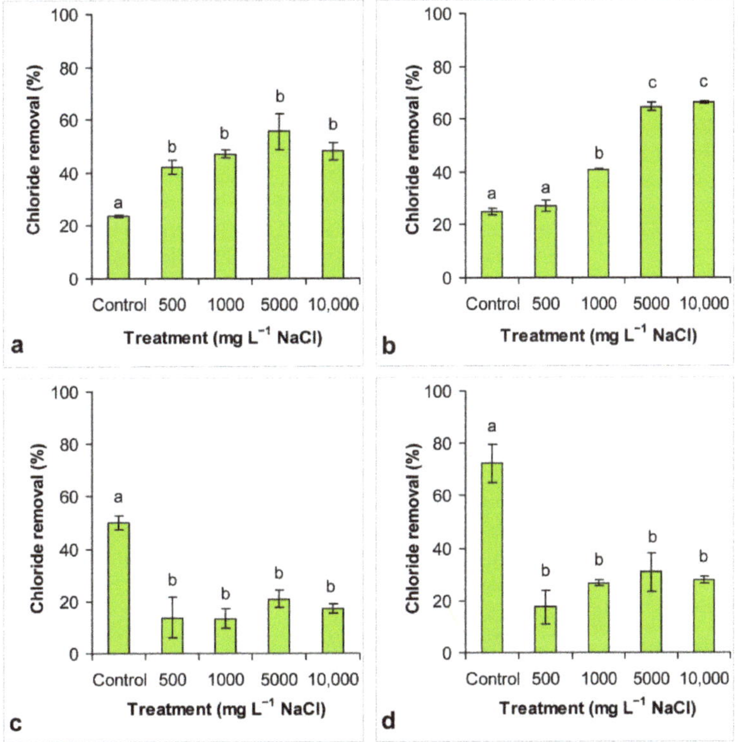

Figure 5. Chloride removal (%) of control and NaCl-treated *Coelastrum morus* cultures in: (**a**) HC-LR; (**b**) HC-HR; (**c**) LC-HR and (**d**) LC-LR media. HC: higher content; LC: lower content; HR: higher ratio; LR: lower ratio. Mean values ($n = 3$) and standard deviations are plotted. Significant differences ($p < 0.05$) are indicated by different lowercase letters.

Comparing the different media, the chloride removal in NaCl treated cultures was significantly higher ($p < 0.05$) in the presence of high nutrient content (HC-LR and HC-HR media). Chloride removal was higher in cultures treated with 500 and 1000 mg L^{-1} NaCl if the N:P ratio was low (HC-LR), but in cultures treated with 5000 and 10,000 mg L^{-1} NaCl, the higher removal was performed, when N:P ratio was higher (HC-HR). In media with lower nutrient content, chloride removal was low in NaCl treated cultures (Table S9). These result suggest that higher nutrient content favors to more efficient chloride removal, and higher N:P ratio has significant role only in the case of higher nutrient content.

4. Discussion

4.1. Growth and Salt Tolerance

Although literature data generally show that higher N:P ratios are better for phytoplankton growth both in the cases of higher [18,19] or lower nutrient contents [36,37], our results did not reveal that higher nutrient content with higher N:P ratio is better for growth and salt tolerance than higher nutrient content with lower ratio, or lower nutrient contents. In control cultures without salt treatment, the highest individual numbers were observed in LC-LR medium, while it was significantly lower in HC-HR medium and it was the worst in HC-LR medium. Despite the intensive research on the field, there are only few studies comparing the effect of different nutrient concentrations with the same N:P ratio. Liu et al. (2011) observed that the cyanobacterium *Microcystis aeruginosa* showed better growth in media with higher nutrient content than in lower ones at similar low (1) N:P ratio [38]. Although the used media were designated as higher and lower nutrient content, neither HC-LR nor LC-LR media used in our study can be considered as limiting from nutrient concentration point of view studying wastewaters. Growth tendencies in our experiments suggest that there could be an upper limit of nitrate and phosphate concentrations, at which growth is inhibited. Terry showed in the case of a marine prymnesiophyte that phosphate interacts with nitrate at certain concentrations [39]. Our results clearly show that only ~40% of phosphate is removed from control cultures at N:P = 1.7, independently of the initial concentrations. According to our knowledge, the underlying mechanism is still not clear. The supposed negative interaction between phosphate and nitrate, when their concentrations are close to each other could explain the observed growth characteristics, but simple physical-chemical features do not support the observations [40]. Exact explanation of the phenomenon requires further, more detailed nutrient uptake studies.

Although there were no clear correlations of salt tolerance, nutrient content and N:P ratio (rejecting our first hypothesis), cultures with the best growth without salt treamtments (controls in LC-LR and HC-HR media) showed the best growth also in all cases of NaCl treatments. However, EC_{50} values showed a different picture: they were not calculable within the first week and it was the highest on day 14 in HC-LR medium, in which the weakest growth was observed. Our results also revealed that higher nutrient content favors to withstand against the negative effects of high chloride concentrations, especially in the case of longer exposure times. These observations are in accordance with the results of Park et al. about another coenobial green alga, *Scenedesmus quadricauda* [41]. It seems that similarly to *S. quadricauda* and other coenobial green algae (e.g., *Scenedesmus bijugatus* [42]), the studied *Coelastrum morus* requires different N:P ratios for its survival under different salinity levels. It was reported in the case of a cyanobacterium that nitrate as nitrogen source has a protecting role against dissolved salts toxicity [43]. This also could be part of the background of higher salt tolerance of cultures in HC-LR medium, although salt tolerance of cultures in HC-HR medium with similar nitrate content was significantly higher only at the end of the exposition.

Comparing simply the salt tolerance of the studied strain with other *Coelastrum* isolates, it can be said, that there are more tolerant ones: Úbeda et al. [23] and Rauytanaparti et al. [25] both reported growth of *Coelastrum* species under higher salinity conditions than the ones presented here.

4.2. Nutrient Removal

4.2.1. Nitrate

Our results clearly proved that independently the NaCl concentration, the initial amount of nitrate in media was rather important in nitrate removal than N:P ratio: lower nitrate content favored complete nitrate removal (partly rejecting our second hypothesis). However, nitrate removal was impressive also in the cases of high initial nitrate content. Shriwastav et al. observed similar phenomena in the case of *Chlorella sorokiniana* cultured in media with different N:P ratio: they reported efficient nitrate removal in all types of media they used [36]. Nitrate removal also did not changed in the case of a *Chlorella vulgaris* strain cultured in media with different N:P ratio [37].

The extent of nitrate removal is important for wastewater treatment, because high nitrogen content can be a major cause of eutrophication in the aquatic environment [44]. Our results clearly highlighted that despite the different amount of nutrients and N:P ratio, a large amount of nitrate was removed by the green alga *C. morus*, regardless of the used salt concentrations (500–10,000 mg L^{-1} NaCl). Similarly large extent of nitrate removal were observed in cultures of freshwater green algae grown in HC-LR medium at different salt concentrations (500–20,000 mg L^{-1} NaCl) in our previous experiments [20]. It can be concluded that nitrate removal is not significantly affected by the salt content in a wide concentration range in the case of common green algae, only indirectly, through growth inhibition.

4.2.2. Phosphate

Our results pointed a strong and clear relationship between N:P ratio and phosphate removal (proving the second hypothesis from the point of view of phosphate): higher N:P ratio indeed favored phosphate removal independently from the initial phosphate concentration. Although there were higher removed amounts of phosphate from HC-LR and LC-LR medium than from HC-HR and LC-HR medium on phosphate concentration basis (Figure S2), the reason is simply because of the higher initial phosphate concentrations of the formers. The remaining amount of phosphate was also higher in media with higher initial phosphate concentration (HC-LR and LC-LR media, Table S4), so taking the phosphate content of the "effluent" into account, lover initial concentrations and higher N:P ratios are the most favorable conditions. Better phosphorous removal from media with higher N:P ratios also were observed in the case of some other isolates (*Pseudanabaena*, *Cladophora* and *Klebsormidium* isolates [45]; *Chlorella sorokiniana* [36]; *C. vulgaris* [37,46]; *Scenedesmus* sp. [47]). Our results also indicated that salt content does not affect directly the phosphate uptake, only indirectly via growth inhibition, similarly to nitrate removal. Media with different initial nutrient contents and N:P ratios can be ranked almost in the same order in the presence of a certain NaCl concentration. Lagus et al. studied phosphate removal in mesocosms with salinity above 6‰ (6 g L^{-1}) along different nutrient concentrations and N:P ratios [48]. They also observed that salt content barely affected phosphate accumulation, which extent was the best in the presence at higher N:P ratio (5.77). Although there are several chemical methods for phosphate removal [49], the more environmentally friendly biological methods seem to be more difficult to perform than nitrate removal, because of the probable sensitivity of the process on N:P ratio. Results of the present study and literature data suggest that satisfactory phosphorous removal could be achieved even in a wide range of salinity setting lower phosphate concentration (below 50 mg L^{-1}) and higher (at least 5) N:P ratio.

4.3. *Conductivity Reduction and Chloride Removal*

The results show that conductivity reduction and salt removal ability were not in connection with favorable growth conditions (rejecting our third hypothesis). Conductivity reduction and chloride removal were significantly higher in HC-LR medium, in which lower growth was observed compared to the other media, especially at lower (500 and 1000 mg/L) NaCl concentrations. It suggests that individuals inhibited in cell division, are able to accumulate high amounts of ions, which is a well-known phenomenon: both freshwater and marine algae are able to accumulate chloride and sodium ions in vacuoles [50]. It is also known that nitrogen source influences the accumulation of chloride [51], although the understanding of the context of initial nitrate concentration, chloride accumulation and the above mentioned possible protecting effect of higher nitrate content against high salt concentrations [47] definitely requires further investigations.

Comparing the conductivity reducing and chloride removal ability of the studied *C. morus* to other green algae, it seems that the isolate can be characterized more remarkable abilities regarding these processes, than other green algae previously studied in our laboratory [20] or by other authors [52].

5. Conclusions

Our results pointed out that slight differences in nitrate and phosphate content of the culturing media (ranging within the "nutrient rich" category) result in differences in the N:P ratio could lead significantly different algal growth characteristics. The results suggest that high nitrate content (above 100 mg L^{-1}) with a similarly high phosphate concentration (resulting low N:P ratio) is not the most favorable situation for the growth of the green alga *C. morus*. The studied isolate can be considered as a halotolerant species, showing remarkable growth up to 1000 mg L^{-1} NaCl. It seems that higher nutrient content contribute to higher halotolerance. In the case of nitrate and phosphate uptake, it can be concluded that these physiological processes are not directly affected by NaCl content in the studied range, as a further approval of the halotolerance of the species. Significant amount of nitrate was removed in media with different nutrient contents and N:P ratios along different salt concentrations. High N:P ratios favor phosphate removal, which is more inhibited by increasing NaCl concentration than nitrate uptake. Conductivity reduction and chloride removal were not in strong connection with favorable growth conditions, similarly to salt tolerance. *C. morus* is able to remove chloride ions regardless of nutrient content and N:P ratio. However, higher nutrient content seemed to be more favorable for the process. Overall, with a relatively higher nutrient content and a favorable (5 or higher) N:P ratio, a common green algae species such as *C. morus* could be able to significantly improve wastewater quality. Examining the composition of the biomass produced in the process can also shed light on the advantages of the species's further uses.

Supplementary Materials: The following are available online at https://www.mdpi.com/article/10.3390/en14082112/s1, Table S1: Composition of the used culturing media. Table S2: Nitrate content values (mg L^{-1}) measured at the start (day 0) and at the end (day 14) of the experiments, Table S3: Extent of nitrate removal (%) from the different media with different NaCl contents, Table S4: Phosphate content values (mg L^{-1}) measured at the start (day 0) and at the end (day 14) of the experiments, Table S5: Extent of phosphate removal (%) from the different media with different NaCl contents, Table S6: Conductivity values (mg L^{-1}) measured at the start (day 0) and at the end (day 14) of the experiments, Table S7: Extent of conductivity reduction (%) from the different media with different NaCl contents, Table S8: Chloride content values (mg L^{-1}) measured at the start (day 0) and at the end (day 14) of the experiments, Table S9: Extent of chloride removal (%) from the different media with different NaCl contents. Figure S1: Nitrate removal (mg L^{-1}) from the different media with different NaCl contents; Figure S2: Phosphate removal (mg L^{-1}) from the different media with different NaCl contents; Figure S3: Conductivity reduction (μS cm^{-1}) in the different media with different NaCl contents; Figure S4: Chloride removal (mg L^{-1}) from the different media with different NaCl contents.

Author Contributions: Conceptualization, I.B.; methodology, A.F., I.B.; formal analysis, A.F., K.M. investigation, A.F., K.M.; resources, I.B.; data curation, A.F., K.M. and I.B.; writing—original draft, A.F. and K.M.; writing—review and editing, I.B., V.B.-B.; visualization, A.F., K.M., I.B., V.B.-B.; project administration, I.B.; supervision, I.B.; funding acquisition, I.B. All authors have read and agreed to the published version of the manuscript.

Funding: This research was funded by the National Research Development and Innovation Office NKFIH FK 131 917 Grant (István Bácsi). Project no. TKP2020-IKA-04 has been implemented with the support provided from the National Research Development and Innovation Fund of Hungary, financed under the 2020-4.1.1-TKP2020 funding scheme.

Institutional Review Board Statement: Not applicable.

Informed Consent Statement: Not applicable.

Data Availability Statement: The data that support the findings of this study are available from the corresponding author upon reasonable request.

Conflicts of Interest: Authors declare no conflict of interest. The funders had no role in the design of the study; in the collection, analyses, or interpretation of data; in the writing of the manuscript, or in the decision to publish the results.

References

1. Edwards, A.C.; Withers, P.J.A. Linking phosphorus sources to impacts in different types of water body. *Soil Use Manag.* **2007**, *23*, 133–143. [CrossRef]
2. Abdel-Raouf, N.; Al-Homaidan, A.A.; Ibraheem, I.B.M. Microalgae and wastewater treatment. *Saudi. J. Biol. Sci.* **2012**, *19*, 257–275. [CrossRef]
3. Fernandes, T.V.; Suárez-Muñoz, M.; Trebuch, L.M.; Verbraak, P.J.; Van de Waal, D.B. Toward an ecologically optimized N:P recovery from wastewater by microalgae. *Front. Microbiol.* **2017**, *8*, 1742. [CrossRef] [PubMed]
4. Rani, S.; Gunjyal, N.; Ojha, C.S.P.; Asce, F.; Singh, R.P. Review of challenges for algae-based wastewater treatment: Strain selection, wastewater characteristics, abiotic, and biotic factors. *J. Hazard. Toxic Radioact. Waste* **2021**, *25*, 03120004. [CrossRef]
5. Ebrahimian, A.; Kariminia, H.R.; Vosoughi, M. Lipid production in mixotrophic cultivation of *Chlorella vulgaris* in a mixture of primary and secondary municipal wastewater. *Renew. Energy* **2014**, *71*, 502–508. [CrossRef]
6. Zhou, W.; Li, Y.; Min, M.; Hu, B.; Chen, P.; Ruan, R. Local bioprospecting for high-lipid producing microalgal strains to be grown on concentrated municipal wastewater for biofuel production. *Bioresour. Technol.* **2011**, *102*, 6909–6919. [CrossRef] [PubMed]
7. Oswald, W.J. Micro-algae and waste-water treatment. In *Micro-Algal Biotechnology*; Borowitzka, M.A., Borowitzka, L.J., Eds.; Cambridge University Press: New York, NY, USA, 1988; pp. 305–328.
8. Muñoz, R.; Jacinto, M.; Guieysse, B.; Mattiasson, B. Combined carbon and nitrogen removal from acetonitrile using algal–bacterial bioreactors. *Appl. Microbiol. Biotechnol.* **2005**, *67*, 699–707. [CrossRef]
9. Perales-Vela, H.V.; Peña-Castro, J.M.; Cañizares-Villanueva, R.O. Heavy metal detoxification in eukaryotic microalgae. *Chemosphere* **2006**, *64*, 1–10. [CrossRef] [PubMed]
10. Wang, Y.; Liu, J.; Kang, D.; Wu, C.; Wu, Y. Removal of pharmaceuticals and personal care products from wastewater using algae-based technologies: A review. *Rev. Environ. Sci. Biotechnol.* **2017**, *16*, 717–735. [CrossRef]
11. Safi, C.; Zebib, B.; Merah, O.; Pontalier, P.Y.; Vaca-Garcia, C. Morphology, composition, production, processing and applications of *Chlorella vulgaris*: A review. *Renew. Sustain. Energy Rev.* **2014**, *35*, 265–278. [CrossRef]
12. Zhai, J.; Li, X.; Li, W.; Rahaman, M.H.; Zhao, Y.; Wei, B.; Wei, H. Optimization of biomass production and nutrients removal by *Spirulina platensis* from municipal wastewater. *Ecol. Eng.* **2017**, *108*, 83–92. [CrossRef]
13. Singh, P.; Jain, K.; Desai, C.; Tiwari, O.; Madamwar, D. Microbial community dynamics of extremophiles/extreme environment. In *Microbial Diversity in the Genomic Era*; Das, S., Dash, H.R., Eds.; Academic Press: Cambridge, MA, USA, 2019; pp. 323–332.
14. Ventosa, A.; de la Haba, R.R.; Sánchez-Porro, C.; Papke, R.T. Microbial diversity of hypersaline environments: A metagenomic approach. *Curr. Opin. Microbiol.* **2015**, *25*, 80–87. [CrossRef] [PubMed]
15. Redfield, A.C.; Ketchum, B.H.; Richards, F.A. The influence of organisms on the composition of seawater. In *The Sea*, 2nd ed.; Hill, M.H., Ed.; Wiley: Hoboken, NJ, USA, 1963.
16. Martiny, A.C.; Vrugt, J.A.; Lomas, M.W. Concentrations and ratios of particulate organic carbon, nitrogen, and phosphorus in the global ocean. *Sci. Data* **2014**, *1*, 140048. [CrossRef]
17. Schreurs, H. Cyanobacterial Dominance, Relation to Eutrophication and Lake Morphology. Ph.D. Thesis, University of Amsterdam, Amsterdam, The Netherlands, 1992.
18. Mayers, J.J.; Flynn, K.J.; Shields, R.J. Influence of the N:P supply ratioon biomass productivity and time-resolved changes in elemental and bulk biochemical composition of *Nannochloropsis* sp. *Bioresour. Technol.* **2014**, *169*, 588–595. [CrossRef] [PubMed]
19. Rasdi, N.W.; Qin, J. Effect of N:P ratio on growth and chemical composition of *Nannochloropsis oculata* and *Tisochrysis lutea*. *J. Appl. Psychol.* **2015**, *27*, 2221–2230. [CrossRef]
20. Figler, A.; B-Béres, V.; Dobronoki, D.; Márton, K.; Nagy, S.A.; Bácsi, I. Salt tolerance and desalination abilities of nine common green microalgae isolates. *Water* **2019**, *11*, 2527. [CrossRef]
21. Valdez-Ojeda, R.; Gonzalez-Munoz, M.; Us-Vazquez, R.; Narvaez-Zapata, J.; Chavarria-Hernandez, J.C.; Lopez-Adrian, S.; Barahona-Perez, F.; Toledano-Thompson, T.; Garduno-Solorzano, G.; Medrano, R.M.E.G. Characterization of five fresh water microalgae with potential for biodiesel production. *Algal Res.* **2014**, *7*, 33–44. [CrossRef]
22. Yang, Z.H.; Zhao, Y.; Liu, Z.Y.; Liu, C.F.; Hu, Z.P.; Hou, Y.Y. A Mathematical Model of Neutral Lipid Content in terms of Initial Nitrogen Concentration and Validation in *Coelastrum* sp. HA-1 and Application in *Chlorella sorokiniana*. *BioMed Res. Int.* **2017**, 9253020. Available online: https://www.hindawi.com/journals/bmri/2017/9253020/ (accessed on 8 April 2021). [CrossRef]
23. Úbeda, B.; Galvez, J.A.; Michel, M.; Bartual, A. Microalgae cultivation in urban wastewater: *Coelastrum* cf. *pseudomicroporum* as a novel carotenoid source and a potential microalgae harvesting tool. *Bioresour. Technol.* **2016**, *228*, 210–217. [CrossRef]
24. Mousavi, S.; Najafpour, G.D.; Mohammadi, M.; Seifi, M.H. Cultivation of newly isolated microalgae *Coelastrum* sp. in wastewater for simultaneous CO_2 fixation, lipid production and wastewater treatment. *Bioprocess Biosyst. Eng.* **2018**, *41*, 519–530. [CrossRef]
25. Rauytanapanit, M.; Janchot, K.; Kusolkumbot, P.; Sirisattha, S.; Waditee-Sirisattha, R.; Praneenararat, T. Nutrient Deprivation-Associated Changes in Green Microalga *Coelastrum* sp. TISTR 9501RE Enhanced Potent Antioxidant Carotenoids. *Mar. Drugs* **2019**, *17*, 328. [CrossRef]

26. Valdez-Ojeda, R.A.; Serrano-Vazquez, M.G.D.; Toledano-Thompson, T.; Chavarria-Hernandez, J.C.; Barahona-Perez, L.F. Effect of media composition and culture time on the lipid profile of the green microalga *Coelastrum sp.* and its suitability for Biofuel Production. *Bioenergy Res.* **2021**, *14*, 241–253. [CrossRef]
27. Tharek, A.; Yahya, A.; Salleh, M.M.; Jamaluddin, H.; Yoshizaki, S.; Dolah, R.; Hara, H.; Iwamoto, K.; Mohamad, S.E. Improvement of Astaxanthin Production in *Coelastrum sp.* by Optimization Using Taguchi Method. *Appl. Food. Biotechnol.* **2020**, *7*, 205–214.
28. CCAP Media Recipes a. Available online: https://www.ccap.ac.uk/media/documents/BB.pdf (accessed on 5 February 2021).
29. CCAP Media Recipes b. Available online: https://www.ccap.ac.uk/media/documents/JM.pdf (accessed on 5 February 2021).
30. European Standard EN 15204: Water Quality—Guidance Sandard on the Enumeration of Phytoplankton Using Inverted Microscopy (Utermöhl Technique). Available online: https://standards.iteh.ai/catalog/standards/cen/d9020a71-2bd3-478f-867f-46ea3a0c0620/en-15204-2006 (accessed on 24 May 2019).
31. Hungarian Standard MSZ 1484-13: 2009. Water Quality. Part 13: Determination of Nitrate and Nitrite Content by Spectrophotometric Method. Available online: http://www.mszt.hu/web/guest/webaruhaz (accessed on 24 May 2019).
32. Hungarian Standard MSZ EN ISO 6878: 2004. Water Quality. Determination of Phosphorus. Ammonium Molybdate Spectrometric Method (ISO 6878:2004). Available online: https://www.iso.org/standard/36917.html (accessed on 24 May 2019).
33. Németh, J. *Methods of Biological Water Qualification*; Institute of Environmental Management, Environmental Protection Information Service: Budapest, Hungary, 1998.
34. Zar, H. *Biostatistical Analysis*, 3rd ed.; Prentice-Hall International: Hoboken, NJ, USA, 1996.
35. Hammer, O.; Harper, D.A.T.; Ryan, P.D. PAST: Paleontological statistics software package for education and data analysis. *Palaeontol. Electron.* **2001**, *2001*, 9.
36. Shriwastav, A.; Gupta, S.K.; Ansari, F.A.; Rawat, I.; Bux, B. Adaptability of growth and nutrient uptake potential of *Chlorella sorokiniana* with variable nutrient loading. *Bioresour. Technol.* **2014**, *174*, 60–66. [CrossRef]
37. Choi, H.J.; Lee, S.M. Effect of the N/P ratio on biomass productivity and nutrient removal from municipal wastewater. *Bioprocess Biosyst. Eng.* **2015**, *38*, 761–766. [CrossRef]
38. Liu, Y.; Li, L.; Jia, R. The Optimum Resource Ratio (N:P) for the Growth of *Microcystis aeruginosa* with Abundant Nutrients. *Procedia Environ. Sci.* **2011**, *10*, 2134–2140. [CrossRef]
39. Terry, K.L. Nitrate and phosphate uptake interactions in a marine prymnesiophyte. *J. Phycol.* **1982**, *18*, 79–86. [CrossRef]
40. Berkessa, Y.W.; Mereta, S.T.; Feyisa, F.F. Simultaneous removal of nitrate and phosphate from wastewater using solid waste from factory. *Appl. Water Sci.* **2019**, *9*, 28. [CrossRef]
41. Park, M.H.; Park, C.H.; Sim, Y.B.; Hwang, S.J. Response of *Scenedesmus quadricauda* (Chlorophyceae) to Salt Stress Considering Nutrient Enrichment and Intracellular Proline Accumulation. *Int. J. Environ. Res. Public Health* **2020**, *17*, 3624. [CrossRef] [PubMed]
42. Mohapatra, P.K.; Dash, R.C.; Panda, S.S.; Mishra, R.K.; Mohanty, R.C. Effects of nutrients at different salinities on growth of the freshwater green alga *Scenedesmus bijugatus* in water of Narendra Pond, Puri, Orissa. *Int. Rev. Hvdrobiol.* **1998**, *83*, 297–304. [CrossRef]
43. Rai, A.K.; Abraham, G. Relationship of combined nitrogen sources to salt tolerance in freshwater cyanobacterium *Anabaena doliolum*. *J. Appl. Microbiol.* **1995**, *78*, 501–506.
44. Collos, Y.; Vaquer, A.; Souchu, P. Acclimation of nitrate uptake by phytoplankton to high substrate levels. *J. Phycol.* **2005**, *41*, 466–478. [CrossRef]
45. Liu, J.; Vyverman, W. Differences in nutrient uptake capacity of the benthic filamentous algae *Cladophora* sp., *Klebsormidium* sp. and *Pseudanabaena* sp. under varying N/P conditions. *Bioresour. Technol.* **2015**, *179*, 234–242. [CrossRef] [PubMed]
46. Alketife, A.M.; Judd, S.; Znad, H. Synergistic effects and optimization of nitrogen and phosphorus concentrations on the growth and nutrient uptake of a freshwater *Chlorella vulgaris*. *Environ. Technol.* **2017**, *38*, 94–102. [CrossRef] [PubMed]
47. Arora, N.; Laurens, L.M.L.; Sweeney, N.; Pruthi, V.; Poluri, K.M.; Pienkos, P.T. Elucidating the unique physiological responses of halotolerant *Scenedesmus* sp. cultivated in sea water for biofuel production. *Algal Res.* **2019**, *37*, 260–268. [CrossRef]
48. Lagus, A.; Suomela, J.; Weithoff, G.; Heikkila, K.; Helminen, H.; Sipura, J. Species-specific differences in phytoplankton responses to N and P enrichments and the N:P ratio in the Archipelago Sea, northern Baltic Sea. *J. Plankton Res.* **2004**, *26*, 779–798. [CrossRef]
49. Ruzhitskaya, O.; Gogina, E. Methods for Removing of Phosphates from Wastewater. *MATEC Web Conf.* **2017**, *106*, 07006. [CrossRef]
50. Kirst, G.O. Salinity tolerance of eukaryotic marine algae. *Annu. Rev. Plant Physiol.* **1989**, *41*, 21–53. [CrossRef]
51. Raven, J.A. Chloride: Essential micronutrient and multifunctional beneficial ion. *J. Exp. Bot.* **2017**, *68*, 359–367. [CrossRef] [PubMed]
52. Sahle-Demessie, E.; Aly Hassan, A.; El Badawy, A. Bio-desalination of brackish and seawater using halophytic algae. *Desalination* **2019**, *465*, 104–113. [CrossRef] [PubMed]

Article

Optimizing Docosahexaenoic Acid (DHA) Production by *Schizochytrium* sp. Grown on Waste Glycerol

Natalia Kujawska [1], Szymon Talbierz [1], Marcin Dębowski [2,*], Joanna Kazimierowicz [3] and Marcin Zieliński [2]

[1] InnovaTree Sp. z o.o., 81-451 Gdynia, Poland; natalia.kujawska@innovatree.pl (N.K.); szymon.talbierz@innovatree.pl (S.T.)
[2] Department of Environment Engineering, Faculty of Geoengineering, University of Warmia and Mazury in Olsztyn, 10-720 Olsztyn, Poland; marcin.zielinski@uwm.edu.pl
[3] Department of Water Supply and Sewage Systems, Faculty of Civil Engineering and Environmental Sciences, Bialystok University of Technology, 15-351 Bialystok, Poland; j.kazimierowicz@pb.edu.pl
* Correspondence: marcin.debowski@uwm.edu.pl

Abstract: The aim of this study was to optimize biomass and docosahexaenoic acid (DHA) production by *Schizochytrium* sp. grown on waste glycerol as an organic carbon source. Parameters having a significant effect on biomass and DHA yields were screened using the fractional Plackett–Burman design and the response surface methodology (RSM). *Schizochytrium* sp. growth was most significantly influenced by crude glycerin concentration in the growth medium (150 g/dm^3), process temperature (27 °C), oxygen in the bioreactor (49.99% v/v), and the concentration of peptone as a source of nitrogen (9.99 g/dm^3). The process parameter values identified as optimal for producing high DHA concentrations in the biomass were as follows: glycerin concentration 149.99 g/dm^3, temperature 26 °C, oxygen concentration 30% (v/v), and peptone concentration 2.21 g/dm^3. The dry cell weight (DCW) obtained under actual laboratory conditions was 66.69 ± 0.66 g/dm^3, i.e., 1.27% lower than the predicted value. The DHA concentration obtained in the actual culture was at 17.25 ± 0.33 g/dm^3, which was 3.03% lower than the predicted value. The results obtained suggest that a two-step culture system should be employed, with the first phase focused on high production of *Schizochytrium* sp. biomass, and the second focused on increasing DHA concentration in the cells.

Keywords: docosahexaenoic acid; *Schizochytrium* sp.; crude glycerin; optimization; Plackett–Burman design; response surface methodology

Citation: Kujawska, N.; Talbierz, S.; Dębowski, M.; Kazimierowicz, J.; Zieliński, M. Optimizing Docosahexaenoic Acid (DHA) Production by *Schizochytrium* sp. Grown on Waste Glycerol. *Energies* **2021**, *14*, 1685. https://doi.org/10.3390/en14061685

Academic Editor: José Carlos Magalhães Pires

Received: 8 February 2021
Accepted: 15 March 2021
Published: 18 March 2021

1. Introduction

The properties of microalgae make them a useful resource for environmental engineering technologies, including wastewater treatment, bio-sequestration of carbon dioxide, manufacture of biofuels, and sorption of contaminants [1,2]. Microalgal biomass is also a source of value-added products useful in medicine, pharmaceuticals, fertilizer industry, animal feed industry, and the food sector [3,4]. The potential of the widespread industrial use of microalgae is limited by the lack of technologies that are process-efficient, simple in terms of design/technology, cost-effective to build and operate, and environmentally friendly [5]. One of the major drivers of operating cost for microalgal biomass production systems are the chemical components of the growth medium [6]. Therefore, there is a real need to seek, develop, and optimize methods that could serve as a competitive alternative to the current solutions. One of the most promising and encouraging options is found in the development of technologies that use waste substrates as the main growth medium ingredient [7,8], a method consistent with the idea of the circular economy and the principles of an integrated biorefinery approach [9]. With the use of biorefinery complexity index (BCI) as an indicator of technical and economic risk, algal- and waste-based bio-refinery platforms are considered to be one of the most promising approaches for producing fuel, food, animal feed, food supplements, fertilizers, and pharmaceuticals [10].

The use of waste glycerol as a carbon source for the heterotrophic cultivation of microalgae seems to be one of the viable trajectories for exploring such integrated technologies. The supply of waste glycerol is steadily growing due to the increasing global production of biodiesel, spurred by requirements for blending specific percentages of biocomponents with conventional fuels [11]. Global biodiesel production totals about 41.3 billion liters, a process that involves the co-production of waste glycerol at over 12% of total esters produced, regardless of the catalyst or technological process used [12,13]. Crude glycerol is the primary by-product in the biodiesel industry. The utilization of the glycerol becomes an urgent issue for the biodiesel business for two reasons; the environment and cost reduction. Large-scale biodiesel producers can choose to upgrade glycerol and move it towards the chemical products market. However, the refining of crude glycerol has a high cost. For small to medium scale biodiesel companies this is very problematic [14,15]. Therefore, there is a real need to seek new technologies for harnessing and neutralizing glycerol waste, as well as refining existing ones [15].

Glycerol has been thermochemically converted to dipropylene glycol [16] and hydroxyacetone [17]. Other processes have also been explored, such as the reformation of glycerin to produce hydrogen and synthesis gas [18], production of epichlorohydrin [19], etherification [20], and hydrogenation [21]. Glycerin can serve as a source of carbon in biochemical processes that produce 1,3-propanediol colorants or omega-3 fatty acids [22]. Waste glycerin is also commonly used in energy production as a substrate for fermentative methane production [23]. By harnessing specific strains of bacteria and synthesizing new enzymes, new conversion pathways become available, including ethanol production (*Saccharomyces cerevisiae* bacteria) or β-carotene production (*Blakeslea trispora*) [24,25].

One little-explored approach to utilizing the waste glycerol fraction and converting it into value-added biocomponents is to use the biomass of *Schizochytrium* sp. heterotrophic microalgae. These algae accumulate large quantities of docosahexaenoic acid (DHA) in their cells, making them a prime resource for use in the food, pharmaceutical, and animal feed industries [26,27]. DHA is an unsaturated fatty acid belonging to the Omega-3 group, which is an important structural component of cell membranes in some tissues of the human body, such as in the phospholipids that make up the neurons of the cortex of the brain and the retina of the eye [28]. DHA suppresses inflammatory responses, plays an important neuroprotective role, and prevents neuronal damage and apoptosis [29]. It has been proven that DHA protects against the development of arterial hypertension and plays a large role in the proper development of the brain in newborns [30]. DHA also increases calcium absorption, helps maintain normal levels of "bad" and "good" cholesterol in the body, and supports the immune system [31]. It is necessary to supply DHA with food [32], and currently its main source are vegetable oils and fats from fish meat [33]. Due to the growing awareness of consumers, the demand for acids from the Omega-3 group is still growing [34]. Therefore, there is a justified need to search for alternative methods for their production, which will be environmentally friendly, and justified in terms of technology and economics. *Schizochytrium* sp. biomass has been shown to grow on a broad range of carbon sources [35], substantiating efforts to develop efficient methods of growing it on waste glycerol as a carbon source. Studies to date have examined the growth of *Thraustochytriacae* microalgae on waste materials, such as breadcrumbs [36], spent brewer's yeast [37], empty palm fruit bunches [38], coconut water [39], okara powder [40], beer and potato processing residues [41], and sweet sorghum juice [42].

To ensure optimal *Schizochytrium* sp. growth rate and enable culture scale-up, multiple variables must be considered and fine-tuned. They include physicochemical parameters, the nutritional value of the waste material, and the presence of growth inhibitors, as well as the availability and cost of the waste material [43]. Considering the commercially and environmentally informed need to implement technologies for integrated waste neutralization, microalgal biomass production, energy recovery, and extraction of value-added substances, it is necessary to pursue research on optimizing technological parameters to achieve high performance [44].

The aim of this research was to use a fractional Plackett–Burman design and the response surface methodology (RSM) to optimize docosahexaenoic acid (DHA) production by heterotrophic *Schizochytrium* sp., with waste glycerol used as a source of organic carbon.

2. Materials and Methods

2.1. General Design

The study aimed to identify and quantify the key process parameter values that affect the growth and DHA production of *Schizochytrium* sp. cultures, using glycerol as an external carbon source. The values were determined through a series of experiments structured using a Plackett–Burman design. After screening the most significant parameters having an effect on the performance, the parameter values resulting in highest dry cell weight (DCW) and DHA levels were determined using the response surface methodology. The modeled parameter values and predicted performance were then experimentally verified using a batch *Schizochytrium* sp. culture.

2.2. Materials

The study used *Schizochytrium* sp., a strain of single-cell heterotrophic microalgae from the family *Thraustochytriaceae*. The inoculum was obtained from the ATCC (American Type Culture Collection). *Schizochytrium* sp. cells were maintained in sterile agar slants of ATCC790By+ medium containing 2% agar (*w*/*w*). The cells were spread into new agar slants on a monthly basis. To obtain enough inoculum for the experiment, the cells were transferred from the agar slants to 50 cm^3 Erlenmeyer flasks, containing 15 cm^3 of the agarless ATCC790By+ liquid medium with a pH of 6.5, with the composition given in Tables 1 and 2.

Table 1. Composition of ATCC790 By+ culture medium used in the experiment.

Component	Unit	Concentration
Glucose	[g/dm^3)	5.0
Yeast extract	[g/dm^3)	1.0
Peptone	[g/dm^3)	1.0
Artificial seawater	[dm^3)	1.0

Table 2. Profile of artificial seawater.

Component	Unit	Concentration
$(NH_4)_2SO_4$	(g/dm^3)	1.0
KH_2PO_4	(g/dm^3)	3.0
Na_2SO_4	(g/dm^3)	12.0
$MgSO_4$	(g/dm^3)	5.0
K_2SO_4	(g/dm^3)	7.0
KCl	(g/dm^3)	2.0
	Trace elements	
$CaCl_2$	(mg/dm^3)	50
$MnCl_2$	(mg/dm^3)	5.2
$ZnSO_4$	(mg/dm^3)	5.2
$CuSO_4$	(mg/dm^3)	0.8
Na_2MoO_4	(mg/dm^3)	0.016
$NiSO_4$	(mg/dm^3)	0.8
$FeSO_4$	(mg/dm^3)	0.01
$CoCl_2$	(mg/dm^3)	0.066
thiamine	(mg/dm^3)	0.76
vitamin B_{12}	(mg/dm^3)	1.2
vitamin B_5 (calcium salt of pantothenic acid)	(mg/dm^3)	25.6

The medium was autoclave-sterilized before use (Systec V-95 autoclave, parameters: 121 °C, 15 min, 2 bars, automatic demineralized water feed for steam generation was used). The flasks were agitated in a temperature-controlled orbital shaker (Excella E24R, New Brunswick/Eppendorf) set to 170 rpm at 25 °C. After 120 h, 10 cm^3 of the culture (*Schizochytrium* sp. inoculum) was transferred into a 250 cm^3 Erlenmeyer flask containing 90 cm^3 of fresh ATCC790By+. The resultant 100 cm^3 culture was inoculated into 500 cm^3 bioreactors (with optimization performed using the Plackett–Burman design and response surface methodology) and a Biostat B Twin (Sartorius Stedim) bioreactor with a working capacity of 2.0 dm^3 (for the verification step).

Crude glycerin sourced from the PKN Orlen Południe S.A. plant in Trzebinia (Poland) was used as the substrate and the sole carbon source in the cultures. During the production of biodiesel, pressed rapeseed oil is filtered and transesterified with the use of NaOH catalyst in the amount of 1.2% to the weight of the oil (process temperature is 25 °C, reaction time 45 min). The ester and glycerol phases are separated by gravity or by centrifugation and the methanol is recovered from both phases by distillation. The substrate contained at least 80% (*w*/*w*) glycerol, up to 5% (*w*/*w*) ash, up to 6% (*w*/*w*) MONG (matter organic non glycerol), and trace water, according to its safety data sheet. The liquid had pH = 5, a light-brown coloration, and a characteristic odor.

2.3. Experimental Equipment

Batch photobioreactors (Erlenmeyer flasks) with an active volume of 500 cm^3 were used for the optimization step. The flasks were agitated in a temperature-controlled orbital shaker (Excella E24R, New Brunswick/Eppendorf) set to 170 rpm. During the validation step, the *Schizochytrium* sp. biomass was grown in a Biostat B Twin (Sartorius Stedim) bioreactor with a working capacity of 2 dm^3. The bioreactor was fitted with acid, base, antifoaming agents, and organic substrate pumps. The system had a gas module for monitoring dissolved oxygen (DO), a system for stabilizing pH based on injecting acid or base via peristaltic pumps, and a temperature measurement/stabilization system. The vessels were stirred by a six-bladed Rushton turbine (53 mm diameter).

2.4. Analytical Methods

The crude glycerin concentration in the culture medium was determined by pre-centrifugation (8000× *g*, 4 min, 10 °C; UNIVERSAL 320 R centrifuge, Hettich, Westphalia, Germany). The supernatant was then filtered (pore size = 0.2 mm), and the filtrate was assayed for glycerol levels using a Glycerol GK Assay Kit (Megazyme). The test involved phosphorylating the glycerol with adenosine 5′-triphosphate (ATP), with the reaction product, adenosine 5′-diphosphate (ADP) then used to further phosphorylate d-glucose, which oxidizes producing nicotinamide adenine dinucleotide (NADH). The concentration of NADH was measured spectrophotometrically (Multiskan GO Microplate, Thermo Scientific, Waltham, MA, USA) at a wavelength of 340 nm.

The dry cell weight (DCW) of the microalgae was determined according to the method described by Chang et al., (2013) [45]. It was done by transferring a 50 cm^3 sample of the culture to a pre-weighed centrifuge tube, which was then centrifuged (8000 g for 15 min, UNIVERSAL 320 R centrifuge, Hettich). The supernatant was discarded and the concentrated biomass was washed twice with distilled water, then dried at 60 °C for 12 h in a moisture balance (MAR, Radwag, Radom, Poland) to stabilize the biomass.

The lipid content of the biomass was determined by adding 7 cm^3 of a 20% hydrochloric acid solution to 1.0 g of freeze-dried biomass (ALPHA 1-4 LD plus freeze dryer, Christ), which was then placed in a water bath (GFL 1003) at 75 °C for 40 min. The sample was treated with 20 cm^3 of *n*-hexane to extract the lipids and placed in a vacuum evaporator (Hei-VAP Advantage G3, Heidolph, Schwabach, Germany) to evaporate the solvent. The lipid content of the sample was measured gravimetrically. The determination of fatty acids in the microalgal biomass was done using a modified direct transmethylation process by Grayburn (1992) [46]. The freeze-dried microalgal biomass (ALPHA 1-4 LD plus, Christ,

Bethlehem, Palestine) of 20–100 mg was transferred to a reaction vial, enriched with 2 mg of triheptadecanoylglycerol as an internal standard, and spiked with 2 cm^3 of a 1% solution of H_2SO_4 in methanol. The vial was then thoroughly mixed and heated to a temperature of 80 °C for 2 h (MKR 13 blockthermostat–Ditabis, Pforzheim, Germany). After cooling, 2 cm^3 of chloroform and 1 cm^3 of distilled water were added. The vial was mixed at 1250 rpm (Vortex Reax top, Heidolph), and centrifuged at 1500× *g* (UNIVERSAL 320 R, Hettich). The organic phase containing fatty acid methyl esters (FAME) was harvested and analyzed by chromatography. A Clarus 680 GC (Perkin Elmer, Waltham, MA, USA) gas chromatograph was used for FAME analysis, with helium used as the carrier gas. The column temperature was raised from 150 °C to 250 °C at 10 °C/min, then kept at 250 °C for 10 min. The injector was kept at 275 °C with an injection volume of 1 µL. Detection was made using a flame ionization detector (FID) at 280 °C. The peak areas were identified by comparing their retention times with those of standard mixtures.

2.5. Optimization Design

A Plackett–Burman based experimental design was used to screen the factors having significant effects on the DCW increase and DHA production (Design-Expert software by Stat-Ease Inc.). The evaluated parameters were as follows: temperature (°C), initial growth medium pH, volumetric air flow rate ($L_{area}/min{\cdot}L_{cont{\cdot}}$), oxygen in the growth medium (%), initial inoculum level (% *v*/*v*), concentration of crude glycerin (g/dm^3), salinity (psu), concentration of yeast extract (g/dm^3), turbine speed (rpm), and concentration of peptone (g/dm^3). A design matrix of experiments with specific parameters was generated using the method. Parameters that significantly affected the values of: the rate of DHA production by microalgae—r_{DHA} (g/$dm^3{\cdot}h$) and the growth rate of the microalgal biomass—r_{DCW} (g/$dm^3{\cdot}h$), were identified in the course of the study.

Each independent variable was investigated at a low (−) and a high (+) level. The values of parameters for both levels were selected through the analysis of the available literature data. The low levels (−) of a variable always corresponded to the lowest parameter value at which *Schizochytrium* sp. growth and DHA production were possible. A high (+) level of a variable was taken as the lowest parameter value at which these parameters were not inhibited. Each experiment was conducted in triplicate.

Table 3 shows a list of the analyzed parameters and their values for each of the levels. Twelve experiments were established, as shown in Table 4. To determine the significant level for each parameter, as prescribed by the Analytical Methods Committee method, an additional "dummy" variable, "d_1", was added to the design matrix. The resultant effects of dummy variables reflected the standard error of the experiments, which was used to derive the significant level for each of the parameters. Parameters at $p < 0.10$ were taken as significant factors, which were further optimized using response surface methodology. For the experiments designed using the response surface methodology matrix, the levels of non-significant parameters were maintained at −1. The effects of each variable ($E_{(xi)}$), the significant (P) levels, and the F-test results for the obtained data are presented herein.

Table 3. Ranges of variables used in the Plackett–Burman design experiments.

Variable	Unit	Variable Designation	Low (−1)	High (+1)
Temperature	(°C)	X_1	25	28
Initial pH	-	X_2	6.5	7.5
Oxygen percentage	(%)	X_3	30	50
Volumetric air flow rate	($L_{air}/min{\cdot}L_{react}$)	X_4	0.3	1.0
Initial level (percent by volume)	(% *v*/*v*)	X_5	10	30
Concentration of crude glycerin	(g/dm^3)	X_6	75	150
Salinity	(psu)	X_7	17.5	35
Concentration of yeast extract	(g/dm^3)	X_8	0.4	10.0
Turbine speed	(rpm)	X_9	175	1000
Concentration of peptone	(g/dm^3)	X_{10}	2	5

Table 4. The Plackett–Burman design of the experiments.

Experiment No.	X_1	X_2	X_3	X_4	X_5	X_6	X_7	X_8	X_9	X_{10}	d_1
1	+1	+1	−1	−1	+1	+1	+1	+1	−1	+1	−1
2	−1	+1	+1	−1	−1	+1	+1	+1	+1	−1	+1
3	+1	−1	+1	+1	−1	−1	+1	+1	+1	+1	−1
4	−1	+1	−1	+1	+1	−1	−1	+1	+1	+1	+1
5	+1	−1	+1	−1	+1	+1	−1	−1	+1	+1	+1
6	+1	+1	−1	+1	−1	+1	+1	−1	−1	+1	+1
7	+1	+1	+1	−1	+1	−1	+1	+1	−1	−1	+1
8	+1	+1	+1	+1	−1	+1	−1	+1	+1	−1	−1
9	−1	+1	+1	+1	+1	−1	+1	−1	+1	+1	−1
10	−1	−1	+1	+1	+1	+1	−1	+1	−1	+1	+1
11	+1	−1	−1	+1	+1	+1	+1	−1	+1	−1	+1
12	−1	−1	−1	−1	−1	−1	−1	−1	−1	−1	−1

The effect of each variable ($E_{(xi)}$) was derived using the equation:

$$E_{(xi)} = \frac{2\cdot(\sum M_{i+} - \sum M_{i-})}{N}. \quad (1)$$

where $E_{(xi)}$ is the effect of the variable, M_{i+} is theDCW or DHA concentration for the high (+) level of the variable, M_{i-} is the DCW or DHA concentration for the low (−) level of the variable, and N is the number of runs.

In order to determine how significant the effect of each parameter was on the technological process, ANOVA-derived calculations were used, i.e., the sum of squares of the effects (SS) of each parameter (with the exception of the dummy variable). The parameter was calculated according to the equation:

$$SS = \frac{N\cdot {E_{(xi)}}^2}{4}. \quad (2)$$

where SS is the sum of the squares of the effects of each variable, $E_{(xi)}$ is the effect of the variable, and N is the number of runs.

The most significant parameters affecting the DCW and DHA concentration were screened, followed by the determination of the parameter values resulting in the highest DCW and DHA levels. This was achieved by using a statistical response surface methodology to formulate a list of experiments with Design-Expert software by Stat-Ease Inc. The method involved testing the parameters at five levels: −2, −1, 0, 1, and 2, and establishing an experimental design with central and axial points (Table 5). Other parameters previously identified as non-significant were maintained at a constant level. The experimental design matrix consisted of a 2^4 full factor design combined with six central points, and eight axial points, where one variable was set at an extreme level while the others were set at the central point level (Table 6).

Table 5. The Plackett–Burman design of the experiments.

Variable	Unit	Level				
		−2	−1	0	+1	+2
Temperature	(°C)	23.5	25.0	26.5	28.0	29.5
Oxygen percentage concentration	(%)	20	30	40	50	60
Concentration of crude glycerin	(g/dm^3)	37.5	75.0	112.5	150.0	187.5
Concentration of peptone	(g/dm^3)	0.5	2.0	3.5	5.0	6.5

Table 6. Experimental matrix designed using response surface methodology, with coded levels of significant parameters.

Experiment No.	Temperature (°C)	Concentration of Crude Glycerin (g/dm^3)	Oxygen Concentration (%)	Peptone Concentration (g/dm^3)
1	−1	−1	−1	−1
2	−1	−1	−1	1
3	−1	−1	1	−1
4	−1	−1	1	1
5	−1	1	−1	−1
6	−1	1	−1	1
7	−1	1	1	−1
8	−1	1	1	1
9	1	−1	−1	−1
10	1	−1	−1	1
11	1	−1	1	−1
12	1	−1	1	1
13	1	1	−1	−1
14	1	1	−1	1
15	1	1	1	−1
16	1	1	1	1
17	−2	0	0	0
18	2	0	0	0
19	0	−2	0	0
20	0	2	0	0
21	0	0	−2	0
22	0	0	2	0
23	0	0	0	−2
24	0	0	0	2
25	0	0	0	0
26	0	0	0	0
27	0	0	0	0
28	0	0	0	0
29	0	0	0	0
30	0	0	0	0

Based on the experimental results, the DCW and DHA values were correlated by the second-order polynomial equation:

$$Y = \beta_0 + \Sigma\beta_i x_i + \Sigma\beta_i x_i^2 + \Sigma\beta_{ij} x_i x_j \quad (3)$$

where Y is the predicted response of the design, β is the coefficient of the equation, and x_i and x_j are coded levels of parameters i and j, respectively.

The variable matrix equation was generated using Design-Expert software by Stat-Ease Inc. The significance level for the model was confirmed with an F-test. The selected cultural conditions resulting in highest DCW and DHA levels were verified through two experiments. The first experiment was to compare the experimental data with the predicted DCW levels, the other was to compare the experimental data with the predicted DHA concentrations.

2.6. Statistical Analysis

Each experimental variant was conducted in triplicate. The statistical analysis of experimental results was conducted using the STATISTICA 13.1 PL package. The hypothesis concerning the normality of distribution of each analyzed variable was verified using a W Shapiro–Wilk test. One-way analysis of variance (ANOVA) was conducted to determine differences between variables. Homogeneity of variance in groups was determined using a Levene test. The Tukey (HSD) test was applied to determine the significance of differences between the analyzed variables. In the tests, results were considered significant at $\alpha = 0.05$.

3. Results

3.1. Screening Significant Culture Factors According to the Plackett–Burman Design

The dry cell weight (DCW) of *Schizochytrium* sp. microalgae varied between 16.1 and 65.4 g/dm^3, DHA concentration ranged from 3.6 to 14.7 g/dm^3, depending on the culture parameters. Mean DCW and DHA concentrations were 35.72 ± 4.36 g/dm^3 and 8.09 ± 0.62 g/dm^3, respectively (Figure 1). The lowest DCW and DHA concentrations were observed in experiment 12 with all culture parameters at low (−1) levels. Conversely, the highest DCW and DHA concentrations were observed in experiment 8, which had high (+1) levels of parameters designated X_1, X_2, X_3, X_4, X_6, X_8, and X_9, and low (−1) levels of parameters X_5, X_7, and X_{10}. Therefore, an inoculum level of 10% (v/v), salinity of 17.5 psu, and peptone concentration of 2 g/dm^3 were sufficient to obtain the highest levels of *Schizochytrium* sp. biomass and DHA across the experiments. A screening of physicochemical cultural parameters was performed with Design-Expert software by Stat-Ease Inc. Table 7 shows the significance levels for those parameters that were highly correlated with DCW level and DHA concentration ($p < 0.10$), namely: crude glycerin concentration, temperature, oxygen concentration, and peptone concentration. The selected parameters were optimized using the response surface methodology.

Table 7. Significance (p values) of the effects of factors coded using the Plackett–Burman design on the dry cell weight (DCW) and docosahexaenoic acid (DHA) levels (values with $p < 0.10$ were denoted in italics).

Parameter	X_1	X_2	X_3	X_4	X_5	X_6	X_7	X_8	X_9	X_{10}
P-levels for DCW	*0.088*	0.133	*0.098*	0.955	0.564	*0.079*	0.224	0.205	0.102	*0.099*
P-levels for DHA	*0.089*	0.130	*0.099*	0.525	0.719	*0.076*	0.220	0.220	0.100	*0.097*

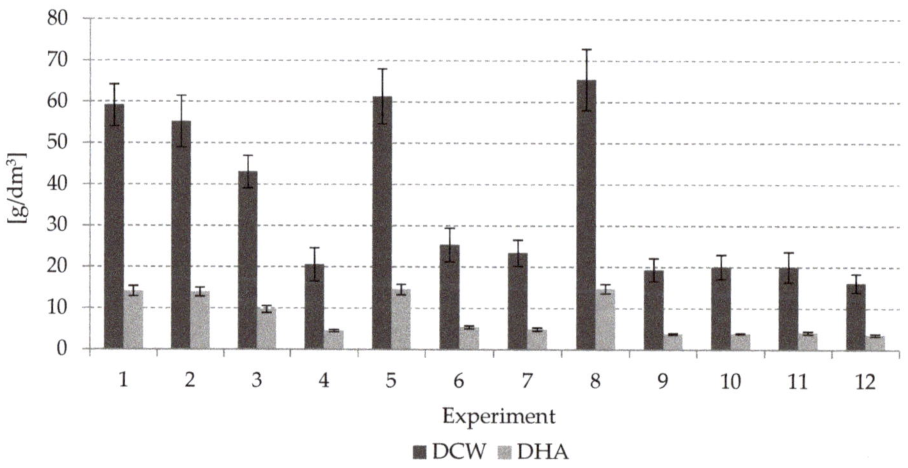

Figure 1. Biomass dry cell weight (DCW) and DHA concentrations obtained in the experimental series designed using the Plackett–Burman method.

3.2. Quantification of Culture Parameter Values Using the Response Surface Methodology

The objective of this step was to define those values of the *Schizochytrium* sp. batch culture parameters that resulted in the highest DCW and DHA levels. The results are presented in Figure 2. The DCW of *Schizochytrium* sp. microalgae varied between 38.9 and 65.4 g/dm^3, whereas DHA levels in the culture ranged from 8.43 to 16.00 g/dm^3, depending on the culture parameters (Figure 2). The lowest DCW and DHA levels were almost 2.5 times higher than the lowest values obtained in the first part of this experimental step. Mean DCW and DHA concentrations were 56.07 ± 4.68 g/dm^3 and 13.49 ± 1.72 g/dm^3, respectively, and were 1.6 times higher than the mean Plackett–Burman design results. The highest biomass concentration was recorded for experiment 20, with the crude glycerin concentration set at level +2 (187.5 g/dm^3) and other significant parameters at 0 (oxygen concentration 40%, peptone concentration 3.5 g/dm^3, temperature 26.5 °C). The highest increase in DHA levels (experiment 12) occurred with the crude glycerin concentration set at level −1 (75 g/dm^3) and other parameters at +1 (temperature 28 °C, oxygen concentration 50%, and peptone concentration 5 g/dm^3). This means that the concentrations of cell biomass and DHA were most significantly influenced by the same parameters, though at different values.

It was shown that the parameters having a statistically significant ($p < 0.10$) effect on DCW and DHA levels were: crude glycerin concentration, temperature, oxygen concentration, and peptone concentration (Table 8). The data analysis was followed by the determination of coefficients for the second-order polynomial equation, indicating the correlation of the most significant variables and their effect on DCW/DHA levels (Table 9). These values of coefficients for the second-order polynomial equation served as the basis for deriving the predicted DCW/DHA levels in relation to the significant parameters screened by the response surface methodology. The predicted parameter values for the biomass-optimized culture were DCW 67.55 g/dm^3 and DHA 15.53 g/dm^3; whereas the predicted parameter values for the DHA-optimized culture were DHA 17.79 g/dm^3 and DCW 63.59 g/dm^3.

The DCW estimation model was characterized by an estimation error on the level of ±4.075 g/dm^3, and reflected 89.45% of changes in the process (coefficient of determination R2 = 0.8945). The DHA estimation model reflected 90.92% of changes in the process (coefficient of determination R2 = 0.9092), with an estimation error at the level of ±2.061 g/dm^3. The ANOVA calculation shows that the p-value was 0.0035 for the DCW estimation model and 0.0010 for the DHA estimation model, which means the relationship between the independent variables and the response values investigated in this experiment were significant and the scheme was reliable.

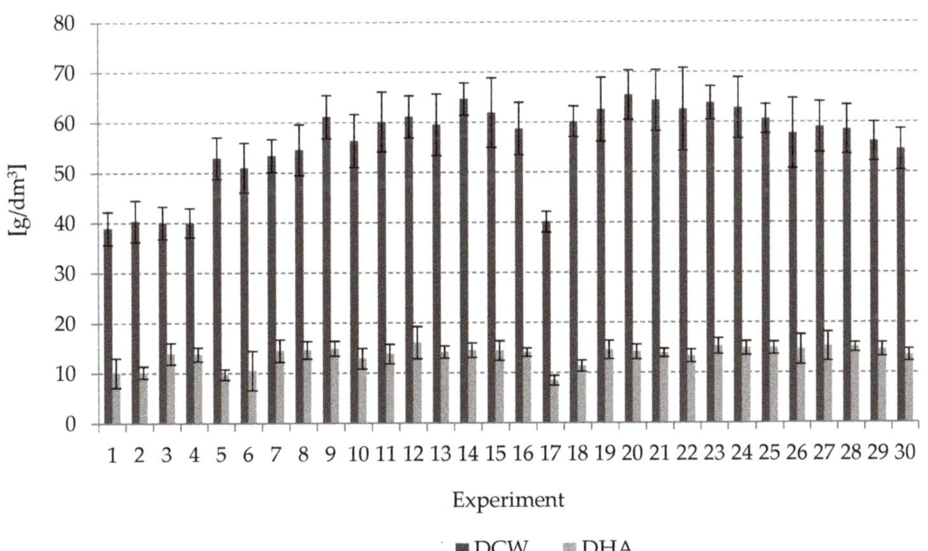

Figure 2. Biomass dry cell weight (DCW) and DHA concentrations obtained in the experimental series designed using the response surface methodology.

Table 8. Effect of variables on DCW and DHA concentrations, with related statistical test results.

Variable	DCW			DHA		
	$E_{(xi)}$	*F-Test*	*P*	$E_{(Xi)}$	*F-Test*	*P*
Temperature	27.73	53.26	*0.0867*	6.283	50.6	*0.0889*
Initial pH	17.97	22.35	0.133	4.25	23.15	0.13
Concentration of oxygen	24.43	41.34	*0.0982*	5.617	40.43	*0.0993*
Volumetric air flow rate	−0.27	0.0005	0.955	−0.817	0.855	0.525
Inoculum level	3.10	0.666	0.564	0.417	0.222	0.719
Concentration of crude glycerin	30.67	65.13	*0.0785*	7.35	69.23	*0.0761*
Salinity	10.33	7.395	0.224	2.45	7.693	0.22
Concentration of peptone	24.10	40.22	*0.0996*	5.717	41.88	*0.0976*
Turbine speed	23.47	38.136	0.1022	5.55	39.4763	0.10048
Concentration of yeast extract	11.40	9	0.205	2.45	7.693	0.22

Table 9. Second-order polynomial equation coefficient values, with related F-test and P-test results.

Coefficient	Variable	DCW			DHA		
		Coefficient Value	*F*-Test	*P*-Level	Coefficient Value	*F*-Test	*P*-Level
ß0	Constant	57.73	-	-	14.62	-	-
ß1	A—Temperature	6.35	38.14	<0.0001	0.98	20.13	0.0004
ß2	B—Concentration of crude glycerin	2.69	6.84	0.0195	0.0029	0.000179	0.9895
ß3	C—Concentration of oxygen	0.062	0.0037	0.9523	0.70	10.44	0.0056
ß4	D—Concentration of peptone	−0.14	0.018	0.8953	0.011	0.00266	0.9595
ß11	A^2	−3.04	10.01	0.0064	−1.19	33.99	<0.0001
ß22	B^2	0.42	0.19	0.6670	−0.075	0.14	0.7175
ß33	C^2	0.28	0.088	0.7714	−0.26	1.60	0.2253
ß44	D^4	0.26	0.073	0.7909	0.100	0.24	0.6309
ß12	AB	−2.91	5.33	0.0356	−0.11	0.18	0.6808
ß13	AC	−0.28	0.050	0.8262	−0.89	11.09	0.0046
ß14	AD	−0.16	0.015	0.9028	−0.039	0.022	0.8846
ß23	BC	−0.26	0.041	0.8414	−0.047	0.031	0.8828
ß24	BD	0.22	0.030	0.8643	0.047	0.031	0.8628
ß34	CD	−0.056	0.0019	0.9649	0.17	0.40	0.5379
Model		-	4.43	0.0035	-	5.62	0.0010

The correlation between the DCW and DHA values and the parameter values shows that increasing temperature and crude glycerin concentration (up to 27.25 °C and 150 g/dm^3, respectively) caused the DCW to increase as well. The opposite was found to be true at higher values (Figure 3). Raising the oxygen concentration (up to a value of 30%) and temperature (up to a value of 26.37 °C) increased DHA concentration in the biomass. As previously, the trend was reversed at higher values (Figure 4).

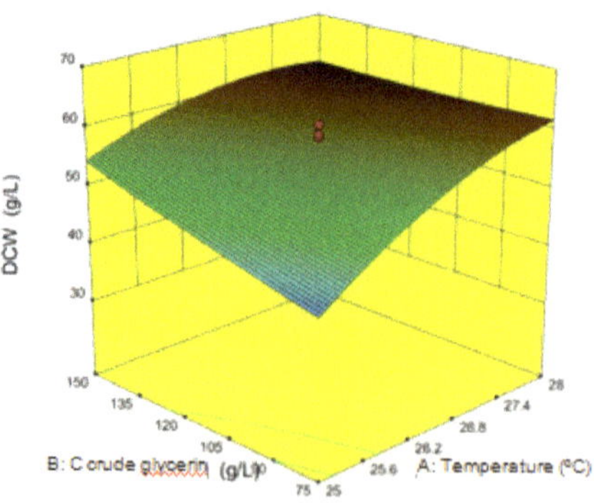

Figure 3. Correlation between the DCW levels and crude glycerin concentration (red points indicate the maximum DCW values).

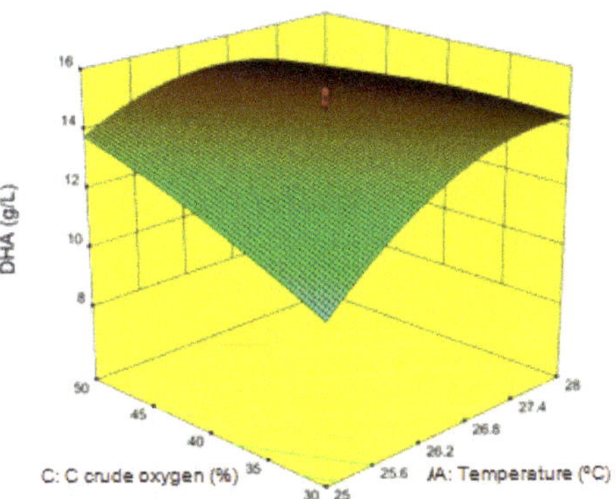

Figure 4. Correlation between the DHA concentration and temperature/oxygen (red points indicate the maximum DHA values).

3.3. Verification of Process Parameter Optimization

Optimization using the response surface methodology (RSM) showed that a culture optimized for DCW concentration could yield DCW levels of 67.55 g/dm^3 (Figure 5). The DCW levels obtained under actual laboratory conditions were 66.69 ± 0.66 g/dm^3, being 1.27% lower than the predicted value. When cell DHA levels are compared between the DHA-optimized culture (17.25 ± 0.33 g/dm^3) and the value predicted using RSM (17.79 g/dm^3), the former is found to be lower by 3.03% (Figure 5).

The culture optimized for high DCW levels produced r_{DCW} of 0.56 ± 0.005 g/dm^3·h and r_{DHA} of 0.12 ± 0.006 g/dm^3·h. While the r_{DCW} value was the same as predicted using RSM (0.56 g/dm^3·h), the r_{DHA} was 7.7% lower than predicted (0.13 g/dm^3·h). The culture optimized for DHA concentration resulted in a r_{DCW} value of 0.51 ± 0.014 g/dm^3·h, whereas the r_{DHA} equaled 0.14 ± 0.003 g/dm^3·h. In this case, r_{DCW} deviated from the predicted value by 3.77%, whereas the r_{DHA} was 6.67% lower (Figure 5).

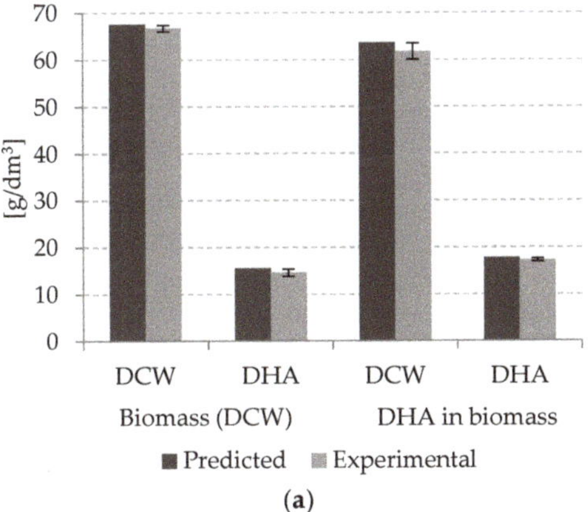

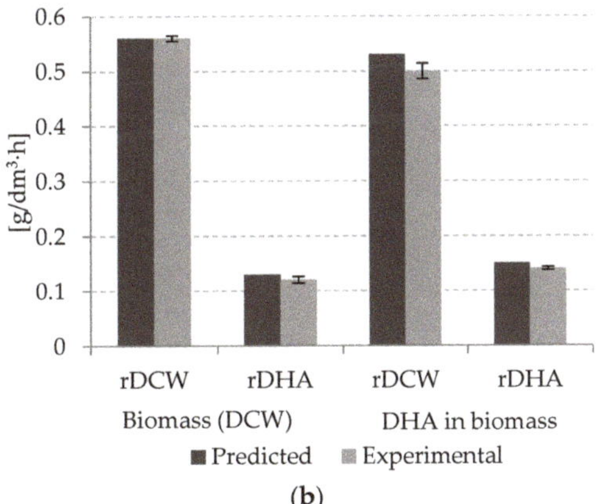

Figure 5. Comparison of (**a**) obtained levels of biomass (DCW) and DHA; and (**b**) biomass growth rate and microalgal DHA production with the parameters set using the response surface methodology and experimental data (by culture purpose).

The experiments confirmed that the significant culture parameters must be adjusted differently depending on whether the culture is intended to produce high biomass growth or high DHA yields. With cultural parameters set for high biomass growth (temperature 27 °C, crude glycerin concentration 150 g/dm^3, oxygen concentration 49.99%, and peptone concentration 9.99 g/dm^3), the dry cell weight was 66.69 ± 0.66 g/dm^3, and the DHA level was 14.60 ± 0.76 g/dm^3. On the other hand, cultural conditions optimized for higher cell DHA levels (temperature 26 °C, crude glycerin concentration 149.99 g/dm^3, oxygen concentration 30%, and peptone concentration 2.21 g/dm^3) produced DCW levels of 61.73 ± 1.70 g/dm^3 and DHA levels of 17.25 ± 0.33 g/dm^3 (Figure 5). These results suggest that a two-step culture system should be employed, with the culture process divided into two phases. The first phase would target a higher cell concentration, while the second, higher DHA yields from the biomass.

4. Discussion

The present experiments are part of larger pioneer research efforts on the use of waste organic substrates to grow heterotrophic microalgae *Schizochytrium* sp. for industrial purposes [47,48]. As these microalgae are able to grow on a variety of carbon sources, they may reasonably be grown on waste glycerol [49]. The biomass of *Schizochytrium* sp. heterotrophic microalgae contains high amounts of DHA and no heavy metals (e.g., mercury), making it highly suitable for use in food supplements, food products, and animal feed [50,51]. To ensure an optimum growth rate of *Schizochytrium* sp. biomass and, above all, to enable scale-up of production for commercial purposes, a variety of factors need to be optimized and controlled for, including physicochemical parameters of the culture, properties of the waste substrate, presence of potential growth inhibitors, substrate availability and cost, culture productivity, investment costs, and the choice of the microalgae strain [52]. Therefore, more advanced research is needed on this issue.

The present study showed that DCW and DHA levels in *Schizochytrium* sp. microalgae are most significantly influenced by: temperature, crude glycerin concentration in the medium, oxygen concentration, and peptone concentration in the bioreactor. A temperature of 26 °C favored DHA production in the cells, whereas higher temperatures promoted biomass growth. These findings are corroborated by the authors of [53,54], who reported that temperature was the primary driver of cell concentration in the culture medium and the accumulation of bioactive compounds in the biomass. The optimal value for producing the highest biomass levels varies with the microalgae strain and its source environment. Many authors [55–58], have noted that fatty acid production in *Chlorella minutissima*, *Pythium irregulare*, and *Crypthecodinium cohnii* cells consistently occurred at low temperatures, as opposed to temperature levels most favorable to biomass growth in process systems. According to [59,60], the increased PUFA content in the investigated microalgae cells at low culture temperatures was linked to increased cell membrane elasticity in the microorganisms, triggered as a defense mechanism. The increase might also be attributed to the higher availability of intracellular molecular oxygen at low temperatures, leading to the activation of oxygen-dependent enzymes that desaturate and elongate PUFAs in cells [61,62].

The present study demonstrated that a crude glycerin concentration of 150 g/dm^3 led to the highest DCW levels at 67.55 g/dm^3, as well as an increase of DHA concentration in *Schizochytrium* sp. cells to 17.25 g/dm^3. Comparable results were obtained by Huang et al. (2012). Their study results indicated that at 100 g/dm^3 glycerol, *Aurantiochytrium limacinum* SR21 achieved a biomass at 61.76 g/dm^3 and DHA concentration at 20.3 g/dm^3 [63]. A lower production was observed by Chi et al. (2007), according to whom the optimal range of crude glycerol concentration is lower to support the growth of *Schizochytrium limacinum* algae and DHA production, and amounts to 75–100 g/dm^3. A highest DHA yield of 4.91 g/dm^3 with 22.1 g/dm^3 DCW was obtained [50]. Moreover, Scott et al. (2011) achieved less efficient results. Using 64.7 g/dm^3 of crude glycerol, they achieved 31.66 g/dm^3 biomass of *Thraustochytrium* sp. ONC T18 and 4.41 g/dm^3 DHA [64]. Other researchers have also noted how the type and level of the external carbon source affected microalgal cultures [65,66]. These authors reported that the carbon source affected biomass growth and might influence PUFA synthesis. Glycerol-based cultures produced higher yields than processes based on glucose, coconut oil, brewery wastes, or wastewater from soymilk production. *Thraustochytriacae* microalgae biomass (*Y*) grown on crude (waste) glycerol was found to produce higher DHA yields (166–550 mg/g) than a pure glycerol-based culture (110–223 mg/g) [50,67]. It is worth noting, however, that the final cell DHA levels are mainly determined by the cell growth phase at the time of DHA extraction.

Previous scientific reports have optimized the oxygen level in microalgae cultures for the production of PUFAs, including DHA. The study described in patent [68] proposed the induction of higher DHA concentrations by lowering oxygen in the growth medium. Oxygen limitation caused a decrease in monounsaturated fatty acids and activated oxygen-dependent PUFA synthase, thus increasing cell DHA and DPA [69,70]. A study by Chi et al. (2009) [71] showed that cultures grown at 10% oxygen exhibited low microalgal growth in favor of an increasing concentration of fatty acids in the biomass, whereas the reverse was true at 50% oxygen. These findings are corroborated by the present results, where the 30% oxygen concentration was found to favor DHA production in the cells, (DCW = 63.59 g/dm^3, C_{DHA} = 17.79 g/dm^3), and the highest biomass concentration was obtained when the system was oxygen-saturated to 50% (DCW = 67.55 g/dm^3, C_{DHA} = 15.53 g/dm^3). This can be explained by the fact that biomass growth necessitates the production of large quantities of primary metabolites, such as nucleic acids, enzymes, and proteins. The production of these compounds requires high oxygen in the process system [71,72].

The dry mass of *Schizochytrium* sp. microalgae contains 14–20% nitrogen (*w*/*w*), mainly incorporated into the structure of proteins and nucleic acids [53,73]. In order to meet cellular nitrogen demand, growth mediums are spiked with various nitrogen sources, such as corn steep, peptone, ammonium sulfate, or mixtures thereof [74,75]. Other alternatives to cheap nitrogen sources include fish waste hydrolysate, silkworm larvae, and wheat bran extract [76]. It is important to note that different nitrogen sources are preferred for different strains to ensure rapid cell growth and the accumulation of metabolites (such as DHA) in technological processes [77,78]. The present study confirms that a 9.99 g/dm^3 dose of peptone is required for an increased concentration of *Schizochytrium* sp. cells in the growth medium, whereas only 2.21 g/dm^3 is needed at the lipid (DHA) accumulation stage. This observation supports the thesis that the generation of new cells and biomass requires a supply of primary metabolite components, such as enzymes, proteins, or nucleic acids. The nitrogen demand is less pronounced in cultures designed for secondary metabolite (Omega-3 acids) synthesis [71,79]. It is worth noting that peptone is a complex nitrogen source, meaning that it contains proteins, peptides, and free amino acids, while being low in carbohydrates, lipids, inorganic ions, vitamins, and growth factors. In addition to its primary function of supplying nitrogen to cell biomass, peptone also promotes overall cell development, granting an advantage over industrial nitrogen sources [80,81].

5. Conclusions

The experiments presented in this study showed the high potential of using the waste fraction of glycerol from biodiesel production to produce biomass and DHA by *Schizochytrium* sp. microalgae. In the course of designing and executing the study, as well as analyzing the results of optimization and experimental work, the theses made prior to the experiments were verified.

In the course of the study, parameters having a significant effect on *Schizochytrium* sp. cell growth and DHA accumulation were screened using the Plackett–Burman design and quantified using the response surface methodology.

The study has shown that the growth and DHA production of *Schizochytrium* sp. microalgae was most significantly influenced by the following physicochemical parameters: crude glycerin concentration in the growth medium (150 g/dm^3), process temperature (27 °C), oxygen concentration in the bioreactor (49.99% *v*/*v*), and peptone concentration as a source of nitrogen (9.99 g/dm^3). Other parameters, such as initial pH, volumetric air flow rate, inoculum level, salinity, turbine speed, and concentration of yeast extract, were not statistically significant and were therefore eliminated from further analyses.

The optimization methods used in the study identified the following physicochemical parameter values as optimal for producing high DHA concentrations in the biomass:

temperature 26 °C, crude glycerin concentration 149.99 g/dm^3, oxygen concentration 30% (v/v), and peptone concentration 2.21 g/dm^3.

Laboratory verification produced DCW levels of 66.69 ± 0.66 g/dm^3, i.e., 1.27% lower than the predicted value. When cell DHA levels are compared between the DHA-optimized culture (17.25 ± 0.33 g/dm^3) and the value predicted using the response surface methodology (17.79 g/dm^3), the former is found to be lower by 3.03%.

The results suggest that a two-step culture system should be employed, with the culture process divided into two phases. The first phase would target higher cell concentration, while the second, higher DHA yields from the biomass.

Author Contributions: Conceptualization, N.K., S.T. and M.D.; Data curation, M.D.; Formal analysis, M.D. and M.Z.; Funding acquisition, M.D., J.K. and M.Z.; Investigation, N.K., S.T. and M.D.; Methodology, N.K. and S.T.; Project administration, M.Z.; Resources, J.K.; Software, N.K. and S.T.; Validation, M.Z.; Visualization, M.D. and J.K.; Writing—original draft, M.D. and J.K.; Writing—review & editing, M.D., J.K. and M.Z. All authors have read and agreed to the published version of the manuscript.

Funding: The manuscript was supported by "Project financially co-supported by Minister of Science and Higher Education in the range of the program entitled "Regional Initiative of Excellence" for the years 2019–2022, Project No. 010/RID/2018/19, amount of funding 12,000,000 PLN" and the work WZ/WBiIŚ/2/2019, funded by the Ministry of Science and Higher Education.

Institutional Review Board Statement: Not applicable.

Informed Consent Statement: Not applicable.

Conflicts of Interest: The authors declare no conflict of interest.

Abbreviations

DHA	docosahexaenoic acid
DCW	dry cell weight
RSM	response surface methodology
ATCC	American Type Culture Collection
MONG	Matter Organic Non Glycerol
DO	dissolved oxygen
ATP	adenosine 5′-triphosphate
ADP	adenosine 5′-diphosphate
NADH	nicotinamide adenine dinucleotide
FAME	fatty acid methyl esters
FID	flame ionization detector
r_{DHA}	the rate of DHA production by microalgae
r_{DCW}	the growth rate of the microalgal biomass
$E_{(xi)}$	effects of each variable
M_{i+}	DCW or DHA concentration for the high (+) level of the variable
M_{i-}	DCW or DHA concentration for the low (−) level of the variable
N	number of runs
SS	sum of the squares of the effects of each variable
Y	the predicted response of the design
ß	coefficient of the equation
x_i, x_j	coded levels of parameters i and j

References

1. Stiles, W.A.V.; Styles, D.; Chapman, S.P.; Esteves, S.; Bywater, A.; Melville, L.; Silkina, A.; Lupatsch, I.; Fuentes, C.; Lovitt, R.; et al. Using microalgae in the circular economy to valorise anaerobic digestate: Challenges and opportunities. *Bioresour. Technol.* **2018**, *267*, 732–742. [CrossRef]
2. SundarRajan, P.; Gopinath, K.P.; Greetham, D.; Antonysamy, A.J. A review on cleaner production of biofuel feedstock from integrated CO2 sequestration and wastewater treatment system. *J. Clean. Prod.* **2019**, *210*, 445–458. [CrossRef]
3. Barsanti, L.; Gualtieri, P. Is exploitation of microalgae economically and energetically sustainable? *Algal Res.* **2018**, *31*, 107–115. [CrossRef]

4. Bhalamurugan, G.L.; Valerie, O.; Mark, L. Valuable bioproducts obtained from microalgal biomass and their commercial applications: A review. *Environ. Eng. Res.* **2018**, *23*, 229–241. [CrossRef]
5. Vassilev, S.V.; Vassileva, C.G. Composition, properties and challenges of algae biomass for biofuel application: An overview. *Fuel* **2016**, *181*, 1–33. [CrossRef]
6. Okoro, V.; Azimov, U.; Munoz, J.; Hernandez, H.H.; Phan, A.N. Microalgae cultivation and harvesting: Growth performance and use of flocculants—A review. *Renew. Sustain. Energy Rev.* **2019**, *115*, 109364. [CrossRef]
7. Ende, S.; Noke, A. Heterotrophic microalgae production on food waste and by-products. *J. Appl. Phycol.* **2019**, *31*, 1565–1571. [CrossRef]
8. Osundeko, O.; Ansolia, P.; Gupta, S.K.; Bag, P.; Bajhaiya, A.K. Promises and Challenges of Growing Microalgae in Wastewater. In *Water Conservation, Recycling and Reuse: Issues and Challenges*; Singh, R., Kolok, A., Bartelt-Hunt, S., Eds.; Springer: Singapore, 2019. [CrossRef]
9. Silkina, A.; Zacharof, M.P.; Ginnever, N.E.; Gerardo, M.; Lovitt, R.W. Testing the Waste Based Biorefinery Concept: Pilot Scale Cultivation of Microalgal Species on Spent Anaerobic Digestate Fluids. *Waste Biomass Valor* **2020**, *11*, 3883–3896. [CrossRef]
10. Chandra, R.; Iqbal, H.M.N.; Vishal, G.; Lee, H.S.; Nagra, S. Algal biorefinery: A sustainable approach to valorize algal-based biomass towards multiple product recovery. *Bioresour. Technol.* **2019**, *278*, 346–359. [CrossRef]
11. Liu, X.; Zhu, F.; Zhang, R.; Zhao, L.; Qi, J. Recent progress on biodiesel production from municipal sewage sludge. *Renew. Sustain. Energy Rev.* **2021**, *135*. [CrossRef]
12. Sondhi, S.; Kumar, P. Biodiesel from sewage sludge: An alternative to diesel. In *Recent Trends in Biotechnology*; MedDocs: Punjab, India, 2020; pp. 1–6.
13. Renewables 2019 Global Status Report [Chapter 3]. 2019. Available online: https://www.ren21.net/gsr-2019 (accessed on 6 December 2020).
14. Manara, P.; Zabaniotou, A. Co-valorization of crude glycerol waste streams with conventional and/or renewable fuels for power generation and industrial symbiosis perspectives. *Waste Biomass Valoriz.* **2016**, *7*, 135–150. [CrossRef]
15. Nomanbhay, S.; Ong, M.Y.; Chew, K.W.; Show, P.L.; Chen, W.H. Organic carbonate production utilizing crude glycerol derived as by-product of biodiesel production: A review. *Energies* **2020**, *13*, 1483. [CrossRef]
16. Seretis, A.; Tsiakaras, P. Hydrogenolysis of glycerol to propylene glycol by in situ produced hydrogen from aqueous phase reforming of glycerol over SiO_2–Al_2O_3 supported nickel catalyst. *Fuel Process. Technol.* **2016**, *142*, 135–146. [CrossRef]
17. Kaur, J.; Sarma, A.K.; Jha, M.K.; Gera, P. Valorisation of crude glycerol to value-added products: Perspectives of process technology, economics and environmental issues. *Biotechnol. Rep.* **2020**, *27*. [CrossRef]
18. Dang, C.; Wu, S.; Cao, Y.; Wang, H.; Peng, F.; Yu, H. Co-production of high quality hydrogen and synthesis gas via sorption-enhanced steam reforming of glycerol coupled with methane reforming of carbonates. *Chem. Eng. J.* **2019**, *360*, 47–53. [CrossRef]
19. Okhlopkova, E.A.; Serafimov, L.A.; Frolkova, A.V. Methods of Preparing Epichlorohydrin. *Theor. Found. Chem. Eng.* **2019**, *53*, 864–870. [CrossRef]
20. Chiosso, M.E.; Casella, M.L.; Merlo, A.B. Synthesis and catalytic evaluation of acidic carbons in the etherification of glycerol obtained from biodiesel production. *Catal. Today* **2020**. [CrossRef]
21. Bouriakova, A.; Mendes, P.S.F.; Katryniok, B.; De Clercq, J.; Thybaut, J.W. Co-metal induced stabilization of alumina-supported copper: Impact on the hydrogenolysis of glycerol to 1,2-propanediol. *Catal. Commun.* **2020**, *146*. [CrossRef]
22. Wang, X.L.; Zhou, J.J.; Shen, J.T.; Zheng, Y.F.; Sun, Y.; Xiu, Z.L. Sequential fed-batch fermentation of 1,3-propanediol from glycerol by *Clostridium butyricum* DL07. *Appl. Microbiol. Biotechnol.* **2020**, *104*, 9179–9191. [CrossRef] [PubMed]
23. Yue, L.; Cheng, J.; Hua, J.; Dong, H.; Zhou, J.; Li, Y.-Y. Improving fermentative methane production of glycerol trioleate and food waste pretreated with ozone through two-stage dark hydrogen fermentation and anaerobic digestion. *Energy Convers. Manag.* **2020**, *203*. [CrossRef]
24. Yu, K.O.; Jung, J.; Kim, S.W.; Park, C.H.; Han, S.O. Synthesis of FAEEs from glycerol in engineered *Saccharomyces cerevisiae* using endogenously produced ethanol by heterologous expression of an unspecific bacterial acyltransferase. *Biotechnol. Bioeng.* **2011**, *109*, 110–115. [CrossRef] [PubMed]
25. Bindea, M.; Rusu, B.; Rusu, A.; Trif, M.; Leopold, L.F.; Dulf, F.; Vodnar, D.C. Valorification of crude glycerol for pure fractions of docosahexaenoic acid and β-carotene production by using *Schizochytrium limacinum* and *Blakeslea trispora*. *Microb. Cell Fact.* **2018**, *17*, 97. [CrossRef]
26. Pyle, D.; Garcia, R.; Wen, Z. Producing docosahexaenoic acid (DHA)-rich algae from biodiesel- derived crude glycerol: Effects of impurities on DHA production and algal biomass composition. *J. Agric. Food Chem.* **2008**, *56*, 3933–3939. [CrossRef] [PubMed]
27. Quilodrán, B.; Cortinez, G.; Bravo, A.; Silva, D. Characterization and comparison of lipid and PUFA production by native thraustochytrid strains using complex carbon sources. *Heliyon* **2020**, *6*. [CrossRef] [PubMed]
28. Zárate, R.; el Jaber-Vazdekis, N.; Tejera, N.; Pérez, J.A.; Rodríguez, C. Significance of long chain polyunsaturated fatty acids in human health. *Clin. Transl. Med.* **2017**, *6*. [CrossRef]
29. Yamagata, K. Dietary docosahexaenoic acid inhibits neurodegeneration and prevents stroke. *J. Neurosci. Res.* **2020**. [CrossRef]
30. Watanabe, Y.; Tatsuno, I. Omega-3 polyunsaturated fatty acids focusing on eicosapentaenoic acid and docosahexaenoic acid in the prevention of cardiovascular diseases: A review of the state-of-the-art. *Expert Rev. Clin. Pharmacol.* **2021**. [CrossRef] [PubMed]

31. Hei, A. Fish as a Functional Food in Human Health, Diseases and Well-Being. Current status of researches in fish and fisheries, today and tomorrow printers and publishers. In Proceedings of the 107th Indian Science Congress, Banglore, India, 3–7 January 2020.
32. Colombo, S.M.; Rodgers, T.F.M.; Diamond, M.L.; Bazinet, R.P.; Arts, M.T. Projected declines in global DHA availability for human consumption as a result of global warming. *Ambio* **2020**, *49*, 865–880. [CrossRef]
33. Yadav, A.K.; Rossi, W.; Habte-Tsion, H.-M.; Kumar, V. Impacts of dietary eicosapentaenoic acid (EPA) and docosahexaenoic acid (DHA) level and ratio on the growth, fatty acids composition and hepatic-antioxidant status of largemouth bass (Micropterus salmoides). *Aquaculture* **2020**, *529*. [CrossRef]
34. Tan, K.; Ma, H.; Li, S.; Zheng, H. Bivalves as future source of sustainable natural omega-3 polyunsaturated fatty acids. *Food Chem.* **2020**. [CrossRef]
35. Wang, Q.; Sen, B.; Liu, X.; He, Y.; Xie, Y.; Wang, G. Enhanced saturated fatty acids accumulation in cultures of newly-isolated strains of *Schizochytrium* sp. and *Thraustochytriidae* sp. for large-scale biodiesel production. *Sci. Total Environ.* **2018**, *631–632*, 994–1004. [CrossRef]
36. Thyagarajan, T.; Puri, M.; Vongsvivut, J.; Barrow, C. Evaluation of bread crumbs as a potential carbon source for the growth of thraustochytrid species for oil and omega-3. *Nutrients* **2014**, *6*, 2104–2114. [CrossRef]
37. Ryu, B.; Kim, K.; Kim, J.; Han, J.; Yang, J. Use of organic waste from brewery industry for high-density cultivation of docosahexaenoic acid-rich microalga Aurantochytrium sp. KRS101. *Bioresour. Technol.* **2012**, *129*, 351–359. [CrossRef] [PubMed]
38. Hong, W.; Yu, A.; Heo, S.; Oh, B.; Kim, C.; Sohn, J.; Yang, J.W.; Kondo, A.; Seo, J.W. Production of lipids containing high levels of docosahexaenoic acid from empty palm fruit bunches by Aurantiochytrium sp. KRS101. *Bioprocess Biosyst. Eng.* **2013**, *36*, 959–963. [CrossRef] [PubMed]
39. Unagul, P.; Assantachai, C.; Phadungruengluij, S.; Suphantharika, M.; Tanticharoen, M.; Verduyn, C. Coconut water as a medium additive for the production of docosahexaenoic acid (C22:6 n3) by Schizochytrium mangrovei Sk-02. *Bioresour. Technol.* **2007**, *98*, 281–287. [CrossRef] [PubMed]
40. Fan, K.W.; Chen, F.; Jones, E.B.G.; Vrijmoed, L.L.P. Eicosapentaenoic and docosahexaenoic acids production by and okara-utilizing potential of thraustochytrids. *J. Ind. Microbiol. Biotechnol.* **2001**, *27*, 199–202. [CrossRef] [PubMed]
41. Quilodran, B.; Hinzpeter, I.; Quiroz, A.; Shene, C. Evaluation of liquid residues from beer and potato processing for the production of docosahexaenoic acid (C22:6n-3, DHA) by native thraustochytrid strains. *World J. Microbiol. Biotechnol.* **2009**, *25*, 2121–2128. [CrossRef]
42. Liang, Y.; Sarkany, N.; Cui, Y.; Yesuf, J.; Trushenski, J.; Blackburn, J.W. Use of sweet sorghum juice for lipid production by *Schizochytrium limacinum* SR21. *Bioresour. Technol.* **2010**, *101*, 3623–3627. [CrossRef]
43. Bwapwa, J.K.; Anandraj, A.; Trois, C. Possibilities for conversion of microalgae oil into aviation fuel: A review. *Renew. Sustain. Energy Rev.* **2017**, *80*, 1345–1354. [CrossRef]
44. Nazir, Y.; Shuib, S.; Kalil, M.S.; Song, Y.; Hamid, A.A. Optimization of Culture Conditions for Enhanced Growth, Lipid and Docosahexaenoic Acid (DHA) Production of *Aurantiochytrium* SW1 by Response Surface Methodology. *Sci. Rep.* **2018**, *8*, 8909. [CrossRef]
45. Chang, G.; Gao, N.; Tian, G.; Wu, Q.; Chang, M.; Wang, X. Improvement of docosahexaenoic acid production on glycerol by *Schizochytrium* sp. S31 with constantly high oxygen transfer coefficient. *Bioresour. Technol.* **2013**, *142*, 400–406. [CrossRef]
46. Grayburn, W.; Collins, G.; Hildebrand, D. Fatty acid alteration by a h-9 desaturase in transgenic tabacco tissue. *Biotechnology* **1992**, *10*, 675–678.
47. Hakim, A. Potential od heterotrophic microalgae (*Schizochytrium* sp.) as a source of DHA. *Squalen* **2012**, *7*, 29–38.
48. Lee, C.; Nichols, C.; Blackburn, S.; Dunstan, G.; Koutoulis, A.; Nichols, P. Comparison of *Thraustochytrids: Aurantochytrium* sp., *Schizochytrium* sp., *Thraustochytrium* sp. and *Ulkenia* sp. for production of biodiesel long-chain omega-3 oils and exopolysaccharide. *Mar. Biotechnol.* **2014**, *16*, 396–411.
49. Mata, T.; Santosa, J.; Mendesa, A.; Caetanoa, N.; Martinsc, A. Sustainability Evaluation of Biodiesel Produced from Microalgae *Chlamydomonas* sp. grown in brewery wastewater. *Chem. Eng. Trans.* **2014**, *37*, 823–828.
50. Chi, Z.; Pyle, D.; Wen, Z.; Frear, C.; Chen, S. A laboratory study of producing docosahexaenoic acid form biodiesel-waste glycerol by microalgal fermentation. *Process Biochem.* **2007**, *42*, 1537–1545. [CrossRef]
51. Christien, E.; Matthias, P.; Maria, B.; Lolke, S. *Microalgae-Based Products for the Food and Feed Sector: An Outlook for Europe*; European Commission, Joint Research Centre, Institute for Prospective Technological Studies: Luxemburg, 2014.
52. Diwan, B.; Parkhey, P.; Gupta, P. From agro-industrial wastes to single cell oils: A step towards prospective biorefinery. *Folia Microbiol.* **2018**, *63*, 547–568. [CrossRef] [PubMed]
53. Wen, Z.; Chen, F. Application of statistically-based experimental designs for the optimization of eicosapentaenoic acid production by the diatom Nitschia laevis. *Biotechnol. Bioeng.* **2001**, *75*, 159–169. [CrossRef] [PubMed]
54. Fagundes, M.B.; Alvarez-Rivera, G.; Vendruscolo, R.G.; Voss, M.; Arrojo da Silva, P.; Barin, J.S.; Jacob-Lopes, E.; Zepka, L.Q.; Wagner, R. Green microsaponification-based methodology followed by gas chromatography for cyanobacterial sterol and squalene determination. *Talanta* **2020**, 121793. [CrossRef]
55. Jiang, Y.; Chen, F. Effects of temperature and temperature shift on doc-osahexaenoic acid production by the marine microalga Crypthecodinium cohnii. *J. Am. Oil Chem. Soc.* **2000**, *77*, 613–617. [CrossRef]

56. Seto, A.; Wang, H.; Hesseltine, C. Culture conditions affect eicosapentaenoic acid content of *Chlorella minutissima*. *J. Am. Oil Chem. Soc.* **1984**, *61*, 892–894. [CrossRef]
57. Stinson, E.; Kwoczak, R.; Kurantz, M. Effect of culture conditions on production of eicosapentaenoic acid by *Pythium irregulare*. *J. Ind. Microbiol.* **1991**, *8*, 171–178. [CrossRef]
58. Sun, X.-M.; Ren, L.-J.; Zhao, Q.-Y.; Ji, X.-J.; Huang, H. Enhancement of lipid accumulation in microalgae by metabolic engineering. Biochim. Biophys. Acta-Mol. *Cell Biol. Lipids* **2019**, *1864*, 552–566. [CrossRef]
59. Richmond, A. Mico alga culture. *Crit. Rev. Biotechnol.* **1986**, *4*, 368–438.
60. Wang, Y.; He, B.; Sun, Z.; Chen, Y.-F. Chemically enhanced lipid production from microalgae under low sub-optimal temperature. *Algal Res.* **2016**, *16*, 20–27. [CrossRef]
61. Higashiyama, K.; Murakami, K.; Tsujimura, H.; Matsumot, N.; Fujikawa, S. Effects of dissolved oxygen on the morphology of an arachidonic acid production by *Mortierella alpina* 1S-4. *Biotechnol. Bioeng.* **1999**, *63*, 442–448. [CrossRef]
62. Shi, K.; Gao, Z.; Shi, T.-Q.; Song, P.; Ren, L.-J.; Huang, H.; Ji, X.-J. Reactive Oxygen Species-Mediated Cellular Stress Response and Lipid Accumulation in Oleaginous Microorganisms: The State of the Art and Future Perspectives. *Front. Microbiol.* **2017**, *8*. [CrossRef]
63. Huang, T.Y.; Lu, W.C.; Chu, I.M. A fermentation strategy for producing docosahexaenoic acid in *Aurantiochytrium limacinum* SR21 and increasing C22:6 proportions in total fatty acid. *Bioresour. Technol.* **2012**, *123*, 8–14. [CrossRef] [PubMed]
64. Scott, S.D.; Armenta, R.E.; Berryman, K.T.; Norman, A.W. Use of raw glycerol to produce oil rich in polyunsaturated fatty acids by a thraustochytrid. *Enzym. Microb. Technol.* **2011**, *48*, 267–272. [CrossRef] [PubMed]
65. Ethier, S.; Woisard, K.; Vaughan, D.; Wen, Z. Continous culture of the microalgae *Schizochytrium* limacinum na biodiesel-derived crude glycerol for producing docosahexaenoic acid. *Bioresour. Technol.* **2010**, *102*, 88–93. [CrossRef]
66. Lage, S.; Kudahettige, N.P.; Ferro, L.; Matsakas, L.; Funk, C.; Rova, U.; Gentili, F.G. Microalgae Cultivation for the Biotransformation of Birch Wood Hydrolysate and Dairy Effluent. *Catalysts* **2019**, *9*, 150. [CrossRef]
67. Pooksawang, N.; Nangtharat, S.; Yunchalard, S. Batch fermentation of marine microalgae using glycerol obtained from biodiesel plant for docosahexaenoic acid (DHA) production. In Proceedings of the 3rd International Conference on Fermentation Technology for Value-Added Agriculture Production, Khon Kaen, Thailand, 26–28 August 2009; Volume 26, pp. 1–7.
68. Bailey, R.; DiMasi, D.; Hansen, J.; Mirrasoul, P.; Ruecker, C.; Kaneko, T.; Barclay, W. Enhanced Production of Lipids Containing Polyenoic Fatty Acid by Very High Density Cultures of Eukaryotic Microbes in Fermenters. U.S. Patent 20010046691, 19 August 2003.
69. Jakobsen, A.; Aasen, I.; Strom, A. Endogenously synthesized (-)-proto-quercitol and glycine betaine are principal compatible solutes fo *Schizochytrium* sp. strain S8 (ATCC 20889) and three new isolates of phylogenetically related thraustochytrids. *Appl. Environ. Microbiol.* **2007**, *73*, 5848–5856. [CrossRef]
70. Morabito, C.; Bournaud, C.; Maes, C.; Schuler, M.; Aiese Cigliano, R.; Dellero, Y.; Marechal, E.; Amato, A.; Rebeille, F. The lipid metabolism in thraustochytrids. *Prog. Lipid Res.* **2019**, *76*, 101007. [CrossRef]
71. Chi, Z.; Liu, Y.; Frear, C.; Chen, S. Study of a two-stage growth of DHA-producing marine algae *Schizochytrium limacinum* SR21 with shifting dissolved oxygen level. *Appl. Microbiol. Biotechnol.* **2009**, *81*, 1141–1148. [CrossRef] [PubMed]
72. Coronado-Reyes, J.A.; Salazar-Torres, J.A.; Juárez-Campos, B.; González-Hernández, J.C. *Chlorella vulgaris*, a microalgae important to be used in Biotechnology: A review. *Food Sci. Technol.* **2020**. [CrossRef]
73. Chen, G.; Fan, K.-W.; Lu, F.-P.; Li, Q.; Aki, T.; Chen, F.; Jiang, Y. Optimization of nitrogen source for enhanced production of squalene from thraustochytrid *Aurantiochytrium* sp. *New Biotechnol.* **2010**, *27*, 382–389. [CrossRef]
74. Yaguchi, T. Production of high yields of docosahexaenoic acid by *Schizochytrium* sp. strain SR21. *J. Am. Oil Chem. Soc.* **1997**, *74*, 1431–1434. [CrossRef]
75. Rawoof, S.A.A.; Kumar, P.S.; Vo, D.-V.N.; Devaraj, K.; Mani, Y.; Devaraj, T.; Subramanian, S. Production of optically pure lactic acid by microbial fermentation: A review. *Environ. Chem. Lett.* **2020**. [CrossRef]
76. Kiros, H.; Zong, J.P.; Li, D.X.; Liu, C.; Lu, X.H. Anaerobic co-digestion process for biogas production: Progress, challenges and perspectives. *Renew. Sust. Energy Rev.* **2017**, *76*, 1485–1496.
77. Fan, K.; Chen, F. Production of high value products by marine microalgae thraustochytrids. In *Bioprocessing for Value-Added Products from Renewable Resources: New Technologies and Applications*; Yang, S., Ed.; Elsevier: Amsterdam, The Netherlands, 2006; pp. 293–324.
78. Chen, X.; He, Y.; Ye, H.; Xie, Y.; Sen, B.; Jiao, N.; Wang, G. Different carbon and nitrogen sources regulated docosahexaenoic acid (DHA) production of *Thraustochytriidae* sp. PKU#SW8 through a fully functional polyunsaturated fatty acid (PUFA) synthase gene (pfaB). *Bioresour. Technol.* **2020**, *318*, 124273. [CrossRef]
79. Li, S.; Hu, Z.; Yang, X.; Li, Y. Effect of Nitrogen Sources on Omega-3 Polyunsaturated Fatty Acid Biosynthesis and Gene Expression in *Thraustochytriidae* sp. *Mar. Drugs* **2020**, *18*, 612. [CrossRef] [PubMed]
80. Orak, T.; Caglar, O.; Ortucu, S.; Ozkan, H.; Taskin, M. Chicken feather peptone: A new alternative nitrogen source for pigment production by Monascus purpureus. *J. Biotechnol.* **2018**, *271*, 56–62. [CrossRef] [PubMed]
81. Xiong, Z. *Fermentation: Process and Theory*; China Medical Science and Technology Press: Beijing, China, 1995.

Article

An Easily Accessible Microfluidic Chip for High-Throughput Microalgae Screening for Biofuel Production

Shubhanvit Mishra [1], Yi-Ju Liu [2], Chi-Shuo Chen [3,*] and Da-Jeng Yao [1,4,5,*]

1 Institute of Nano Engineering and MicroSystems, National Tsing Hua University, Hsinchu 30013, Taiwan; shubhanvit@gmail.com
2 Food Industry Research and Development Institute, Hsinchu 300193, Taiwan; lyj20@firdi.org.tw
3 Department of Biomedical Engineering and Environmental Sciences, National Tsing Hua University, Hsinchu 30013, Taiwan
4 Department of Power Mechanical Engineering, National Tsing Hua University, Hsinchu 30013, Taiwan
5 Department of Engineering and System Science, National Tsing Hua University, Hsinchu 30013, Taiwan
* Correspondence: chen.cs@mx.nthu.edu.tw (C.-S.C.); djyao@mx.nthu.edu.tw (D.-J.Y.)

Citation: Mishra, S.; Liu, Y.-J.; Chen, C.-S.; Yao, D.-J. An Easily Accessible Microfluidic Chip for High-Throughput Microalgae Screening for Biofuel Production. *Energies* **2021**, *14*, 1817. https://doi.org/10.3390/en14071817

Academic Editor: José Carlos Magalhães Pires

Received: 28 February 2021
Accepted: 20 March 2021
Published: 24 March 2021

Abstract: Microalgae are important green energy resources. With high efficiency in fixing carbon dioxide, microalgae are broadly applied for biofuel production. Integrating various cultivation parameters, we applied ultraviolet (UV) mutagenesis, one of the most common approaches, to induce genomic mutation in microalgae and thus enhance the production of lipid content, but the screening process is convoluted and labor-intensive. In this study, we aimed to develop an accessible microfluidic platform to optimize the biofuel production of microalgae. Instead of traditional lithography, we designed hanging-drop microfluidic chips that were fabricated using a cheap computer numerical control (CNC) micro-milling technique. On each chip, we cultured in parallel *Botryococcus braunii*, one of the most common freshwater microalgae for biofuel production, in sets of ten separated hanging drops (~30 μL each); we monitored their growth in each individual drop for more than 14 days. To optimize the culturing conditions, using drops of varied diameter, we first identified the influence of cell density on algae growth and lipid production. After introducing UV-induced random mutations, we quantified the lipid content of the microalgae *in situ*; the optimized UV-C dosage was determined accordingly. In comparison with wild-type *B. braunii*, the results showed increased biomass growth (137%) and lipid content (149%) of the microalgae mutated with the desired UV process. Moreover, we showed a capacity to modulate the illumination on an addressed chip area. In summary, without using an external pump system, we developed a hanging-drop microfluidic system for long-term microalgae culturing, which can be easily operated using laboratory pipettes. This microfluidic system is expected to facilitate microalgae mutation breeding, and to be applied for algae cultivation optimization.

Keywords: biofuel; microfluidic; microalgae; UV mutagenesis

1. Introduction

Global warming, caused by CO_2 emissions and the depletion of fossil fuels, has necessitated the search for sustainable and renewable energy sources [1]. In the past decade, biofuel has attracted great attention among researchers as a promising renewable energy source as it can be used to replace petrol and diesel while remaining carbon neutral [2,3]. Biodiesel can be produced from oil-containing crops [4]. As oil-containing crops require large land areas, the generation of biodiesel from high-oil-containing crops can, however, be the cause other problems, such as food scarcity [5]. Further, the use of farm land for biofuel production can cause increased emissions of greenhouse gases, which are emitted during the land use changes [6]. Under these circumstances, microalgae can be considered as a promising feedstock for biodiesel production, as they have a greater growth rate and lipid content than agricultural crops [3]. *Botryococcus braunii* is a green alga that is characterized by its ability to produce hydrocarbon-rich oil up to 61% of its dry mass [7–9]. *B. braunii* has attracted substantial interest as a potential species for the production of biofuels [1]; when the extracted hydrocarbon-rich oil from *B. braunii* is cracked, which is a process of catalytic and thermal degradation of long-chain hydrocarbons to form usable shorter-chain hydrocarbons, it yields petrol 67%, aviation fuel 15% and diesel distillate 15%, which proportions are similar to the gasoil fraction of crude oil; it is hence considered to be an effective candidate for third-generation biofuels.

Various microalgae studies have been conducted on culturing the cells in laboratory-scale flasks, open-raceway ponds or closed photobioreactors [10–12]. These studies significantly contribute to the understanding of basic algal biology and the selection of strains for oil production, and explore various culture parameters such as light intensity, light cycle [13], temperature [14], nutrient concentration [15], CO_2, and pH [13]. In the development of any optimization of biofuel production, two limitations can hamper the progress. Firstly, because of intensive labor requirements and high operating costs, high-throughput screening is difficult using conventional photobioreactors [16]. Secondly, in conventional photobioreactors, it is challenging to maintain the culture conditions in a consistent and precise manner [16]. In recent decades, many microfluidic-based techniques have been developed that are capable of providing well-controlled culture conditions to optimize algal growth and lipid productivity [17]. For instance, recent studies have demonstrated the usefulness of devices with microchambers in the study of bacteria, as in the following: algae that are used in a micro-scale bioreactor for the automated culture and density analysis of microorganisms [18]; a fully automatic and programmable microfluidic device that enables parallel culture of cells [19]; the cultivation of microalgae in a droplet-based condition for the rapid screening of algal growth [20–22]; and simple chamber-based microfluidic devices for the quantification of lipid content in microalgae under varied stress conditions [23]. These studies helped to identify optimal conditions for microbial growth [19,20,23].

In this work, we aimed to develop an easily accessible platform for the high-throughput screening of microalgae with rapid growth and a high lipid content. In addition, UV mutagenesis and various modulations of culture parameters, such as cell density and intensity of illumination, were integrated into this system to achieve culture optimization for the production of microalgae biofuel.

2. Materials and Methods

2.1. Microalgae Culture

Botryococcus braunii (AL20016, UTEX 2441) was supplied by Bioresource Collection and Research Centre (BCRC) in the Food Industry Research and Development Institute (FIRDI), Taiwan. It was cultivated in Bold-3N medium, which was prepared with double-distilled water (ddH2O) and various inorganic salts were added as per standard protocol [24]. Both experimental and stock cultures were kept in a biochemical incubator at 20 ± 1 °C, with a 12:12 h light/dark cycle (intensity of 1500 lux) provided by Warm White LED (WISVA OPTOELECTRONICS HH-S60F010-5050-12 WW) with a regulated intensity control switch. The cell cultures were shaken three times a week without aeration. Culture stock was maintained in 20 mL volume, and the cell density reached approximately 1.8×10^5 cells/mL in 14 days.

2.2. Chip Design and Fabrication

Microfluidic hanging-drop devices were made using a computer numerical control (CNC) micro-milling system (Roland MDR-40A) as described elsewhere [25]. The devices were composed of two polystyrene layers (Figure 1A) of 25 × 75 mm. The thickness of the upper layer was 1.5 mm, and it was used to make the inlet and outlet; the lower layer was also made of polystyrene which was 2 mm thick and consisted of a channel (thickness: 1200 µm) with opening wells at the bottom (Figure 1A). The chip used to study the effect of cell density on cell growth consisted of 13 × 4 opening wells with diameters of 1.1, 1.5 and 1.8 mm. For the other tests, the chip consisted of 13 × 4 opening wells of diameter 1.1 mm.

For the assembly of the hanging-drop device, the fabricated layer was first immersed in ethanol solution (75%) and cleaned with an ultrasonic cleaner for 30 min to remove any residual particles formed during fabrication. After cleaning, both the upper and bottom layers were assembled using adhesive tape (0.05 mm, 300LSE, 3 M) which was cut using a laser-cutting instrument (C180II GCC, LaserPro). The outlet was made with the upper part of a pipette tip (1 mL, diameter 4.5 mm, height 5 mm). The outlet was bonded with poly(dimethylsiloxane) (PDMS) and cured at 60 °C for 90 min. The base stand for the chips were also made from PDMS, cast using a laser-cut polystyrene mold, and cured at 90 °C for 60 min. To study the effect of light on the growth, we placed this chip on a holder stand made of black opaque polystyrene sheet (thickness 3 mm) to block the light in the desired region (Figure 1A).

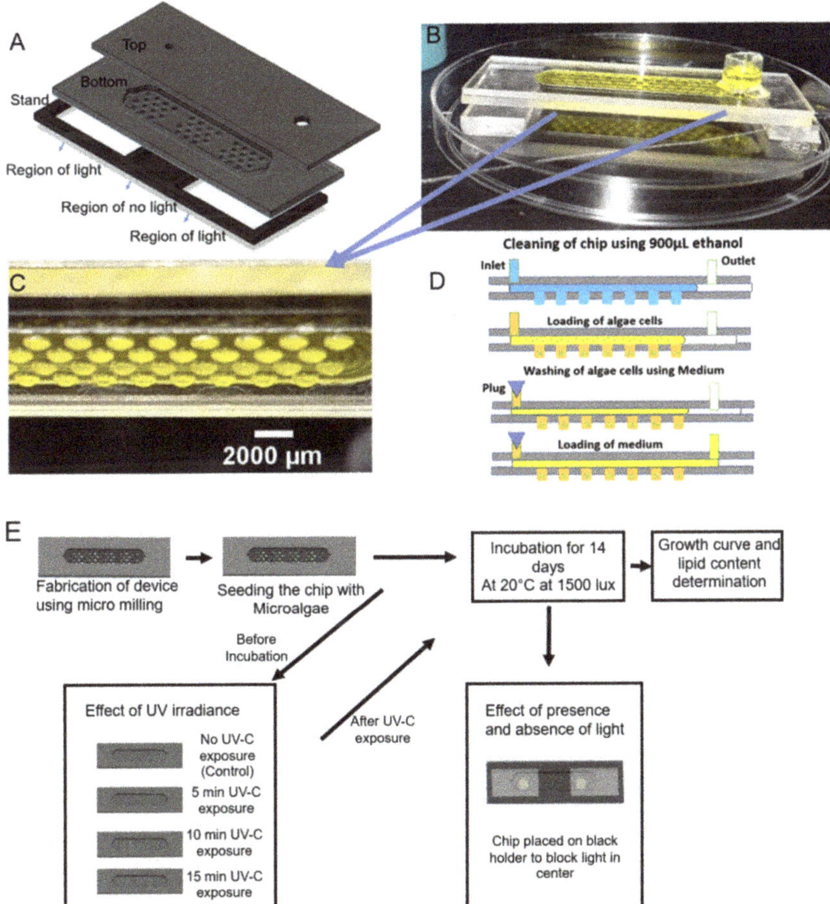

Figure 1. Illustration of the chip design. (**A**) Layout of a chip used to study the effect of light. The chip holder consists of three regions, two of which were exposed to light and one was an unexposed region. (**B**) Image of an actual assembled chip. (**C**) A chip filled with yellow dye shows the formation of a hanging drop formed on the bottom of the chip through the well (scale = 1 mm). (**D**) Schematic illustration of the hanging-drop chip operation. (**E**) Schematic diagram to summarize the experimental design.

2.3. On-Chip Microalgae Culturing

All cell seeding was performed in a sterilized laminar-flow hood. The chips were sterilized with UV-C light for 30 min before use. The chip was filled with ethanol solution (75%) using a standard pipette. Then a culture medium (900 µL) with algal cells (density 17×10^4 cell/mL) was injected through the inlet. The cell suspension uniformly distributed the cells within the microchannel and allowed the cells to fall into the wells to form hanging drops. After the cells settled into the hanging drops for 30 min, the microchannel was washed with fresh medium three times to remove the residual cells that were unable to fall into the well. After the washing step, the channel was filled with fresh medium (950 µL). The chips were then placed in an incubator at 20 ± 1 °C, in warm white light with a 12:12 h light:dark cycle and an intensity of 1500 lux for 14 days.

2.4. Analysis of the Microalgae Growth Rate

The growth of *B. braunii* inside the hanging-drop chip was characterized by measuring the intensity of chlorophyll autofluorescence of cells over time as in the existing literature [26]. The microalgae were imaged using an epi-fluorescent microscope (Eclipse Ti, Nikon Instruments, Inc., Japan) equipped with a filter set (TXRED, excitation at 559 nm, emission at 630 nm) every two days. To measure the growth rate, we selected 20 samples in each chip from a well with a specific diameter. The total intensity as a sum of chlorophyll autofluorescence was measured with image analysis software (Image J, NIH, USA). Large images (Figure 2A) of each chip (field area 13 mm × 40 mm) were acquired using a microscope with an automatic stage and stitched together afterward.

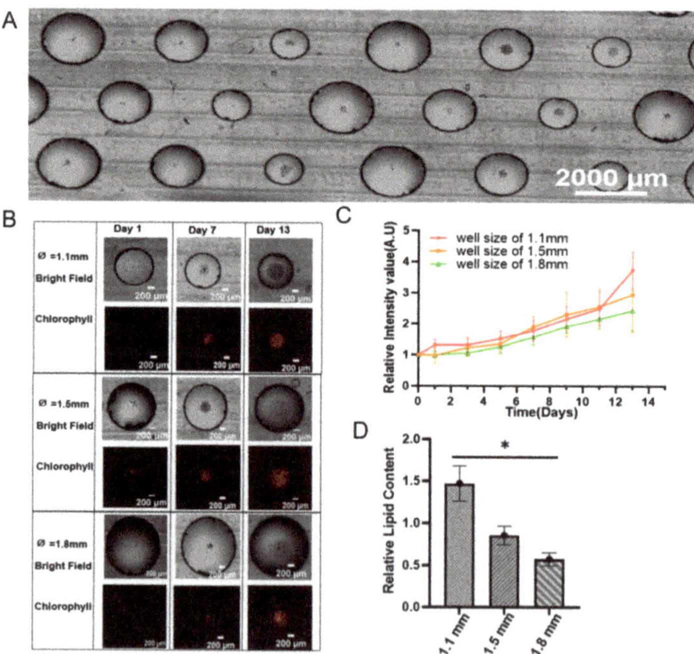

Figure 2. The microalgae cultured in the chip with three well diameters. (**A**) Bright-field image showing the formation of algal cell aggregated on a chip within the wells of sizes 1.1, 1.5 and 1.8 mm on its seventh day of culture. (**B**) Representative time-lapse bright field and chlorophyll (λ = 645 nm) images showing the increased cell count (day 1, day 7 and day 13); the cell aggregations were observed. (**C**) Growth curve measured using the value of total chlorophyll intensity at wavelength 645 nm. (**D**) Quantification analysis of lipid content in wells of varied diameter (* indicated p-value < 0.05, n = 54).

2.5. Microalgae Lipid Quantification

To analyze and to quantify the lipid content, we utilized fluorescence staining (Nile red) [16]. Nile red, a lipid-soluble fluorescent dye that binds to neutral lipids, has been shown to stain *B. braunii* oil efficiently in both the extracellular matrix and intracellular oil bodies, and has been used to accurately evaluate the oil content in *B. braunii* [27]. For the lipid quantification, we stained the cells with Nile red fluorescent dye (concentration 10 μg/mL dissolved in dimethyl sulfoxide solution 10%) on the fourteenth day of culture. To quantify the fluorescence, we used a microscope (Eclipse Ti, Nikon Instruments, Inc.) equipped with a filter set (TXRED, excitation at 559 nm, emission at 630 nm); the intensity sum was measured with image analysis software (Image J). The lipid content percentage was calculated by finding the difference between the intensities before and after staining, then dividing the difference by the initial intensity.

3. Results and Discussion

Measurement of Growth Rate In Situ and Quantification of Lipid Content of B. braunii

To study the influence of the cell density on the microalgae, we performed experiments with the hanging-drop chip using wells of varied diameter (1.1, 1.5, 1.8 mm). The growth rate after culture for 14 days in the chip was measured by taking the mean value of the total chlorophyll intensity (n = 54, from three independent experiments). Within the same chip, we managed the seeding density in each well by varying the diameter of the well. After seeding, we noticed that the range of seeding density, based on the total intensity, varied from 2.96 ± 0.327 to 7.08 ± 0.430 (A.U.) according to the increasing diameter of the well. The growth of microalgae was monitored and calculated using the total fluorescent intensity of chlorophyll in each well (Figure 2B,C). On day 14, we found that the total chlorophyll signal in the well with a diameter of 1.8 mm was 4.86 times that in the well with a diameter 1.1 mm and 5.65 times that in the well with a diameter 1.5 mm; once normalized with the initial chlorophyll signal on day 0, we noticed that the growth rate of microalgae in a small hanging drop was greater than that in larger drops (Figure 2C). The signals of chlorophyll are strongly associated with biomass production [28]. Being limited by the volume of the hanging drop, the biomass cannot be directly evaluated as for a large-scale approach [16]; our data indicated a greater biomass production with increased cell density. We also noticed a more rapid formation of microalgae aggregation in the smaller droplets. The lipid content on the fourteenth day of microalgae culture in wells of diameter 1.1 mm showed Nile red signals 1.70 and 2.53 times those from wells of 1.5 mm and 1.8 mm, respectively (Figure 2D). It has also been reported that the fluorescent intensity of cells stained with Nile red and the lipid content in *B. braunii* as determined with a conventional solvent extraction system show a linear relation (R^2 = 0.998) [27].

These results indicate that in our hanging-drop chip, the large cell density improves cell growth and lipid production. In a small hanging drop, the microalgae accumulated quickly at the bottom because of the strong curvature of a small drop. We speculate that the spatial distribution of a microalgae community is associated with the alteration of microalgae physiology. The increased cell growth observed in our study confirms previous findings, which indicated that a large cell density is associated with increasing proliferation, such as an algae bloom in nature [29]. Previous authors also reported that small colonies have advantages over dispersed microalgae, such as greater affinities for light and nutrients, and subsequently greater rates of growth [30]. In a conventional culture flask, the cell density and nutrition limitation are coupled so that the effects of a single factor are not determined easily. Here, with our developed chip, algae of varied cell density were cultured in separated wells under the same experimental conditions of nutrition; the results confirm the influence of cell density on cell growth.

A greater lipid content of microalgae was observed in a 1.1 mm well, which might also be associated with increased cell density. We speculated that cell aggregation may facilitate the formation of biofilm matrix and high lipid content. During photosynthesis, extracellular polymeric substances (EPS) are released by algae, and algae attach together with EPS to form a complicated biofilm community. The growth of biofilm may lead to local nutrient starvation and stimulate lipid accumulation [31]. With our system, we decoupled the influences of ensemble nutrition stress and cell density, and reported greater lipid production of microalgae under high cell density in a local environment. Although further investigation is needed to understand the inter- and intra-cellular mechanisms of the influence of cell density on lipid production, we demonstrated that a chip can be used to monitor microalgae growth with a high throughput and evaluate the lipid content. We conducted tests to compare the growth rate and lipid content with a simultaneous comparison of cell density.

As microalgae in a 1.1 mm well showed the greatest oil content, we conducted a subsequent experiment on our 1.1 mm chip to study the effect of UV-C mutagenesis on the chip. Tests showed that UV irradiation of various algae species improves the cell growth and lipid accumulation of microalgae and results in a high lipid-producing strain [32].

Appropriate UV exposure causes random mutagenesis in microalgae, which is caused mainly by the alteration of the DNA of the organism; the UV exposure induces either the formation of thymine dimers that cause a transition of DNA base pairs, or the deletion of base pairs. This genetic alteration further results in an over-expression of nitrate reductase (NR) in the mutant strain that contributes to an increased growth rate [33]. *B. braunii* were seeded on a chip and irradiated with UV-C light (Philips TUV UV-C TL-D, 36W) for 5, 10 and 15 min on the first day. The growth rate and lipid content were calculated from 60 samples (20 in each chip) and compared to a control chip that was unexposed to UV-C (Figure 3C,D). From the curve of growth rate, we observed on the fourteenth day chlorophyll intensity values of 410.48 $\pm$ 29.7, 561.84 $\pm$ 28.7, 504.84 $\pm$ 28.76, and 381.74 $\pm$ 28.78, respectively, from UV exposure of 5 and 10 min, control, and 15 min. The lipid contents on the fourteenth day of culture were found to be 1.61 $\pm$ 0.31, 2.40 $\pm$ 0.16, 1.45 $\pm$ 0.27, and 1.92 $\pm$ 0.64, respectively (Figure 3B). These results indicate that, in our hanging-drop chip with *B. braunii* algae culture, the 5 min UV-C exposure was considered to be the most efficient, increasing both the growth rate by 36.87% and the lipid content by 49.02%. The greater lipid content might be directly associated with a specific mutagenesis of the microalgae. The greater growth rate caused by mutagenesis might also have led to a deficiency of nutrients, which caused nitrogen starvation and further induced an increased lipid content [34].

Our data also showed that an excessive dose of UV exposure can lead to undesired effects on the microalgae, such as a small growth rate and lipid content (Figure 3C,D). UV-C exposure can degrade the algal photosynthetic apparatus and pigment content; the oxidative stress and genotoxicity effects generated by UV exposure can further trigger cellular lethality and senescence [35]. With our developed system, in addition to screening parameters simultaneously for cell growth and lipid productivity, our microfluidic platform can be applied to study the impact of UV toxicity.

In addition to the procedure for optimization, we could easily select an algae mutant from each individual well. For instance, a scatter plot of 20 samples from one experiment was taken on the fourteenth day; the data present the lipid content of microalgae in different samples, which show an uneven distribution under each mutation condition (Figure 3D). Once microalgae with a high lipid content are determined, in contrast to a sophisticated algae extraction process [36], the desired microalgae for further studies can be easily extracted using a 1 mL pipette without affecting the other strains.

We further conducted an experiment on our modified chip to study the effect of the presence and absence of light intensity (Figure 4). On comparison of the fluorescent intensity value of chlorophyll between the regions that were exposed and unexposed to warm white light (1500 lux for 12/12 h cycle), the results showed an initial intensity value of 264 (A.U.) in the region with light and 255 (A.U.) in the region without light. On the fourteenth day of culture the exposed and unexposed regions with light of the specified intensity showed mean intensity values (in A.U.) of 1466.03 $\pm$ 447.00 and 386.25 $\pm$ 42.19 in regions of the presence and absence of light, respectively (Figure 4C); the lipid content on the fourteenth day of culture was found to be 2.09 ± 0.30786 and 1.68 ± 0.29, respectively. The algae in a region exposed to light showed 2.27 times the cell growth of algae in the unexposed region, which showed slow growth, but in the lipid content result there was only a 19.6% difference in lipid content between exposed and unexposed regions (Figure 4B). The results support previous findings, which indicated that the energy produced during high light intensity was used for cell division instead of being stored in the form of lipids, whereas in the absence of light the cells do not go through photo-oxidation, which results in the production of a greater lipid content [37]. These results indicate that our hanging-drop chip can be used to define the light conditions to optimize oil productivity in microalgae. The illumination intensity in regular bioreactors, however, alters when the microalgal density increases over time, which makes it difficult to apply identical conditions to all algal cells in a large culture system [20]. It is worthy of mention, though, that this device can be used to screen culturing parameters for rapid side-by-side comparisons between growth and oil production; more practical considerations and larger-scale tests are required to apply the outcome further for industrial culturing.

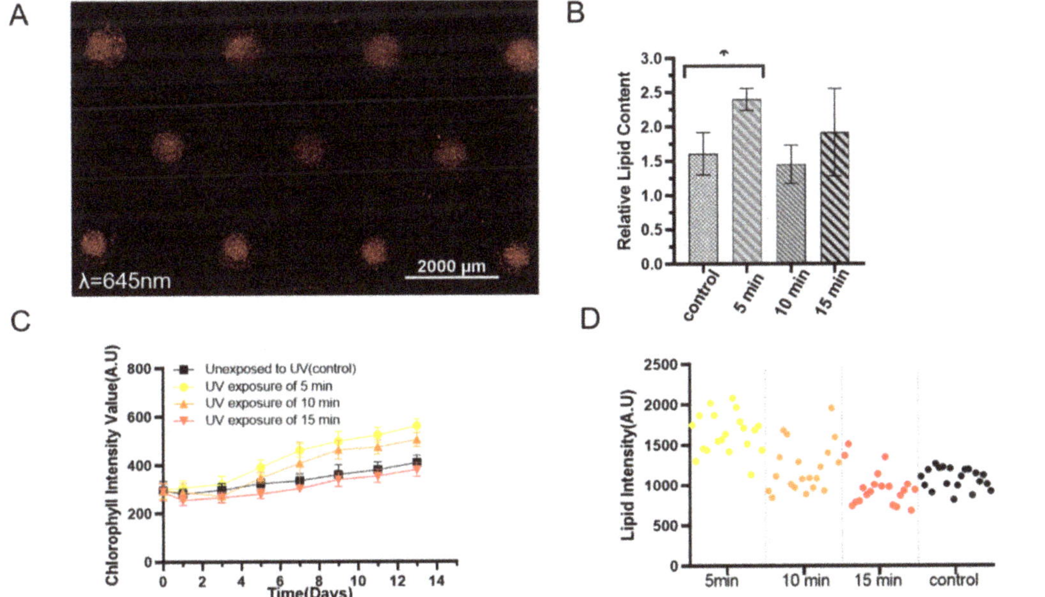

Figure 3. *B. braunii* treated with a UV mutagenesis procedure on a chip, and its lipid content comparison. (**A**) Representative images of algal cells in the chip observed at wavelength 645 nm on its fourteenth day of culture. (**B**) Quantification of lipid content of three UV-C exposures of a chip—5, 10 and 15 min, with unexposed chip as a control (* indicated p-value < 0.05, n = 60). (**C**) Comparison of growth curves of three UV-C chips with exposures 5,10 and 15 min with an unexposed chip as a control (n = 60). (**D**) Scatter plot of the mean intensity of Nile-stained algal cells shows a variation in lipid production with different UV exposure (n = 20 for each UV irradiance condition).

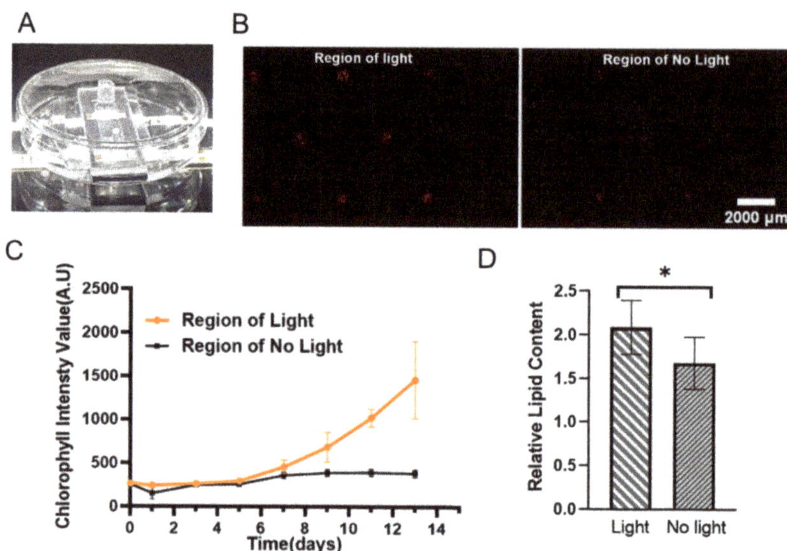

Figure 4. Formation of cell aggregate in the chip with regions of light and no light and the growth curve. (**A**) Experimental setup of chip with LED light source. (**B**) Formation of algal cell aggregate within the wells in the chip, which was observed at wavelength 645 nm with region of presence and absence of light on the fourteenth day of culture. (**C**) Comparison of curves of growth rate in the presence and absence of light ($n = 54$). (**D**) Comparison of lipid content of algae grown in a region of the presence and absence of light (* indicated p-value < 0.05, $n = 54$).

4. Conclusions

In this study, we have developed a hanging-drop microfluidic device for the high-throughput screening of microalgae, based on the growth rate and lipid content. In contrast to previous microfluidic approaches [19,20,23,26,27], we first took advantage of the easily accessible hanging drop as a separate bioreactor for algae culturing. By modulating the design of the hanging-drop chip, we easily customized the culture parameters, such as the cell density and illumination. We found, under our experimental conditions, that the clustering of microalgae was associated with enhanced lipid production. Using our hanging-drop chip, we also demonstrated the process for optimizing the UV dosage for mutagenesis to increase the algae growth rate and lipid content simultaneously. Moreover, integrated with a high-throughput microalgae physiology monitored in situ, we directly selected the desired specimens with a laboratory pipette due to the semi-open nature of the hanging drops. Although more studies are required to expand the outcomes of this study for algae cultivation, based on our obtained results, our easily accessible microfluidic system can accelerate the mutation breeding of microalgae and is expected to facilitate the screening process further for microalgae biofuel production.

Author Contributions: S.M.: Investigation, Writing. Y.-J.L.: Conceptualization, Writing and editing. C.-S.C.: Conceptualization, Writing, Resources. D.-J.Y.: Conceptualization, Writing, Resources. All authors have read and agreed to the published version of the manuscript.

Funding: This research was supported by grants from Ministry of Science and Technology, Taiwan (MOST 108-2221-E-007-091-MY2, MOST 109-2112-M-007-003, MOST 110-2217-E-007-003-MY3. Y.-J. Liu was supported by grant from the Ministry of Economic Affairs (MEA), Taiwan (No. 110-EC-17-A-22-0525).

Institutional Review Board Statement: Not applicable.

Informed Consent Statement: Not applicable.

Data Availability Statement: Data will be available from authors upon request.

Conflicts of Interest: The authors declare no conflict of interest.

References

1. Houghton, J. Global warming. *Rep. Prog. Phys.* **2005**, *68*, 1343. [CrossRef]
2. Chisti, Y. Biodiesel from microalgae. *Biotechnol. Adv.* **2007**, *25*, 294–306. [CrossRef]
3. Heredia-Arroyo, T.; Wei, W.; Hu, B. Oil Accumulation via Heterotrophic/Mixotrophic Chlorella protothecoides. *Appl. Biochem. Biotechnol.* **2010**, *162*, 1978–1995. [CrossRef]
4. Ramli, U.S.; Salas, J.J.; Quant, P.A.; Harwood, J.L. Use of metabolic control analysis to give quantitative information on control of lipid biosynthesis in the important oil crop, Elaeis guineensis (oilpalm). *New Phytol.* **2009**, *184*, 330–339. [CrossRef]
5. Kaenchan, P.; Gheewala, S.H. A review of the water footprint of biofuel crop production in Thailand. *J. Sustain. Energy Environ.* **2013**, *4*, 45–52.
6. Searchinger, T.; Heimlich, R.; Houghton, R.A.; Dong, F.; Elobeid, A.; Fabiosa, J.; Tokgoz, S.; Hayes, D.; Yu, T.-H. Use of U.S. Croplands for Biofuels Increases Greenhouse Gases Through Emissions from Land-Use Change. *Science* **2008**, *319*, 1238–1240. [CrossRef]
7. Banerjee, A.; Sharma, R.; Chisti, Y.; Banerjee, U.C. Botryococcus braunii: A Renewable Source of Hydrocarbons and Other Chemicals. *Crit. Rev. Biotechnol.* **2002**, *22*, 245–279. [CrossRef]
8. Metzger, P.; Casadevall, E.; Pouet, M.; Pouet, Y. Structures of some botryococcenes: Branched hydrocarbons from the b-race of the green alga Botryococcus braunii. *Phytochemistry* **1985**, *24*, 2995–3002. [CrossRef]
9. Metzger, P.; Largeau, C. Botryococcus braunii: A rich source for hydrocarbons and related ether lipids. *Appl. Microbiol. Biotechnol.* **2004**, *66*, 486–496. [CrossRef] [PubMed]
10. Borowitzka, M.A. Microalgae in medicine and human health: A historical perspective. In *Microalgae in Health and Disease Prevention*; Elsevier: Amsterdam, The Netherlands, 2018; pp. 195–210.
11. Yamamoto, S.; Mandokoro, Y.; Nagano, S.; Nagakubo, M.; Atsumi, K.; Watanabe, M.M. Catalytic conversion of Botryococcus braunii oil to diesel fuel under mild reaction conditions. *Environ. Boil. Fishes* **2014**, *26*, 55–64. [CrossRef]
12. Chisti, Y. Biodiesel from microalgae beats bioethanol. *Trends Biotechnol.* **2008**, *26*, 126–131. [CrossRef] [PubMed]
13. Simionato, D.; Basso, S.; Giacometti, G.M.; Morosinotto, T. Optimization of light use efficiency for biofuel production in algae. *Biophys. Chem.* **2013**, *182*, 71–78. [CrossRef] [PubMed]
14. Singh, S.; Singh, P. Effect of temperature and light on the growth of algae species: A review. *Renew. Sustain. Energy Rev.* **2015**, *50*, 431–444. [CrossRef]
15. Juneja, A.; Ceballos, R.M.; Murthy, G.S. Effects of Environmental Factors and Nutrient Availability on the Biochemical Composition of Algae for Biofuels Production: A Review. *Energies* **2013**, *6*, 4607–4638. [CrossRef]
16. Srinivas, R.; Ochs, C. Effect of UV-A Irradiance on Lipid Accumulation in Nannochloropsis oculata. *Photochem. Photobiol.* **2012**, *88*, 684–689. [CrossRef]
17. Au, S.H.; Shih, S.C.C.; Wheeler, A.R. Integrated microbioreactor for culture and analysis of bacteria, algae and yeast. *Biomed. Microdevices* **2010**, *13*, 41–50. [CrossRef]
18. Lee, P.J.; Hung, P.J.; Rao, V.M.; Lee, L.P. Nanoliter scale microbioreactor array for quantitative cell biology. *Biotechnol. Bioeng.* **2006**, *94*, 5–14. [CrossRef]
19. Kim, J.; Taylor, D.; Agrawal, N.; Wang, H.; Kim, H.; Han, A.; Rege, K.; Jayaraman, A. A programmable microfluidic cell array for combinatorial drug screening. *Lab Chip* **2012**, *12*, 1813–1822. [CrossRef] [PubMed]
20. Dewan, A.; Kim, J.; McLean, R.H.; Vanapalli, S.A.; Karim, M.N. Growth kinetics of microalgae in microfluidic static droplet arrays. *Biotechnol. Bioeng.* **2012**, *109*, 2987–2996. [CrossRef] [PubMed]
21. Pan, J.; Stephenson, A.L.; Kazamia, E.; Huck, W.T.S.; Dennis, J.S.; Smith, A.G.; Abell, C. Quantitative tracking of the growth of individual algal cells in microdroplet compartments. *Integr. Biol.* **2011**, *3*, 1043–1051. [CrossRef] [PubMed]
22. Qu, B.; Eu, Y.-J.; Jeong, W.-J.; Kim, D.-P. Droplet electroporation in microfluidics for efficient cell transformation with or without cell wall removal. *Lab Chip* **2012**, *12*, 4483–4488. [CrossRef]
23. Holcomb, R.E.; Mason, L.J.; Reardon, K.F.; Cropek, N.M.; Henry, C.S. Culturing and investigation of stress-induced lipid accumulation in microalgae using a microfluidic device. *Anal. Bioanal. Chem.* **2011**, *400*, 245–253. [CrossRef] [PubMed]
24. Hung, P.J.; Lee, P.J.; Sabounchi, P.; Aghdam, N.; Lin, R.; Lee, L.P. A novel high aspect ratio microfluidic design to provide a stable and uniform microenvironment for cell growth in a high throughput mammalian cell culture array. *Lab Chip* **2005**, *5*, 44–48. [CrossRef] [PubMed]
25. Berges, J.A.; Franklin, D.J.; Harrison, P.J. evolution of an artificial seawater medium: Improvements in enriched seawater, artificial water over the last two decades. *J. Phycol.* **2001**, *37*, 1138–1145. [CrossRef]
26. Wu, H.-W.; Hsiao, Y.-H.; Chen, C.-C.; Yet, S.-F.; Hsu, C.-H. A PDMS-Based Microfluidic Hanging Drop Chip for Embryoid Body Formation. *Molecules* **2016**, *21*, 882. [CrossRef] [PubMed]
27. Cheng, P.; Ji, B.; Gao, L.; Zhang, W.; Wang, J.; Liu, T. The growth, lipid and hydrocarbon production of Botryococcus braunii with attached cultivation. *Bioresour. Technol.* **2013**, *138*, 95–100. [CrossRef]
28. Lee, S.J.; Yoon, B.-D.; Oh, H.-M. Rapid method for the determination of lipid from the green alga Botryococcus braunii. *Biotechnol. Tech.* **1998**, *12*, 553–556. [CrossRef]

29. Vigeolas, H.; Duby, F.; Kaymak, E.; Niessen, G.; Motte, P.; Franck, F.; Remacle, C. Isolation and partial characterization of mutants with elevated lipid content in Chlorella sorokiniana and Scenedesmus obliquus. *J. Biotechnol.* **2012**, *162*, 3–12. [CrossRef]
30. Fransolet, D.; Roberty, S.; Herman, A.-C.; Tonk, L.; Hoegh-Guldberg, O.; Plumier, J.-C. Increased Cell Proliferation and Mucocyte Density in the Sea Anemone Aiptasia pallida Recovering from Bleaching. *PLoS ONE* **2013**, *8*, e65015. [CrossRef] [PubMed]
31. Li, Y.; Gao, K. Photosynthetic physiology and growth as a function of colony size in the cyanobacteriumNostoc sphaeroides. *Eur. J. Phycol.* **2004**, *39*, 9–15. [CrossRef]
32. Schnurr, P.J.; Espie, G.S.; Allen, D.G. Algae biofilm growth and the potential to stimulate lipid accumulation through nutrient starvation. *Bioresour. Technol.* **2013**, *136*, 337–344. [CrossRef] [PubMed]
33. Arora, N.; Yen, H.-W.; Philippidis, G.P. Harnessing the Power of Mutagenesis and Adaptive Laboratory Evolution for High Lipid Production by Oleaginous Microalgae and Yeasts. *Sustain. J. Rec.* **2020**, *12*, 5125. Available online: https://www.mdpi.com/2071-1050/12/12/5125 (accessed on 14 March 2021). [CrossRef]
34. Noorhana; Nigam, S.; Rai, M.P.; Sharma, R. Effect of Nitrogen on Growth and Lipid Content of Chlorella pyrenoidosa. *Am. J. Biochem. Biotechnol.* **2011**, *7*, 124–129. [CrossRef]
35. Borderie, F.; Laurence, A.-S.; Naoufal, R.; Faisl, B.; Geneviève, O.; Dominique, R.; Badr, A.-S. UV–C irradiation as a tool to eradicate algae in caves. *Int. Biodeterior. Biodegrad.* **2011**, *65*, 579–584. [CrossRef]
36. Kim, H.S.; Devarenne, T.P.; Han, A. A high-throughput microfluidic single-cell screening platform capable of selective cell extraction. *Lab Chip* **2015**, *15*, 2467–2475. [CrossRef] [PubMed]
37. Nzayisenga, J.C.; Farge, X.; Groll, S.L.; Sellstedt, A. Effects of light intensity on growth and lipid production in microalgae grown in wastewater. *Biotechnol. Biofuels* **2020**, *13*, 1–8. [CrossRef]

 *energies* 

Article

Operation Regimes: A Comparison Based on *Nannochloropsis oceanica* Biomass and Lipid Productivity

Inês Guerra [1,2], Hugo Pereira [3], Margarida Costa [1,*], Joana T. Silva [1], Tamára Santos [4], João Varela [3,4], Marília Mateus [2] and Joana Silva [1]

1 ALLMICROALGAE Natural Products S.A., R&D Department, Rua 25 de Abril s/n, 2445-413 Pataias, Portugal; ines.funico@gmail.com (I.G.); joanatlfsilva@gmail.com (J.T.S.); joana.g.silva@allmicroalgae.com (J.S.)
2 iBB—Institute for Bioengineering and Biosciences, IST, Universidade de Lisboa, Av. Rovisco Pais, nº1, 1049-001 Lisbon, Portugal; marilia.mateus@tecnico.ulisboa.pt
3 GreenCoLab—Associação Oceano Verde, University of Algarve, Campus de Gambelas, 8005-139 Faro, Portugal; galvaohugo@gmail.com (H.P.); jvarela@ualg.pt (J.V.)
4 CCMAR—Centre of Marine Sciences, University of Algarve, Gambelas, 8005-139 Faro, Portugal; tamarafilipasantos@gmail.com
* Correspondence: costa.anamarg@gmail.com

Abstract: Microalgae are currently considered to be a promising feedstock for biodiesel production. However, significant research efforts are crucial to improve the current biomass and lipid productivities under real outdoor production conditions. In this context, batch, continuous and semi-continuous operation regimes were compared during the Spring/Summer seasons in 2.6 m^3 tubular photobioreactors to select the most suitable one for the production of the oleaginous microalga *Nannochloropsis oceanica*. Results obtained revealed that *N. oceanica* grown using the semi-continuous and continuous operation regimes enabled a 1.5-fold increase in biomass volumetric productivity compared to that cultivated in batch. The lipid productivity was 1.7-fold higher under semi-continuous cultivation than that under a batch operation regime. On the other hand, the semi-continuous and continuous operation regimes spent nearly the double amount of water compared to that of the batch regime. Interestingly, the biochemical profile of produced biomass using the different operation regimes was not affected regarding the contents of proteins, lipids and fatty acids. Overall, these results show that the semi-continuous operation regime is more suitable for the outdoor production of *N. oceanica*, significantly improving the biomass and lipid productivities at large-scale, which is a crucial factor for biodiesel production.

Keywords: tubular photobioreactor; pilot-scale; operation regimes; outdoor cultivation; *Nannochloropsis oceanica*

Citation: Guerra, I.; Pereira, H.; Costa, M.; Silva, J.T.; Santos, T.; Varela, J.; Mateus, M.; Silva, J. Operation Regimes: A Comparison Based on *Nannochloropsis oceanica* Biomass and Lipid Productivity. *Energies* **2021**, *14*, 1542. https://doi.org/10.3390/en14061542

Academic Editor: José Carlos Magalhães Pires

Received: 20 February 2021
Accepted: 8 March 2021
Published: 11 March 2021

1. Introduction

Major climate changes have been observed since 1950, and human impact, due to industrial activity, is one of the main leading causes. The emissions of greenhouse gases, including CO_2, CH_4 and NO_2, have been increasing since the pre-industrial era, and those gases remain in the atmosphere, soil and oceans [1]. This concern emphasizes the importance of fossil fuel replacements, like the ones based on biomass feedstocks for biofuel production [2].

Microalgae are ubiquitous microscopic photosynthetic organisms mainly found in aquatic environments (freshwater and saline), but also on the surface of soil and stone, from deserts to polar sea habitats [3]. These organisms are known to efficiently fix CO_2, through photosynthesis, and convert it into organic matter with an efficiency up to 10-fold faster than terrestrial plants [4]. Additionally, when compared to terrestrial plants, microalgae have the advantage of requiring less area to reach the same amount of biomass, being able to grow in non-arable soil and having the ability to grow using non-potable water, namely in saline or wastewater [2,5].

Nannochloropsis oceanica is a fast growing microalga known to intracellularly accumulate high amounts of lipid, up to 53% of biomass dry weight [6,7], making this species suitable to be used as biodiesel feedstock [8,9]. This small unicellular marine microalga (2–4 µm in diameter) is an ochrophyte, belonging to the class Eustigmatophyceae [10] and is also of considerable interest as a source of polyunsaturated fatty acids, namely for the production of eicosapentaenoic acid for human disease-prevention [8].

Although significant progress has been made in the final microalgae biodiesel properties [11,12], the high cost of culture growth and biomass harvesting and the limited biomass productivity are still major limitations for the use of this rich biomass as a biodiesel feedstock [13]. Several laboratory studies have shown that the operation regime is crucial to significantly increase biomass and lipid productivity in *Nannochloropsis* sp., underlining the importance of testing it in settings closer to an industrial scenario [14,15].

Biomass production of microalgal feedstocks can be achieved using different operation regimes, namely batch, continuous and semi-continuous [16,17]. A batch operation regime consists of introducing all needed nutrients in the bioreactor, being this culture entirely harvested after the production period [17,18]. Under a continuous regime, the medium and all needed nutrients are continuously added to the cultivation system. At the same time, the culture is continuously removed from the system, at the same flow rate [17,18]. Semi-continuous operation regime is a combination between the batch and continuous operations. Usually, a percentage of the cultures (10–50%) is removed when cultivation reaches the mid to late exponential phase and the volume is replaced with fresh medium [18]. The major difference between these operation regimes is the achieved productivity, which is usually higher in the continuous and semi-continuous systems, as they allow maintaining the culture near the maximum growth rate [19]. On the other hand, the susceptibility to contamination is much lower and the accumulation of target substances is usually higher using the batch regime [19,20]. Therefore, identifying the adequate production regime for effective microalgae production considering all the mentioned factors is of the utmost importance. Although there is no overall better production method, the tendency in bioprocessing has been to adopt increasingly more continuous processes [21].

Production of microalgal biomass can occur using the aforementioned operation regimes, in open and closed systems. Open systems are usually more economically viable and can be divided into three major types, including natural water bodies, circular ponds, raceway ponds, and thin layer cascade systems [6,22]. Nevertheless, closed systems are known to display better growth performances since they limit the direct gas exchange, reduce the contaminants in the culture, and better control important physicochemical variables. Closed systems can be classified into three main groups: vertical column, flat panel, and tubular photobioreactors (PBRs) [23,24].

The present work aimed to cultivate *N. oceanica* in pilot-scale tubular PBRs with the outdoor light and temperature conditions, using three operation regimes: batch, continuous and semi-continuous. The main goal was to identify the most suitable operation regime that can ensure the highest biomass productivity, uses fewer resources while providing high-quality biomass with a high lipid percentage for further biodiesel application.

2. Materials and Methods

The outdoor work was performed at the facilities of Allmicroalgae (Pataias, Portugal) between 1 March and 8 July 2019.

2.1. Microalgae Strain and Culture Media

The microalga *Nannochloropsis oceanica* CCAP849/10 was obtained from the culture collection Algae and Protozoa (Oban, Scotland, UK) and is kept at Allmicroalgae culture collection. The culture medium used for growth assays was Guillard's F/2 medium at

0.31 g/L of NO_3^-, supplemented with 12 µM of iron, 30 g/L of NaCl (Salexpor, Coimbra, Portugal) and magnesium-enriched supplementation (Necton, Faro, Portugal).

2.2. Culture Scale-Up

Initially, the cultures were grown in 5 L airlift reactors, in the laboratory. The aeration of these reactors was made by compressed air pre-mixed with 1% CO_2, to maintain the pH below 8.2, sterilized by 0.2 µm filters (Sartorius, Gottingen, Germany). These reactors were maintained under constant irradiance of approximately 700 µmol of photons m^{-2} s^{-1} at room temperature (24 °C). Five of these 5 L reactors were used to inoculate an outdoor 125 L Flat Panel (FP) PBR, which served as inoculum for an 800 L FP. The aeration conditions were similar to the 5 L reactors, although the CO_2 was added using a pulse system that maintained the pH close to 8.2, and the temperature was maintained below 30 °C by an irrigation system. The 800 L FP was later used to inoculate a 2.6 m^3 tubular PBR. This PBRwas subsequently used to inoculate three 2.6 m^3 tubular PBRs used for the assay. In these systems, the agitation of the culture was performed by pumping the culture through the PBR, using centrifugal pumps. The pH was measured in real-time and kept at 8.2 by an automated system that injected CO_2 on demand. The temperature was maintained below 30 °C through an irrigation system.

2.3. Operation Regime Trial

N. oceanica was grown in three tubular PBRs (Figure 1), each being operated in batch, semi-continuous, or continuous operation regime. The horizontal tubular PBRs used in this trial had a serpentine configuration, with a working volume of 2.6 m^3 and an illuminated volume of 1.6 m^3. In order to begin the assay the PBRs were inoculated atan initial biomass concentration of 0.4 g L^{-1} for the batch regime and 1 g L^{-1} for the continuous and semi-continuous regimes.

Figure 1. Pilot-scale tubular photobioreactors (PBRs) used in the trials with 2.6 m^3 of working volume (Allmicroalgae, Pataias, Portugal).

The cultures were allowed to grow for a day to adapt to the new reactor conditions, and then each of the reactors was operated differently, depending on the correspondent operation regime. The culture grown in the batch operation regime was left to grow until the end of the trial. In the semi-continuous regime, the culture was left to grow for two days and on the second day it was diluted to 1 g L^{-1}. This dilution process was repeated every two days until the end of the trial. In the continuous system, the culture was continuously diluted using peristaltic pumps; the flow rate was adjusted daily, to maintain the culture at around 1 g L^{-1}. Simultaneously, the culture was also continuously removed from the PBR, at the same rate. In all conditions, the Guillard's F/2 medium supplemented with iron was added manually when needed, in order to maintain the NO_3^- concentration between 0.12 and 0.31 g L^{-1}. Three replicates of each trial were performed, with a consecutive rotation of the PBR used in an operation regime.

2.4. Growth Assessment

Culture growth was followed daily through optical density (OD) until the culture operated in batch reached the stationary phase. The OD was measured at 540 nm in a UV/Vis spectrophotometer (Zuzi, Seville, Spain) while the dry weight (DW) was obtained through a calibration curve established previously (Figure S1). The DW was determined by filtering a known amount of culture through a 0.7 μm glass microfiber filter (VWR International, Radnor, PA, USA), which was later washed with an equal volume of 35 g L^{-1} ammonium formate (Biochem Chemopharma, Cosne-Cours-sur-Loire, France).

The specific growth rate was calculated for the batch operated cultures through Equation (1), where X_1 and X_2 represent, respectively, the cellular concentration in the beginning and end of the exponential phase and t_2 and t_1 are the times, in days, corresponding to those respective concentrations.

$$\mu\left(\text{day}^{-1}\right) = \frac{ln\frac{X_2}{X_1}}{t_2 - t_1} \quad (1)$$

The volumetric productivity (P) was calculated as the ratio of the sum of the produced biomass in each day considered ($m_{produced}$, g) by the total reactor volume (V_t, L) and whole growth time (t, day), as shown in Equation (2). The daily produced biomass was calculated by multiplying the biomass concentration reached in that day (X_i, g L^{-1}) by the respective removed volume (V_{ri}, L). The only exception was the last day of the whole cycle, day n, in which the entire working volume of the reactor was processed, being the produced biomass calculated by the difference of the final and initial cell concentration of the last day (X_i and X_0, g L^{-1}) multiplied by the total working volume of the last and first day

$$P\left(\text{gL}^{-1}\text{day}^{-1}\right) = \frac{\sum m_{produced}}{V_t t} \quad (2)$$

$$\sum_{i=1}^{n-1} m_{produced} = X_i V_{ri}$$
$$\sum_{i=n}^{n} m_{produced} = (X_i - X_0)V_t \quad (3)$$

Areal biomass productivity (P_a) (Equation (4)) was determined by multiplying the volumetric biomass productivity by the volume of the reactor (V_t) divided by the ground area occupied by the reactor (A, m^2).

$$P_a\left(\text{gm}^{-2}\text{day}^{-1}\right) = \frac{P\, V_t}{A} \quad (4)$$

The photosynthetic efficiency (PE) was determined by the ratio between the increase of the higher heating value (HHV) and the total sun irradiation that reached the reactor (Equation (5)). The outside solar radiation and temperature were measured using a WatchDog 2000 weather station (Spectrum Technologies. Inc., Aurora, IL, USA). The specific HHV (HHV, kJ g^{-1}) was calculated according to a previous correlation reported by [25], present in Equation (6), where C represents the percentage of carbon, H the percentage of hydrogen and N the percentage of nitrogen obtained by the CHN analysis of the final biomass.

$$PE\ (\%) = \frac{HHV * \left(biomass_{final} - biomass_{initial}\right)}{Total\ incident\ radiation} \times 100 \quad (5)$$

$$HHV\ \left(\text{kJ g}^{-1}\right) = -3.393 + 0.507\text{C} - 0.341\text{H} + 0.067\text{N} \quad (6)$$

The nitrogen source concentration was measured at least once a day and added when needed. Nitrates were determined according to Armstrong (1963), modified [26]. Briefly, the collected supernatant was diluted and hydrochloric acid was added at 30 mM. The absorbance of samples was measured spectrophotometrically (4251/50, Zuzi, Seville, Spain) at 220 and 275 nm. The organic matter interference was corrected by subtracting twice the absorbance read at 275 nm from the reading at 220 nm. The final absorbance was compared to a sodium nitrate calibration curve (Figure S2).

2.5. Elemental Analysis

The biomass was collected and centrifuged (Hermle Labortechnik Z300, Wehingen, Germany) at 2050× *g* for 15 min at the end of each culture trial. The resulting pellet was frozen and stored at −18 °C. Before the biochemical analysis was performed, the biomass was freeze-dried (Telstar, Lisbon, Portugal).

The lyophilized biomass was weighed into aluminum vessels, using a precision balance and inserted in a Vario el III (Vario EL, Element Analyser System, GmbH, Hanau, Germany). The CHN composition was determined according to the procedure provided by the manufacturer. The total protein was determined by multiplying the percentage of nitrogen by the factor 6.25 [10].

The total ash content was determined by gravimetric analysis. Samples were weighed before and after being burned in a muffle (J. P. Selecta, Sel horn R9-L, Barcelona, Spain) for 8 h at 550 °C.

The total lipid content was determined following the Bligh and Dyer method (1959) [27] with few modifications [28]. In brief, lipids were extracted using a mixture of distilled water, chloroform, methanol (1.8:2:2 *v*/*v*/*v*) (Fisher Chemical, NH, USA) and an IKA Ultra-Turrax disperser (IKA-Werke GmbH, Staufen, Germany), for homogenization. Afterwards, the mixture was centrifuged at 2000× *g* for 10 min for phase separation. The organic phase was transferred to a clean tube with a Pasteur pipette and later a known volume of the chloroform phase was pipetted to a pre-weighed tube and placed in a dry bath at 60 °C. Lipids were gravimetrically determined after the chloroform evaporation.

The carbohydrate content was determined by subtracting the weight of proteins, lipids and ashes from the total DW of biomass.

2.6. Fatty Acid Profile

Fatty acids were converted into the corresponding fatty acid methyl esters (FAME) according to the protocol of Lepage and Roy (1984) [29], modified by Pereira et al. (2011) [30]. FAME were analyzed in a GC-MS analyzer (Bruker SCION 456/GC, SCION TQ MS, MA, USA) equipped with a ZB-5MS column (length of 30 m, 0.25 mm of internal diameter, 0.25 μm of film thickens, Phenomenex), using helium as the carrier gas. The temperature program was 60 °C for 1 min, an increase of 30 °C per min up to 120 °C, an increase of 5 °C per min up to 250 °C, and a final increase of 20 °C per min up to 300 °C. The temperature in the injector was 300 °C. For the identification and the quantification of FAME five different concentration of Supelco 37 component FAME Mix standard (Sigma-Aldrich, Sintra, Portugal) were analyzed in order to establish 37 different calibration curves. Then the peak area of each component in each sample was compared to the correspondent calibration curve in order to have a quantitative analysis of that specific FAME.

2.7. Statistical Analyses

Statistical analyses were performed using R software (version 3.6.1) through the RStudio IDE (version 1.2.1335). Experimental results are presented with a 95% confidence level. The normal distribution of data was tested through the Shapiro–Wilk test and the data homogeneity was tested through the Bartlett test. The data was then compared using one-way ANOVA, followed by Tukey's multiple comparison tests. The non-homogeneous data was compared with the Kruskal–Wallis test followed by Dunn's test.

3. Results and Discussion

3.1. Growth Performance

N. oceanica was grown under batch, semi-continuous and continuous operation regimes in three 2.6 m^3 horizontal tubular PBRs, during three consecutive replicas, and the temperature and the incident solar radiation were registered in the three independent trials (Figure 2). The average ambient temperature throughout the whole trial was 17.5 $\pm$ 0.8 °C, with a maximum and minimum of 21.1 $\pm$ 2.8 °C and 15.6 $\pm$ 2.3 °C, respectively. The temperature of the culture inside the reactor was usually higher than the ambient temperature, but maintained below 30 °C through the thermoregulation system. The average daily solar radiation throughout the trial was 20.1 $\pm$ 3.5 MJ m^2, with a maximum registered average daily radiation of 23.8 $\pm$ 2.4 MJ m^{-2} and a minimum of 15.4 $\pm$ 1.6 MJ m^{-2}.

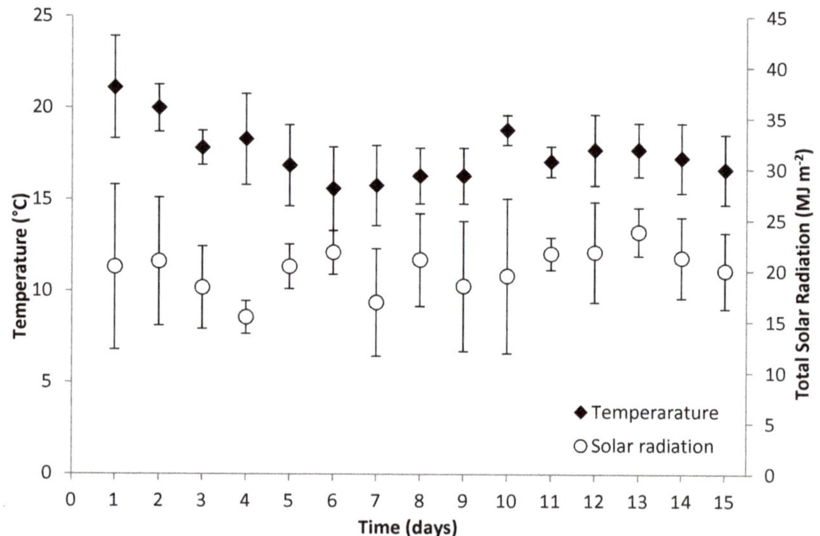

Figure 2. Average daily ambiance temperature and total solar radiation incident in the 2.6 m^3 PBRs, while culturing *Nannochloropsis oceanica*. The presented values are the average values obtained in three biologically independent replicates and the error bars represent the respective standard deviations.

The assay lasted until the batch culture ceased its growth, which happened on the 15th day with a maximum DW of 2.0 g L^{-1} (Figure 3). The batch operated culture presented no lag phase and an exponential phase of 9 days with amaximum specific growth rate of 0.129 day^{-1}, calculated from the 4th to 9th culture day.

The semi-continuous culture was renewed every second day, beginning on the 3rd day and was renovated five more times during each trial. The water volumes spent in each renovation are shown in Figure 4 and ranged between 17.8 $\pm$ 1.8% and 45.4 $\pm$ 1.0% of the total volume, having an average dilution rate of 0.122 $\pm$ 0.011 day^{-1} (Table 1). The culture DW ranged from 1.0 to 1.5 g L^{-1}. The continuous operated culture started to be diluted on the 2nd day, presenting maximum and minimum renovation volumes of 15.5 $\pm$ 0.5% and

12.2 ± 1.4% of total volume, respectively (Figure 4). The average dilution rate throughout the whole trial was 0.140 ± 0.010 day^{-1}. After the second day, until the end of the trial, it was possible to carry a steady-state, with the DW ranging from 1.1 to 1.3 g L^{-1}.

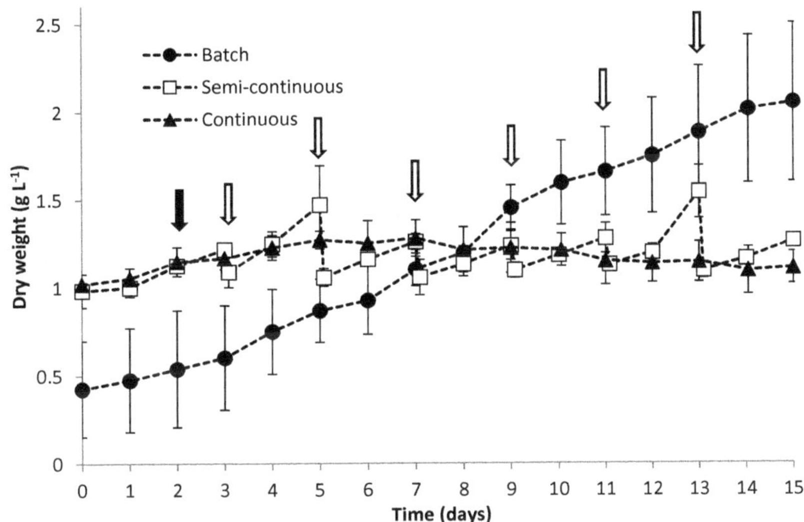

Figure 3. Growth of *Nannochloropsis oceanica* in 2.6 m^3 tubular PBRs using three different operation regimes: batch, semi-continuous and continuous. The values presented are the average of the three independent biological replicates, and the error bars are the respective standard deviations. The black arrow represents the day in which the continuous regimes started to be diluted, and the white arrows represent the moments of medium renewals under the semi-continuous regime.

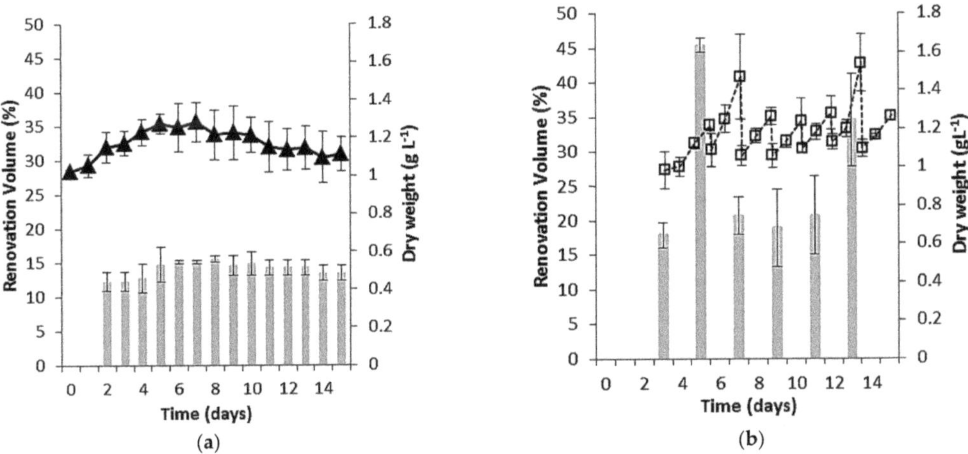

Figure 4. *Nannochloropsis oceanica* growth in 2.6 m^3 PBRs in continuous (**a**) and semi-continuous (**b**) operation regimes. The bars represent the percentage of volume renewed with fresh medium. All the values represent an average of three biologically independent replicates, with the respective standard deviation.

In terms of total water used (Table 1) the batch regime was, by far, the one that needed the least amount of water, more specifically only the initial 2.6 m^3. It was followed by the semi-continuous and continuous operation regimes that used over 2.5 times the water used in the batch regime. Spent water per one kg of produced biomass revealed no significant differences between the semi-continuous and continuous operation systems. However, these values (1.04 ± 0.03 m^3 kg^{-1} and 1.25 ± 0.14 m^3 kg^{-1} respectively) were almost twice those verified for the batch regime (0.62 ± 0.07 m^3 kg^{-1}). Yet, these values do not consider the water spent during cleaning, which would considerably increase the water spent in the batch regime, since the reactor would have to be cleaned at the end of each cycle (±every 15 days). For the other two operation regimes, cultivation would only be discontinued for cleaning if contamination occurred.

Table 1. Specific growth rate, average dilution rate and amount of water spent, either per cycle or per amount of produced biomass, in the growth of *Nannochloropsis oceanica* in 2.6 m^3 tubular PBRs, in three different operation regimes. The values represent the average and standard deviation of three biologically independent replicates. Different letters within the same column represent significantly different values (p-value < 0.05).

Production Regime	Specific Growth Rate (Day^{-1})	Water (m^3/Cycle)	Water (m^3 kg^{-1} Produced Biomass)	Average Dilution Rate (Day^{-1})
Batch	0.129 ± 0.020	2.6 ± 0.0 ^{a}	0.62 ± 0.07 ^{a}	-
Semi-continuous	-	6.7 ± 0.4 ^{ab}	1.04 ± 0.03 ^{b}	0.122 ± 0.011 ^{a}
Continuous	-	7.4 ± 0.1 ^{b}	1.25 ± 0.14 ^{b}	0.140 ± 0.010 ^{a}

The total and maximum volumetric and areal productivities were calculated and the values obtained are presented in Table 2. The semi-continuous and continuous operation regimes reached similar productivity values, with a volumetric productivity of 0.165 ± 0.013 and 0.154 ± 0.021 g L^{-1} day^{-1}, respectively, and areal productivities of 16.3 ± 1.3 and 15.2 ± 2.0 g m^{-2} day^{-1}, respectively. On the other hand, these values were around 1.5-fold higher than the volumetric (0.108 ± 0.01 g L^{-1} day^{-1}) and areal productivity (10.7 ± 1.1 g m^{-2} day^{-1}) obtained by the batch regime. The batch and continuous operation systems presented similar values regarding the maximum volumetric and areal productivities, respectively, 0.333 ± 0.036 g L^{-1} day^{-1} and 0.266 ± 0.028 g L^{-1} day^{-1}, while the semi-continuous showed a significantly higher value of 0.427 ± 0.020 g L^{-1} day^{-1}.

Table 2. Volumetric and areal productivities of biomass during the whole test, and the maximum in a short interval, in the cultivation of *Nannochloropsis oceanica*, for three operation regimes in 2.6 m^3 PBRs. The values represent the average and standard deviation of three biologically independent replicates. Different letters within the same column represent significantly different values (p-value < 0.05).

Production Regime	Volumetric Productivity (g L^{-1} day^{-1})	Maximum Volumetric Productivity (g L^{-1} day^{-1})	Areal Productivity (g m^2 day^{-1})	Maximum Areal Productivity (g m^2 day^{-1})	Photosynthetic Efficiency (%)
Batch	0.108 ± 0.011 ^{a}	0.333 ± 0.03 ^{a}	10.7 ± 1.1 ^{a}	31.6 ± 3.4 ^{a}	0.358 ± 0.016 ^{a}
Semi-continuous	0.165 ± 0.013 ^{b}	0.427 ± 0.020 ^{b}	16.3 ± 1.3 ^{b}	40.4 ± 1.9 ^{b}	0.436 ± 0.043 ^{a}
Continuous	0.154 ± 0.021 ^{b}	0.266 ± 0.028 ^{a}	15.2 ± 2.0 ^{b}	25.2 ± 2.6 ^{a}	0.481 ± 0.073 ^{a}

The registered batch productivity value is in accordance with the 0.15 g L^{-1} day^{-1} reported by Quinn et al. (2012) for flat panel outdoor batch growth of *Nannochloropsis oculata* in the same period of the year (May to June) in the northern hemisphere (Fort Collins, CO, USA) [31].

Ledda et al. (2015) and Chini Zittelli et al. (1999) reported productivities of 0.48 g L^{-1} day^{-1} in a 340 L vertical tubular reactor operated in a semi-continuous regime, with a dilution rate of 0.33 day^{-1}, and 0.56 g L^{-1} day^{-1} for a 36.6 L horizontal tubular reactor operated in the semi-continuous regime, in the same period of the year in the northern hemisphere (Florence, Italy), respectively [32,33]. These values are higher than the 0.165 ± 0.013 g L^{-1} day^{-1} obtained in this study for semi-continuous production of *N. oceanica*. These differences are probably related to the different size and geometries of the

PBRs used in the different works, and the optimized dilution rate used by Ledda et al. (2015). In addition, Ledda et al. (2015) reported a lower maximum areal productivity of 27 g m^{-2} day^{-1}, against 40.4 ± 1.9 g m^{-2} day^{-1} in this study, suggesting that the PBR configuration used by the authors is less efficient than the one used in the present study, in what concerns the occupied area [32]. In a laboratory study, Cai et al. (2013) reported volumetric productivity of 0.068–0.092 g L^{-1} day^{-1} for the batch regime, against 0.087–0.121 g L^{-1} day^{-1} for the semi-continuous operation regime when growing *Nannochloropsis salina* in a 2 L flask reactor and a constant photosynthetic photon flux of approximately 200 µmol $m^{-2}s^{-1}$ with a harvesting frequency of three times a week [14]. However, the productivity values cannot be directly compared with those of the present study, as there is a significant gap in the volumes used. It is important to note, though, that similarly to the present study, Cai et al. (2013) also reported an increase in volumetric productivity when switching the operation regime from batch to semi-continuous [14].

The continuous operation regime presented a volumetric productivity of 0.154 ± 0.021 g L^{-1} day^{-1}, for a dilution rate of 0.14 day^{-1}. This value of global productivity is in the range of the ones reported by Camacho-Rodriguez et al. (2014) for *Nannochloropsis gaditana* in a 2.5 m^3 tubular PBR of 0.12–0.20 g L^{-1} day^{-1} with similar temperature and radiance conditions in Almeria, Spain, for a dilution rate of 0.3 day^{-1} [34]. Despite this, San Pedro et al. (2014) reached maximum volumetric productivity of 0.25 g L^{-1} day^{-1}, for *N. gaditana*, in an outdoor 340 L vertical tubular PBR with 0.1 day^{-1} dilution rate in similar weather conditions [15]. This higher volumetric productivity is probably the consequence of the different configurations and considerably smaller scale used by the author [15].

In what concerns photosynthetic efficiency, there were no significant differences between the 0.358 ± 0.016, 0.436 ± 0.043 and 0.481± 0.073% obtained using batch, semi-continuous and continuous operation regimes, respectively. During the present study, lower photosynthetic efficiency values than the ones reported in the literature were obtained: 0.358–0.481%, in contrast with the 1.2–1.8% obtained during the cultivation of *Nannochloropsis* sp. in an outdoor 0.56 m^3 horizontal tubular reactor [35]. The reason can be attributed to the higher illuminated area in the DeVree et al. (2015) study, 73% [35] against 61%.

The maximum volumetric and maximum areal productivities were higher when using a semi-continuous regime, even though this regime did not present a significantly different global volumetric productivity when comparing with the batch and continuous operation regimes.

3.2. Biochemical Profile

The biomass produced using the different operation regimes was biochemically characterized at the end of the trial (Table 3).

Table 3. Proximate composition, in percentage of total dry weight, of *Nannochloropsis oceanica* grown in 2.6 m^3 tubular PBRs, using different operation regimes. For proteins and lipids, the values represent the average and standard deviation of two biologically independent replicates and two analytical replicates. Different letters within the same column represent significantly different values. Regarding ashes and, consequently, carbohydrates, the presented values are the average of two biological replicates and the minimum and maximum values (p-value < 0.05).

Production Regime	Protein (%)	Lipids (%)	Ash (%)			Carbohydrates (%)		
			Average	Min	Max	Average	Min	Max
Batch	29.6 ± 3.6 [a]	22.0 ± 3.1 [a]	10.7	9.4	11.9	37.8	34.7	40.8
Semi-continuous	28.6 ± 2.8 [a]	24.0 ± 5.3 [a]	14.9	13.0	16.8	32.5	26.4	38.6
Continuous	28.9 ± 2.5 [a]	19.1 ± 2.5 [a]	12.9	12.0	13.8	39.1	36.4	41.8

The macronutrient composition of *N. oceanica* biomass, regarding proteins and lipids, ranged between, 28.6 ± 2.8 to 29.6 ± 3.6%, and 19.1 ± 2.5 to 24.0 ± 5.3% of biomass DW, respectively. No significant differences were observed regarding macronutrients' composition between the biomass produced among the different operation regimes. The global lipid productivity (Table 4) was 22.9, 39.6 and 29.5 mg L^{-1} day^{-1}, for the batch, semi-continuous and continuous operation regimes, respectively; the productivity in the semi-continuous regime was significantly higher than that obtained under batch conditions.

Table 4. Global lipid productivity in the cultivation of *Nannochloropsis oceanica*, for each of the operation regimes in 2.6 m^3 PBRs. The values represent the average and standard deviation of three biologically independent replicates. Different letters within the same column represent significantly different values (p-value < 0.05).

Production Regime	Lipid Productivity (mg L^{-1} day^{-1})
Batch	22.9 ± 3.7 [a]
Semi-continuous	39.6 ± 3.1 [b]
Continuous	29.5 ± 3.9 [ab]

San Pedro et al. (2014) obtained an average protein content of 36.9% in biomass DW of *N. gaditana* grown in 340 L outdoor tubular reactors operated in continuous mode, for dilution rates varying from 0.1 to 0.35 day^{-1} [15]. The same study reported a lipid content of biomass DW within the range 17.7 and 26.7%, values similar to the ones obtained in this work [15]. In the present study, the lipid content was the same in all the operation regimes, which was also verified by Zhang et al. (2014) when growing *Nannochloropsis* sp. in continuous and batch regimes using 2 L glass bubble column PBRs at the laboratory scale [36]. Regarding the lipid productivity, the values in the present study (22.9–39.6 mg L^{-1} day^{-1}) were higher than the ones obtained by Nogueira et al. (2020) for *N. gaditana* grown in an outdoor 100 L vertical column PBR operated in the semi-continuous regime with a dilution rate of 0.5 day^{-1} in winter at Porto Santo, Portugal (7.2–17.82 mg L^{-1} day^{-1}) [37]. The difference can be attributed to the temperature and irradiance differences between seasons since the literature shows a positive correlation between temperature and lipid content in *Nannochloropsis* sp. [38,39]. San Pedro et al. (2014) reported lipid productivity values between 50–60 mg L^{-1} day^{-1} in a 340 L vertical tubular PBR operated in continuous mode using a 0.1 day^{-1} dilution rate under similar weather conditions. The higher values are due to higher biomass productivity achieved by the authors, since the attained lipid content was similar (20–30%) [15]

The lipid content values reached in the present study in a tubular PBR for *N. oceanica* (19.1–24.0%) were higher than the ones obtained by Cunha et al. (2020) (13.2–19.0%), in a raceway reactor using the same strain, in the same location and time of the year [40]. This result highlights that the tubular horizontal PBR produces biomass with higher quality for biofuel than that cultivated in open systems.

The FAME profile represents an important factor for biofuel microalgal biomass applications. The FAME profiles obtained in this study are presented in Table 5; it shows only the fatty acids above 0.50% of total FAME.

The major FAME observed in *N. oceanica* grown in an outdoor 2.6 m^3 tubular reactor were C16:1 and C16:0, together representing more than 60% of total FAME, followed by C18:1 and C20:5 and C14:0. Comparing the three operation regimes, there were only significant differences in the C16:0 and in the C18:0, which were higher in the semi-continuous than in the batch operation regime, not being significantly different from the continuous regime. In terms of saturation ratios, the present work obtained a higher percentage of saturated fatty acids (SFA), around 45% of total FAME, against around 35% of total FAME obtained by San Pedro et al. (2014) in an outdoor tubular reactor [15]. Additionally, in the present study the percentage polyunsaturated fatty acids (PUFA) was lower than that commonly reported in the literature, 10–14% of total FAME, against the 20% of total FAME, reported by San Pedro et al. (2014) [15]. Regarding the saturation degree, the only significant difference was a higher percentage of SFA in the semi-continuous operated culture (46.19 ± 1.64%) when compared to the batch (41.52 ± 1.60%). This difference did not spread to the PUFA percentage and PUFA/SFA ratio, which was not significantly different between the operation regimes, ranging between 10.45–13.73% and 0.23–0.33, respectively.

Table 5. Fatty acid methyl esters (FAME) content and profile, presented in percentage of total FAME and saturation distribution of the FAME profile in three different operation regimes of *Nannochloropsis oceanica* grown in 2.6 m^3 outdoor tubular PBR. The values represent the average and standard deviation of two biologically independent replicates and two analytical replicates. Different letters within the same row represent significantly different values (p-value < 0.05).

FAME	Batch	Semi-Continuous	Continuous
C 14:0 (%)	6.88 ± 0.44 ^{a}	5.99 ± 0.66 ^{a}	6.02 ± 0.39 ^{a}
C 16:1 (%)	32.79 ± 0.55 ^{a}	32.28 ± 0.65 ^{a}	32.28 ± 1.79 ^{a}
C 16:0 (%)	33.58 ± 1.76 ^{a}	38.24 ± 2.10 ^{b}	36.20 ± 0.94 ^{ab}
C 18:2 ω6 (%)	0.75 ± 0.09 ^{a}	0.59 ± 0.59 ^{b}	0.76 ± 0.08 ^{a}
C 18:1 (%)	11.96 ± 2.98 ^{a}	11.07 ± 0.76 ^{a}	12.40 ± 2.46 ^{a}
C 18:0 (%)	1.05 ± 0.26 ^{a}	1.95 ± 0.26 ^{b}	1.59 ± 0.19 ^{ab}
C 20:4 ω6 (%)	1.85 ± 0.48 ^{a}	1.42 ± 0.30 ^{a}	1.46 ± 0.11 ^{a}
C 20:5 ω3 (%)	11.12 ± 3.4 ^{a}	8.44 ± 2.00 ^{a}	9.16 ± 0.08 ^{a}
SFA (%)	41.52 ± 1.60 ^{a}	46.19 ± 1.64 ^{b}	43.80 ± 1.40 ^{ab}
MUFA (%)	44.76 ± 2.43 ^{a}	43.36 ± 0.92 ^{a}	44.68 ± 067 ^{a}
PUFA (%)	13.73 ± 1.00 ^{a}	10.45 ± 2.31 ^{a}	11.52 ± 0.88 ^{a}
PUFA/SFA	0.33 ± 0.11 ^{a}	0.23 ± 0.06 ^{a}	0.26 ± 0.03 ^{a}

Literature reports that a higher percentage of SFA is necessary for a good biodiesel production feedstock as the percentage of SFA is positively correlated with the cold filter plugging point (CFPP) and the cetane number, both important parameters in biodiesel quality [41,42]. Chen et al. (2012) reported a *Nannochloropsis* sp. biomass with a similar FAME profile and SFA content (35%) from which resulted a biodiesel with an HHV comparable to fossil fuels [43].

4. Conclusions

Nannochloropsis oceanica was successfully grown in pilot-scale tubular PBRs in batch, semi-continuous and continuous operation regimes, with the outdoor light and temperature conditions. Regarding biomass productivity, the semi-continuous and continuous regimes achieved higher values when compared to the batch regime. In terms of protein and lipids (19–24%) content as well as in the fatty acid profile there were no significant differences between the three operation regimes. In addition, the semi-continuous operation regime allowed obtaining higher lipid productivities, indicating that it was the most promising operation regime to grow *N. oceanica* as biodiesel feedstock.

Nevertheless, further work in biomass productivity optimization is needed, such as the optimization of dilution rates in the semi-continuous and continuous operation regimes. The ideal period of the day to harvest the biomass also needs to be explored to maximize lipid content and the SFA/PUFA ratio.

Supplementary Materials: The following are available online at https://www.mdpi.com/1996-1073/14/6/1542/s1. Figure S1: Correlation between the optical density at 540 nm and the dry weight of an autotrophic culture of Nannochloropsis oceanica. Figure S2: Correlation between the difference of the optical density at 220 nm and two times the optical density at 275 nm and nitrate concentration in mM. Figure S3: Microscopic picture (400×) of N. oceanica grown in Allmicroalgae's pilot-scale horizontal tubular photobioreactors.

Author Contributions: H.P., M.M., J.V. and J.S. designed the experiments. I.G. performed the experiments. J.T.S. assisted in the outdoor assays. T.S. assisted in the biochemical analyses. I.G. and M.C. wrote the manuscript. All authors have read and agreed to the published version of the manuscript.

Funding: This work has received funding under the project AlgaValor, from the Portugal 2020 program (grant agreement nº POCI-01-0247-FEDER-035234; LISBOA-01-0247-FEDER-035234; ALG-01-0247-FEDER-035234).

Institutional Review Board Statement: Not applicable.

Informed Consent Statement: Not applicable.

Data Availability Statement: Not applicable.

Acknowledgments: The authors greatly thank Allmicroalgae's staff for all the support given throughout this work.

Conflicts of Interest: The authors declare no conflict of interest.

References

1. IPCC. *Climate Change 2014: Synthesys Report. Contribution of Working Groups I, II and III to the Fifth Assessment Report of the Intergovernmental Panel on Climate Change*; Core Writing Team, Pachauri, R.K., Meyer, L.A., Eds.; IPCC: Geneva, Switzerland, 2015.
2. Sajjadi, B.; Chen, W.Y.; Raman, A.A.A.; Ibrahim, S. Microalgae lipid and biomass for biofuel production: A comprehensive review on lipid enhancement strategies and their effects on fatty acid composition. *Renew. Sustain. Energy Rev.* **2018**, *97*, 200–232. [CrossRef]
3. Rastogi, R.P.; Madamwar, D.; Pandey, A. *Algal Green Chemistry: Recent Progress in Biotechnology*; Elsevier: Amsterdam, The Netherlands, 2017; ISBN 9780444640413.
4. Pires, J.C.M.; Alvim-Ferraz, M.C.M.; Martins, F.G.; Simões, M. Carbon dioxide capture from flue gases using microalgae: Engineering aspects and biorefinery concept. *Renew. Sustain. Energy Rev.* **2012**, *16*, 3043–3053. [CrossRef]
5. Salama, E.S.; Kurade, M.B.; Abou-Shanab, R.A.I.; El-Dalatony, M.M.; Yang, I.S.; Min, B.; Jeon, B.H. Recent progress in microalgal biomass production coupled with wastewater treatment for biofuel generation. *Renew. Sustain. Energy Rev.* **2017**, *79*, 1189–1211. [CrossRef]
6. Gupta, P.L.; Lee, S.M.; Choi, H.J. A mini review: Photobioreactors for large scale algal cultivation. *World J. Microbiol. Biotechnol.* **2015**, *31*, 1409–1417. [CrossRef] [PubMed]
7. Ma, Y.; Wang, Z.; Yu, C.; Yin, Y.; Zhou, G. Evaluation of the potential of 9 Nannochloropsis strains for biodiesel production. *Bioresour. Technol.* **2014**, *167*, 503–509. [CrossRef]
8. Ashour, M.; Elshobary, M.E.; El-Shenody, R.; Kamil, A.W.; Abomohra, A.E.F. Evaluation of a native oleaginous marine microalga Nannochloropsis oceanica for dual use in biodiesel production and aquaculture feed. *Biomass Bioenergy* **2019**, *120*, 439–447. [CrossRef]
9. Liu, J.; Song, Y.; Qiu, W. Oleaginous microalgae *Nannochloropsis* as a new model for biofuel production: Review & analysis. *Renew. Sustain. Energy Rev.* **2017**, *72*, 154–162.
10. Richmond, A.; Hu, Q. *Handbook of Microalgal Culture: Applied Phycology and Biotechnology: Second Edition*; Blackwell Publishing: Hoboken, NJ, USA, 2004; ISBN 9781118567166.
11. Pereira, H.; Gangadhar, K.N.; Schulze, P.S.C.; Santos, T.; De Sousa, C.B.; Schueler, L.M.; Custódio, L.; Malcata, F.X.; Gouveia, L.; Varela, J.C.S.; et al. Isolation of a euryhaline microalgal strain, Tetraselmis sp. CTP4, as a robust feedstock for biodiesel production. *Sci. Rep.* **2016**, *6*, 1–11. [CrossRef]
12. Kokkinos, N.; Lazaridou, A.; Stamatis, N.; Orfanidis, S.; Mitropoulos, A.C.; Christoforidis, A.; Nikolaou, N. Biodiesel production from selected microalgae strains and determination of its properties and combustion specific characteristics. *J. Eng. Sci. Technol. Rev.* **2015**, *8*, 1–6. [CrossRef]
13. Borowitzka, M.A.; Moheimani, N.R. Sustainable biofuels from algae. *Mitig. Adapt. Strateg. Glob. Chang.* **2013**, *18*, 13–25. [CrossRef]
14. Cai, T.; Park, S.Y.; Racharaks, R.; Li, Y. Cultivation of Nannochloropsis salina using anaerobic digestion effluent as a nutrient source for biofuel production. *Appl. Energy* **2013**, *108*, 486–492. [CrossRef]
15. San Pedro, A.; Gonzalez-Lopez, C.V.; Acien, F.G.; Molina-Grima, E. Outdoor pilot-scale production of *Nannochloropsis gaditana*: Influence of culture parameters and lipid production rates in tubular photobioreactors. *Bioresour. Technol.* **2014**, *169*, 667–676. [CrossRef] [PubMed]
16. Benvenuti, G.; Bosma, R.; Ji, F.; Lamers, P.; Barbosa, M.J.; Wijffels, R.H. Batch and semi-continuous microalgal TAG production in lab-scale and outdoor photobioreactors. *J. Appl. Phycol.* **2016**, *28*, 3167–3177. [CrossRef] [PubMed]
17. Ganesan, R.; Manigandan, S.; Samuel, M.S.; Shanmuganathan, R.; Brindhadevi, K.; Lan Chi, N.T.; Duc, P.A.; Pugazhendhi, A. A review on prospective production of biofuel from microalgae. *Biotechnol. Rep.* **2020**, *27*, e00509. [CrossRef] [PubMed]
18. Zohri, A.-N.A.; Ragab, S.W.; Mekawi, M.I.; Mostafa, O.A.A. Comparison between batch, fed-batch, semi-continuous and continuous techniques for bio-ethanol production from a mixture of egyptian cane and beet molasses. *Egypt. Sugar J.* **2017**, *9*, 89–111.
19. Lavens, P.; Sorgeloos, P.; Food and Agriculture Organization of the United Nations. Microalgae. In *Manual on the Production and Use of Live Food for Aquaculture*; Food and Agriculture Organization of the United Nations: Roma, Italy, 1996; p. 295. ISBN 9251039348.
20. Brown, M.R.; Garland, C.D.; Jeffrey, S.W.; Jameson, I.D.; Leroi, J.M. The gross and amino acid compositions of batch and semi-continuous cultures of *Isochrysis* sp. (clone T.ISO), *Pavlova lutheri* and *Nannochloropsis oculata*. *J. Appl. Phycol.* **1993**, *5*, 285–296. [CrossRef]
21. Croughan, M.S.; Konstantinov, K.B.; Cooney, C. The future of industrial bioprocessing: Batch or continuous? *Biotechnol. Bioeng.* **2015**, *112*, 648–651. [CrossRef] [PubMed]

22. Bux, F. *Biotechnological Applications of Microalgae: Biodiesel and Value-Added Products*; CRC Press: Boca Raton, FL, USA, 2013; ISBN 9781466515307.
23. Wang, B.; Lan, C.Q.; Horsman, M. Closed photobioreactors for production of microalgal biomasses. *Biotechnol. Adv.* **2012**, *30*, 904–912. [CrossRef]
24. Mata, T.M.; Martins, A.A.; Caetano, N.S. Microalgae for biodiesel production and other applications: A review. *Renew. Sustain. Energy Rev.* **2010**, *14*, 217–232. [CrossRef]
25. Callejón-Ferre, A.J.; Velázquez-Martí, B.; López-Martínez, J.A.; Manzano-Agugliaro, F. Greenhouse crop residues: Energy potential and models for the prediction of their higher heating value. *Renew. Sustain. Energy Rev.* **2011**, *15*, 948–955. [CrossRef]
26. Armstrong, F.A.J. Determination of Nitrate in Water by Ultraviolet Spectrophotometry. *Anal. Chem.* **1963**, *35*, 1292–1294. [CrossRef]
27. Bligh, E.G.; Dyer, W.J. A rapid method of total lipid extraction and purification. *Can. J. Biochem. Physiol.* **1959**, *37*, 911–917. [CrossRef]
28. Guedes, A.C.; Barbosa, C.R.; Amaro, H.M.; Pereira, C.I.; Malcata, F.X. Microalgal and cyanobacterial cell extracts for use as natural antibacterial additives against food pathogens. *Int. J. Food Sci. Technol.* **2011**, *46*, 862–870. [CrossRef]
29. Lepage, G.; Roy, C.C. Improved recovery of fatty acid through direct transesterification without prior extraction or purification. *J. Lipid Res.* **1984**, *25*, 1391–1396. [CrossRef]
30. Pereira, H.; Barreira, L.; Mozes, A.; Florindo, C.; Polo, C.; Duarte, C.V.; Custádio, L.; Varela, J. Microplate-based high throughput screening procedure for the isolation of lipid-rich marine microalgae. *Biotechnol. Biofuels* **2011**, *4*, 61. [CrossRef]
31. Quinn, J.C.; Yates, T.; Douglas, N.; Weyer, K.; Butler, J.; Bradley, T.H.; Lammers, P.J. *Nannochloropsis* production metrics in a scalable outdoor photobioreactor for commercial applications. *Bioresour. Technol.* **2012**, *117*, 164–171. [CrossRef] [PubMed]
32. Ledda, C.; Romero Villegas, G.I.; Adani, F.; Acién Fernández, F.G.; Molina Grima, E. Utilization of centrate from wastewater treatment for the outdoor production of *Nannochloropsis gaditana* biomass at pilot-scale. *Algal Res.* **2015**, *12*, 17–25. [CrossRef]
33. Chini Zittelli, G.; Lavista, F.; Bastianini, A.; Rodolfi, L.; Vincenzini, M.; Tredici, M.R. Production of eicosapentaenoic acid by Nannochloropsis sp. cultures in outdoor tubular photobioreactors. *Prog. Ind. Microbiol.* **1999**, *35*, 299–312.
34. Camacho-Rodríguez, J.; González-Céspedes, A.M.; Cerón-García, M.C.; Fernández-Sevilla, J.M.; Acién-Fernández, F.G.; Molina-Grima, E. A quantitative study of eicosapentaenoic acid (EPA) production by Nannochloropsis gaditana for aquaculture as a function of dilution rate, temperature and average irradiance. *Appl. Microbiol. Biotechnol.* **2014**, *98*, 2429–2440. [CrossRef]
35. De Vree, J.H.; Bosma, R.; Janssen, M.; Barbosa, M.J.; Wijffels, R.H. Comparison of four outdoor pilot-scale photobioreactors. *Biotechnol. Biofuels* **2015**, *8*, 215. [CrossRef]
36. Zhang, D.; Xue, S.; Sun, Z.; Liang, K.; Wang, L.; Zhang, Q.; Cong, W. Investigation of continuous-batch mode of two-stage culture of Nannochloropsis sp. for lipid production. *Bioprocess Biosyst. Eng.* **2014**, *37*, 2073–2082. [CrossRef] [PubMed]
37. Nogueira, N.; Nascimento, F.J.A.; Cunha, C.; Cordeiro, N. Nannochloropsis gaditana grown outdoors in annular photobioreactors: Operation strategies. *Algal Res.* **2020**, *48*, 101913. [CrossRef]
38. Fakhry, E.M.; El Maghraby, D.M. Lipid accumulation in response to nitrogen limitation and variation of temperature in nannochloropsis salina. *Bot. Stud.* **2015**, *56*, 6. [CrossRef]
39. Carneiro, M.; Cicchi, B.; Maia, I.B.; Pereira, H.; Zittelli, G.C.; Varela, J.; Malcata, F.X.; Torzillo, G. Effect of temperature on growth, photosynthesis and biochemical composition of Nannochloropsis oceanica, grown outdoors in tubular photobioreactors. *Algal Res.* **2020**, *49*, 101923. [CrossRef]
40. Cunha, P.; Pereira, H.; Costa, M.; Pereira, J.; Silva, J.T.; Fernandes, N.; Varela, J.; Silva, J.; Simões, M. Nannochloropsis oceanica Cultivation in Pilot-Scale Raceway Ponds—From Design to Cultivation. *Appl. Sci.* **2020**, *10*, 1725. [CrossRef]
41. Golimowski, W.; Berger, W.A.; Pasyniuk, P.; Rzeźnik, W.; Czechlowski, M.; Koniuszy, A. Biofuel parameter dependence on waste fats' fatty acids profile. *Fuel* **2017**, *197*, 482–487. [CrossRef]
42. Wu, H.; Miao, X. Biodiesel quality and biochemical changes of microalgae Chlorella pyrenoidosa and Scenedesmus obliquus in response to nitrate levels. *Bioresour. Technol.* **2014**, *170*, 421–427. [CrossRef]
43. Chen, L.; Liu, T.; Zhang, W.; Chen, X.; Wang, J. Biodiesel production from algae oil high in free fatty acids by two-step catalytic conversion. *Bioresour. Technol.* **2012**, *111*, 208–214. [CrossRef] [PubMed]

Article

Pigments Production, Growth Kinetics, and Bioenergetic Patterns in *Dunaliella tertiolecta* (Chlorophyta) in Response to Different Culture Media

Yanara Alessandra Santana Moura [1], Daniela de Araújo Viana-Marques [2], Ana Lúcia Figueiredo Porto [1], Raquel Pedrosa Bezerra [1] and Attilio Converti [3,*]

1 Department of Morphology and Animal Physiology, Federal Rural University of Pernambuco-UFRPE, Dom Manoel de Medeiros Ave, Recife PE 52171-900, Brazil; yanara.moura@ufrpe.br (Y.A.S.M.); ana.porto@ufrpe.br (A.L.F.P.); raquel.pbezerra@ufrpe.br (R.P.B.)

2 Laboratory of Biotechnology Applied to Infectious and Parasitic Diseases, Biological Science Institute, University of Pernambuco-UPE, Rua Arnóbio Marquês, Recife PE 50100-130, Brazil; daniela.viana@upe.br

3 Department of Civil, Chemical and Environment Engineering, Pole of Chemical Engineering, University of Genoa (UNIGE), Via Opera Pia 15, 16145 Genoa, Italy

* Correspondence: converti@unige.it; Tel.: +39-010-3352593

Received: 25 September 2020; Accepted: 7 October 2020; Published: 14 October 2020

Abstract: This work dealt with the study of growth parameters, pigments production, and bioenergetic aspects of the microalga *Dunaliella tertiolecta* in different culture media. For this purpose, cultures were carried out in Erlenmeyer flasks containing F/2 medium, Bold's Basal medium, or an alternative medium made up of the same constituents of the Bold's Basal medium dissolved in natural seawater instead of distilled water. *D. tertiolecta* reached the highest dry cell concentration (X_{max} = 1223 mgDM·L^{-1}), specific growth rate (μ_{max} = 0.535 d^{-1}), cell productivity (P_X = 102 mgDM·L^{-1}·d^{-1}), and photosynthetic efficiency (PE = 14.54%) in the alternative medium, while the highest contents of carotenoids (52.0 mg·g^{-1}) and chlorophyll (108.0 mg·g^{-1}) in the biomass were obtained in Bold's Basal medium. As for the bioenergetic parameters, the biomass yield on Gibbs energy dissipation was higher and comparable in both seawater-based media. However, the F/2 medium led to the highest values of moles of photons absorbed to produce 1 C-mol of biomass (n_{Ph}), total Gibbs energy absorbed by the photosynthesis (ΔG_a) and released heat (Q), as well as the lowest cell concentration, thus proving to be the least suitable medium for *D. tertiolecta* growth. On the other hand, the highest values of molar development of O_2 and consumption of H^+ and H_2O were obtained in the alternative medium, which also ensured the best kinetic parameters, thereby allowing for the best energy exploitation for cell growth. These results demonstrate that composition of culture medium for microalgae cultivation has different effects on pigments production, growth kinetics, and bioenergetics parameters, which should be taken into consideration for any use of biomass, including as raw material for biofuels production.

Keywords: cell growth; microalgae; chlorophyll; carotenoids; energetic yield

1. Introduction

Species belonging to the *Dunaliella* genus are unicellular and biflagellate green microalgae that grow under high salinity and irradiance conditions [1]. Large-scale *Dunaliella* sp. cultivation has been widely applied because its biomass is a source of valuable compounds such as polysaccharides, lipids, vitamins, proteins, and pigments, mainly carotenoids and chlorophyll, that can be used in several biotechnological applications [2–5]. Therefore, *Dunaliella* sp. have high potential to contribute for a sustainable industry through the generation of high added-value products [6,7].

Microalgae can be cultivated under different conditions, and several factors such as light intensity, photobioreactor type and configuration, mixing regime, temperature, and culture medium composition may influence cell growth and biomass composition [7]. Even though nutrient limitation reduces biomass concentration [8–11], it promotes the accumulation of carotenoids affecting the photosynthetic pathway [12]. Therefore, the composition of culture medium is one of the most important issues to be tackled in commercial production of microalgal biomass.

Microalgae biomass is considered a potential source of biofuels due to its high energy content and for being environmentally friendly compared to fossil fuels [13]. Nevertheless, new production methods have been proposed to improve its feasibility and cost effectiveness [14,15]. Among microalgae, *Dunaliella* sp. has stood out for being a halotolerant unicellular green alga, easily cultivable and resistant to contamination by other species. In addition, *D. tertiolecta* is highly adaptable to abiotic stress and able to accumulate high content of lipids as a potential feedstock to produce biofuels and pigments, especially carotenoids [16–18].

β-Carotene is the best-known compound of this class and has received a lot of attention due to antioxidant, anti-inflammatory, and anticarcinogenic activities [19], with a market value of $1.5 billion in 2017 and an estimated value for 2022 of $2.0 billion [20]. Chlorophyll is another important pigment found in *Dunaliella* sp. biomass, which is commercially used as a natural food coloring agent thanks to its green color and is consumed as nutraceutical due to beneficial properties such as antioxidant, anti-inflammatory, antimutagenic, and antibacterial activities [21,22].

Additionally, culture medium can influence other parameters such as photosynthetic efficiency (*PE*) and microalgal bioenergetics. *PE* is defined as the efficiency by which photons are absorbed, converted into chemical energy and stored as biomass [23], while bioenergetics refers to how energy is managed in living cells [24,25].

The aim of this work was to investigate *D. tertiolecta* growth, pigments production, and bioenergetic parameters in different culture media in order to select the best cultivation conditions to overproduce such added-value compounds. Moreover, within the framework of a biorefinery concept, the lipid fraction of the exhaust biomass will be exploited in the next effort as a raw material for energy production.

2. Materials and Methods

2.1. Microalgal Cultivations

Dunaliella tertiolecta (UTEX 999) was cultivated in three different media, namely the F/2 medium, the Bold's Basal medium and an alternative saline medium. The F/2 medium contained mineral salts, trace elements, and vitamins dissolved in natural seawater previously filtered through membranes with 0.22 μm pore diameter [26]; the Bold's Basal medium is a well-known broth made up of mineral salts dissolved in distilled water [27]; while the alternative medium was composed of the same mineral salts as the Bold's Basal one but using seawater (Table 1) instead of distilled water. The compositions of these culture media are listed in Table 2.

Table 1. Chemical composition of seawater.

Salts	Concentration (mM)
$NaCl$	470
$MgCl_2$	420
$MgSO_4$	640
$CaCl_2$	280
KCl	8
$NaHCO_3$	580

Note: Adapted from Jesus and Maciel Filho [28].

Table 2. Nutrient composition of different culture media. F/2 medium (**A**), Bold's Basal medium (**B**), alternative medium (**C**).

Constituent	Medium		
	A (mM) [a]	B (mM) [b]	C (mM) [a]
$NaNO_3$	0.8823	2.9410	2.9410
$MgSO_4 \cdot 7H_2O$	-	0.6231	0.6231
NaCl	-	0.4277	0.4277
K_2HPO_4	-	0.4305	0.4305
KH_2PO_4	-	1.2860	1.286
$CaCl_2 \cdot 2H_2O$	-	0.1702	0.1702
$ZnSO_4 \cdot 7H_2O$	0.0765	0.0306	0.0306
$MnCl_2 \cdot 4H_2O$	0.9098	0.0072	0.0072
MoO_3	-	0.0047	0.0047
$CuSO_4 \cdot 5H_2O$	0.0392	0.0062	0.0062
$Co(NO_3)_2 \cdot 6H_2O$	-	0.0016	0.0016
H_3BO_3	-	0.1846	0.1846
EDTA	-	0.1710	0.1710
KOH	-	0.5525	0.5525
$FeSO_4 \cdot 7H_2O$	-	0.0125	0.0125
$H_2SO_4 \cdot$(conc)	-	0.0071	0.0071
$NaH_2PO_4 \cdot H_2O$	0.0362	-	-
$Na_2SiO_3 \cdot 9H_2O$	0.1055	-	-
$FeCl_3 \cdot 6H_2O$	0.0116	-	-
$Na_2EDTA \cdot 2H_2O$	0.0116	-	-
$Na_2MoO_4 \cdot 2H_2O$	0.0240	-	-
$CoCl_2 \cdot 6H_2O$	0.0770	-	-
Thiamine HCl (vit. B1)	0.0003	-	-
Biotin (vit. H)	0.0020	-	-
Cyanocobalamin (vit. B12)	0.0003	-	-

[a] diluted in natural seawater. [b] diluted in distilled water.

The microalga was cultivated in 500 mL Erlenmeyer flasks containing 200 mL of the selected culture medium with initial cell concentration of 80 mg·L^{-1} at room temperature, under light intensity of 45 ± 5 μmol photon m^{-2}· s^{-1} and continuous aeration at a flow rate of 2 L·min^{-1} for 13 days. Each culture was carried out in duplicate.

2.2. Analytical Methods

D. tertiolecta cell motility was examined using a light microscope (model Nikon Eclipse E200MV R, Nikon Instruments Inc, Shanghai, China) (magnification 400× *g*). Optical density (*OD*) was determined spectrophotometrically at a wavelength of 680 nm (OD_{680}) by a UV/Visible spectrophotometer using a calibration curve correlating dry biomass concentration to OD_{680}. Biomass concentration was measured daily along cultivations, and values were given as the average of triplicate determinations (n = 3) with standard deviation. The Equation (1) was derived from the standard curve:

$$\text{Cell concentration } (g_{DM} \cdot L^{-1}) = 0.003\ OD_{680} + 0.098 \quad (1)$$

The contents of chlorophylls and carotenoids were quantified by the method described by Lichtenthaler [29]. Briefly, *D. tertiolecta* dry biomass was mixed with acetone and centrifuged at 500× *g* for 5 min to recover the pigment-containing supernatant, whose absorbance (Abs) was measured at 647, 663, and 470 nm. Chlorophyll and carotenoid contents were calculated using the following equations:

$$\text{Chlorophyll a: } C_a = [(12.25\ Abs_{663\ nm}) - (2.79\ Abs_{647\ nm})] \quad (2)$$

$$\text{Chlorophyll b: } C_b = [(21.50\ Abs_{647\ nm}) - (5.10\ Abs_{663\ nm})] \quad (3)$$

$$\text{Carotenoids: } [(1000 \text{ Abs}_{470\text{ nm}}) - (1.82\ C_a) - (85.02\ C_b)]/198 \tag{4}$$

2.3. Growth Parameters

Cell productivity (P_X) was calculated by dividing the variation in cell concentration ($X_{max} - X_i$) by the time of cultivation (T_c), according to the equation:

$$P_X = \frac{X_{max} - X_i}{T_c} \tag{5}$$

where X_{max} is the maximum cell concentration and X_i the initial cell concentration.

Maximum specific growth rate (μ_{max}) was calculated as:

$$\mu_{max} = \frac{1}{\Delta t} \ln \frac{X_j}{X_{j-1}} \tag{6}$$

where X_j and X_{j-1} are cell concentrations at the end and the beginning of each time interval ($\Delta t = 1$ day).

2.4. Photosynthetic Efficiency

Photosynthetic efficiency (PE) was determined by converting the photosynthetic photon flux density (PPFD) to photosynthetic active radiation (PAR). The input of PAR (IPAR) into the Erlenmeyer was obtained by multiplying PAR by the illuminated surface (m^2). Therefore, PE was calculated by the equation:

$$PE = \frac{r_G H_G}{\text{IPAR}} \times 100 \tag{7}$$

where r_G is the maximum daily growth rate ($g_{DM} \cdot d^{-1}$) and H_G = 21.01 $kJ \cdot g_{DM}^{-1}$ dry biomass enthalpy [30].

2.5. Bioenergetic Parameters

The Gibbs energy dissipation for cell growth and maintenance ($1/Y_{GX}$) was estimated by the equation [31]:

$$\frac{1}{Y_{GX}} = \frac{1}{Y_{GX}^{max}} + \frac{m_G}{\mu} \tag{8}$$

in which μ is the specific growth rate, $1/Y_{GX}^{max}$. is the portion of $1/Y_{GX}$ referred only to growth in photoautotrophic cultivation using CO_2 as a carbon source (986 $kJ \cdot C\text{-}mol_{DM}^{-1}$) and m_G is the specific rate of Gibbs energy dissipation for cell maintenance (7.12 kJ $C\text{-}mol_{DM}^{-1} \cdot h^{-1}$) [31].

Knowing $1/Y_{GX}$, the average molar energy of photons at λ = 580 nm (Δg_{Ph} = 206.2 $kJ \cdot mol^{-1}$) and the Gibbs energies of formation under biological standard conditions of the compounds involved in the growth ($\Delta g_{f'_i}$), it was possible to estimate the moles of photons to sustain the autotrophic growth of 1 C-mol of biomass (n_{Ph}) by the equation:

$$n_{Ph} = (\Delta g_{f_{HCO_3^-}} + \Delta g_{f_{NO3^-}} + \Delta g_{f_{H_2O}} + \Delta g_{f_{O2}} + \Delta g_{f_H}{}^{+} + \Delta g_{f_X} + 1/Y_{GX})/{-\Delta g_{Ph}} \tag{9}$$

The average molar energy associated with the absorption of one mol of photons involved in the photosynthetic event was defined as [32]:

$$\Delta g_{Ph} = \frac{hcN_A}{\lambda} \tag{10}$$

where h = $6.626 \cdot 10^{-34}$ J·s is the Planck constant, c = $2.996 \cdot 10^{8}$ $m \cdot s^{-1}$ the light velocity, N_A = 6.023×10^{23} mol^{-1} the Avogadro number.

The molar Gibbs energies of formation of compounds involved in the growth are, under biological standard conditions, $\Delta g_{f'HCO3^-} = -587.2$ $kJ \cdot mol^{-1}$, $\Delta g_{f'_{NO3}^-} = -111.4$ $kJ \cdot mol^{-1}$, $\Delta g_{f'_{H_2O}} = -237.3$ $kJ \cdot mol^{-1}$,

$\Delta g_{f'_{O2}} = 0$, $\Delta g_{f'_{H^+}} = -39.87$ kJ·mol^{-1}, and $\Delta g_{f'_X} = -67.0$ kJ·mol^{-1} [31]. These values differ negligibly from those under experimental conditions [32], with exception of $\Delta g_{fH}{}^+$, which was recalculated at the actual pH using the well-known equation of Gibbs:

$$\Delta g_{f_{H^+}} = \Delta g_{f'_{H^+}} + RT \ln \frac{10^{-pH}}{10^{-7}} \quad (11)$$

where T and R are the absolute temperature and the ideal gas constant, respectively.

The molar rates of O_2 production (qO_2), H^+ consumption ($qH+$) and H_2O consumption or formation (qH_2O) occurring during photosynthesis were calculated, according to Torre et al. [32], by multiplying cell concentration expressed in C-mol$_{DM}$ L^{-1} by their respective stoichiometric coefficients expressed in mol C-mol$_{DM}{}^{-1}$, using the experimental biomass elemental composition reported for *D. tertiolecta* by Kim et al. [33], and by dividing by T_c.

The total Gibbs energy absorbed by the photosynthesis (ΔG_a), estimated by multiplying the moles of photons (n_{Ph}) by their average molar energy (Δg_{Ph}), was considered to be equal to the sum of the energy fixed by the photosystems to increase its own enthalpic content (ΔH), that recovered as ATP (ΔG_{ATP}) and the released heat (Q), according to the equation:

$$\Delta G_a = n_{Ph}\, \Delta g_{Ph} = -\Delta G_{ATP} - \Delta H - Q \quad (12)$$

It should be noticed that ΔG_a, contrary to ΔG_{ATP}, ΔH and Q, conventionally assumes negative values being an energy entering the system.

2.6. Statistical Analysis

Cultures were done in duplicate, while pigments extraction from biomass was performed in triplicate. Results were expressed as means ± standard deviations (SD) and compared by one-way analysis of variance (ANOVA) with a confidence interval of 95%.

3. Results and Discussion

3.1. Cell Growth Profile

As shown in Figure 1, *Dunaliella tertiolecta* cultivation was simultaneously investigated in the three selected culture media, and its growth profile was followed for 13 days. One can see that, in F/2 medium, the exponential growth phase started after about 1 day of cultivation, and cell density achieved a value of 452 mg$_{DM}$·L^{-1} on the fifth day, after which the microalga entered the stationary growth phase that continued until the end of cultivation. This may have been the result of phosphorus limitation, which has been reported to significantly affect the growth of *Dunaliella* sp. [34,35]. Kumar et al. [36] observed the same growth profile for the *D. tertiolecta* CCAP 19/27 strain in the same medium, where it reached a maximum cell concentration of around 700 mg$_{DM}$·L^{-1} after 8 days of cultivation.

When grown in Bold's Basal medium, *D. tertiolecta* underwent a three-day lag phase likely due to the need to adapt itself to this synthetic medium, which is so different from its natural environment consisting of brackish or marine water [37]. Contrariwise, *Dunaliella salina* exhibited cell growth without any lag phase in a modified Bold's Basal medium [38], possibly because it can adapt itself to more habitat types than *D. tertiolecta*, so that it is found in many different environments [37]. Nonetheless, after the lag phase, *D. tertiolecta* showed quicker growth in this medium rather than in the F/2 one, achieving higher cell concentration (607 mg$_{DM}$·L^{-1}) at the end of cultivation.

On the other hand, *D. tertiolecta* grew in the alternative medium without undergoing any lag phase as previously observed in the F/2 one, also prepared in seawater (Figure 1). The achievement of a much higher cell concentration (>1200 mg$_{DM}$·L^{-1}) as a result of an extension of the exponential growth phase suggests a synergistic effect of typical nutrients of seawater and some peculiar components of the Bold's Basal medium such as H_3BO_3 and EDTA in significant levels.

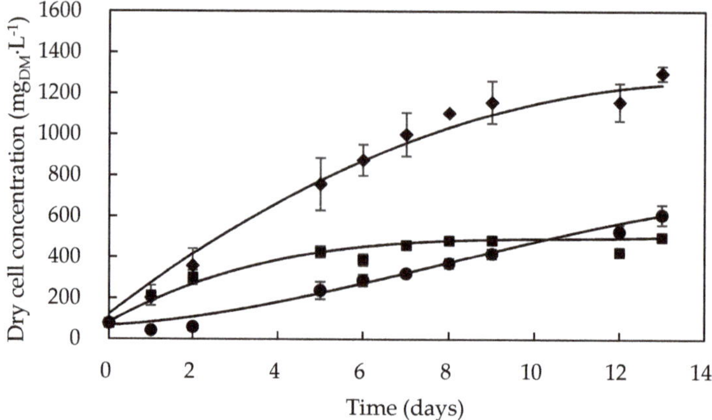

Figure 1. Growth profiles of *Dunaliella tertiolecta* (UTEX 999) cultivated in different culture media: F/2 medium (■), Bold's Basal medium (•), Alternative medium (◆).

To shed more light on these findings, mobility of *D. tertiolecta* cells was examined at the end of cultivations in the three different culture media as a rough, qualitative index of cell viability. Cells maintained in F/2 and alternative media showed normal motility, while that of cells cultivated in the Bold's Basal one was significantly reduced. Considering the limited contents of S, O, Mg, Ca, and K in such a synthetic medium, this observation confirms the essential role of some seawater component as nutrient, whose shortage may have reduced the viability and consequently the motility of microalgal cells [39,40].

This influence of medium composition was confirmed by visual examination of the color of cell cultures. It can be seen in Figure 2 that only the culture grown in the alternative medium had an intense green color, while those grown either in the Bold's Basal or the F/2 medium were yellowish in color, which confirms the above-supposed occurrence of some nutrient-limitation [41,42].

Figure 2. Color of *D. tertiolecta* (UTEX 999) cultivations in different culture media: F/2 medium (**A**), Bold's Basal medium (**B**), Alternative medium (**C**).

3.2. Kinetic Parameters

Kinetic parameters were greatly influenced by the culture medium composition, as the culture grown in the alternative medium showed almost twice the average values of maximum biomass concentration (X_{max} = 1223 ± 91 mgDM·L^{-1}), maximum specific growth rate (μ_{max} = 0.535 d^{-1}), and biomass productivity (P_X = 102 mgDM·L^{-1}·d^{-1}) found in the Bold's Basal medium, while intermediate results were observed in the F/2 one (Table 3). This result suggests that *D. tertiolecta* was able to quickly adapt itself to the composition of the first medium due to its high carbon (C), nitrogen (N) and phosphorus (P) levels. F/2 and alternative media contained bicarbonate as quickly-metabolizable C source provided by seawater (Table 1), whereas the only C source in the Bold's Basal medium was the CO_2 contained in air (about 400 ppm), which was probably the main reason for the poor performance of *D. tertiolecta* in it. This behavior appears to be typical of microalgae, in that Yeh et al. [43] observed a significant increase in biomass concentration (from 0.15 to 0.6 g·L^{-1}) and specific growth rate (from 0.5 to 1.5 day^{-1}) when 1.2 g·L^{-1} of $NaHCO_3$ was added to a *Chlorella vulgaris* autotrophic culture.

Table 3. Growth parameters and pigments production by *Dunaliella tertiolecta* (UTEX 999) cultivated in different culture media.

Medium	X_{max} (mgDM·L^{-1})	P_X (mgDM·L^{-1}·d^{-1})	μ_{max} (d^{-1})	PE_{max} (%)	Carotenoids (mg·g $^{-1}$)	Chlorophyll (mg·g^{-1})
F/2	452 ± 28 ^{a}	75	0.500	9.14	18.2 ± 2.1 ^{a}	65.7 ± 4.8 ^{a}
Bold's Basal	567± 33 ^{b}	47	0.269	5.13	52.0 ± 7.2 ^{b}	162.6 ± 15.7 ^{b}
Alternative	1223 ± 91 ^{c}	102	0.535	14.54	33.4 ± 5.7 ^{c}	108.0 ± 11.3 ^{c}

X_{max} = maximum cell concentration, P_X = biomass volumetric productivity, μ_{max} = maximum specific growth rate, PE_{max} = photosynthetic efficiency. Data expressed as means ± standard deviations of duplicate experiments. ^{a,b,c} Different superscript letters indicate statistically significant differences ($p < 0.05$).

The lower N content of the F/2 medium (0.8823 mM instead of 2.9442 mM) (Table 2) led to lower X_{max} value (452 ± 28 mgDM·L^{-1}) than in Bold's Basal (567 ± 33 mgDM·L^{-1}) or alternative (1223 ± 91 mgDM·L^{-1}) medium. Similarly, Chen et al. [44] observed that, in a medium containing 23 mM $NaNO_3$, *D. tertiolecta* grew up to an OD_{680} of 2.8, whereas, when this salt was 10-fold more diluted, its growth was strongly affected, thereby confirming how this microalga is sensitive to N availability. Moreover, the P content was higher in F/2 and alternative media compared to the Bold's Basal one (Table 2); it has been reported that depletion of this element inhibited *D. salina* growth, stopped cell duplication, and reduced the photosynthetic rate [45]. Resuming, C, N, and P are the fundamental elements for microalgae growth, and their higher availability in the alternative medium probably contributed to improve the kinetic parameters of *D. tertiolecta* growth.

Tammam et al. [46] reported that both *D. salina* and *D. tertiolecta* grew better at high (2.5 to 4.0 M NaCl) rather than at low (0.05 to 1.0 M NaCl) salt concentrations. Similarly, in the present work *D. tertiolecta* grew better in the alternative medium containing higher salt concentrations. In addition, as suggested by Katz and Pick [47], the *Dunaliella* genus has a singular capacity to remove Na^+ ions in hypersaline environments through a redox-driven sodium pump, and this transfer process results in enhanced photosynthetic CO_2 uptake [48]. Some *Dunaliella* species, such as *D. salina*, are able to survive in media containing NaCl in concentration ranging from about 0.05 to 5.5 M [49].

Likewise, Jiang et al. [50] observed that N-depletion significantly reduced *D. tertiolecta* growth, because under nitrogen limitation the microalga slowed its cell division, redirecting the flow of carbon from the formation of proteins and chlorophyll to those of carbohydrates and lipids. Chlorophyll is in fact an easily accessible N-rich compound, which is used as an intracellular nitrogen pool to hold up cell growth and biomass production as the N source in the medium becomes the limiting factor [51]. Therefore, we can infer that N limitation in the F/2 medium was responsible for marked decreases in both the chlorophyll content of *D. tertiolecta* biomass and its final X_{max} value.

On the other hand, in Bold's Basal medium, cells showed lower biomass volumetric productivity (P_X = 47 mgDM·L^{-1}·d^{-1}) and maximum specific growth rate (μ_{max} = 0.269 d^{-1}) than in the other two media, which suggests that the overall salt concentration and, then, the medium osmolarity may have also played a key role in *D. tertiolecta* growth. For instance, under limitation by sulfur (S), which is one of the most abundant elements of seawater, the microalga *Chlamydomonas reinhardtii* showed inhibition of both cell division and photosynthesis [52–54]. A similar sulfur limitation may have occurred with *D. tertiolecta* in Bold's Basal medium prepared using distilled water. Considering all these growth parameters together, it can be said that the alternative medium was the most suitable for *D. tertiolecta* growth, thanks to the simultaneous presence of seawater and Bold's Basal medium components.

Table 3 also lists the values of the maximum photosynthetic efficiency (PE_{max}), which was significantly influenced by the medium composition. The highest value of this parameter was observed in the alternative medium (14.54%), whereas in the F/2 and Bold's Basal ones it was only 9.14 and 5.13%, respectively. These results may have also been due to different availability of nutrients in the selected media. Particularly, the absence of seawater affected cell division and photosynthesis in the Bold's Bold medium, while the N-limitation responsible for the reduced chlorophyll content of biomass was the likely reason for the poor performance of the F/2 one. According to Srinivasan et al. [55], who investigated the combined effect of sodium bicarbonate and macronutrient starvation stress on *D. salina* V-101 physiological and biochemical responses, observed that the photosynthesis efficiency decreased in all N-, P-, or S-deficient cultures compared to the control containing all nutrients. A similar reduction of *PE* accompanied by lipid content increase was observed by Gao et al. [56] in an unspecified strain of *D. salina* under conditions of complete nutrient deprivation. This means that nitrogen depletion in general affects the synthesis of proteins, including those involved in the reaction centers, and results in a decrease of chlorophyll content reducing the functioning of photosystem II (PSII) [55]. *PE* values (7.25%) close to those observed in the F/2 (9.14%) and Bold's Basal (5.13%) media were reported for *Chlorella sorokiniana* cultivation in a conical helical tubular photobioreactor [57].

3.3. Pigments Production

Contents of chlorophyll and total carotenoids were determined for *D. tertiolecta* in the three selected culture media (Table 3). Likely due to the early supposed N-limitation in the F/2 medium, the contents of carotenoids (18.2 ± 2.1 $mg \cdot g^{-1}$) and chlorophyll (65.7 ± 4.8 $mg \cdot g^{-1}$) in biomass were significantly lower than those detected in cells cultured in the alternative medium (33.4 ± 5.7 and 108.0 ± 11.3 $mg \cdot g^{-1}$, respectively) and especially in the Bold's Basal one (52.0 ± 7.2 and 162.6 ± 15.7 $mg \cdot g^{-1}$, respectively). As described by Li et al. [51], when the N source runs out in the medium chlorophyll is used as an internal N source to sustain cell division, until its content drops to a critical value below which growth ceases. Likewise, Lai et al. [58] observed that, under N-limited conditions, the chlorophyll content of *Dunaliella viridis* progressively decreased causing cell growth to stop. Furthermore, it was described that, although the photosynthesis can continue even under N-limited conditions, biomass has smaller contents of N-rich components and accessory pigments such as carotenoids but higher contents of energy-rich components such as lipids and sugars [59].

On the other hand, literature reports suggest that the highest contents of both types of pigments in biomass cultured in the Bold's Basal medium may be ascribed to its low S content. In fact, as already mentioned, the alternative medium is constituted by seawater that has high S content [40], which reduces the production of pigments. For example, when cultured under S deprivation, *C. reinhardtii* showed an increase in the fluorescence yield of chlorophyll [60], and *D. salina* a 20% higher carotenoid content compared to non-limited culture [61]. Similarly to our results in Bold's Basal medium, Volgusheva et al. [54] observed for *C. reinhardtii* an increase in chlorophyll content accompanied by a decay in growth kinetic parameters.

Finally, the carotenoid and chlorophyll contents in biomass cultured in alternative medium were around 35% lower than in Bold's Basal medium; this was probably due to the higher S concentration in the former medium, consistently with the previously-discussed results of Antal et al. [60].

As for the influence of medium salinity on the production of pigments, although *Dunaliella* is usually considered a highly salt-tolerant genus, some studies have shown that halophilicity can vary widely from one species to another and even from one strain to another [62]. For example, maximal carotenoid production by *D. tertiolecta* occurred at lower NaCl concentration (0.7 M) than by *D. salina* (2.0 M) [63]. β-Carotene accumulation triggered by reactive oxygen species has also been found to defend *Dunaliella* cells from the adverse effects of high salinity [64]. In another study, carotenoid production by *D. salina* increased continuously with increasing salinity [65]. Based on these results, we can infer that, due to the moderate halophilicity of *D. tertiolecta*, the NaCl concentration in the seawater used to prepare the alternative medium was excessive for the purpose of carotenoids' production, thus leading to a value of their content significantly lower than that obtained in Bold's Basal medium.

3.4. Bioenergetic Parameters

Figure 3 shows the time course of biomass yield on Gibbs energy dissipation (Y_{GX}), which corresponds to the reciprocal of the previously defined Gibbs energy dissipation for cell growth and maintenance. As a rule, this bioenergetic parameter is the highest at the beginning of cultivation, when cell concentration is low and there is a high metabolic rate favoring cell growth and maintenance. During cultivation, there was a progressive decrease in Y_{GX}, accompanied by a reduction in the specific growth rate, probably due to the onset of growth-limiting conditions resulting from the exhaustion of some nutrients or from shading. The same profile was observed by Silva et al. [66], Sassano et al. [67] and Torre et al. [32] in *Arthrospira (Spirulina) platensis* cultivations carried out in different photobioreactor configurations, using different light intensities and N-sources in the culture medium.

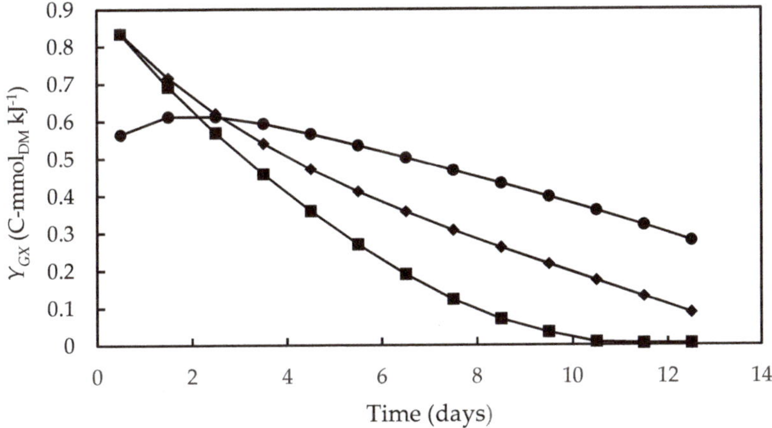

Figure 3. Biomass yield on Gibbs energy dissipation during cultivations of *D. tertiolecta* carried out in different culture media. F/2 medium (■), Bold's Basal medium (•), Alternative medium (♦).

The highest value of Y_{GX} (0.83 C-mmol$_{DM}$·kJ^{-1}), observed in both seawater-based F/2 and alternative media, is close to the maximum Y_{GX} values reported by Silva et al. [66] (0.7–0.9 C-mmol$_{DM}$·kJ^{-1}) and by Sassano et al. [67] (0.8–1.0 C-mmol$_{DM}$·kJ^{-1}) in bioenergetic studies on *A. platensis*. The significantly lower Y_{GX} value obtained in the Bold's Basal medium (0.56 C-mmol$_{DM}$·kJ^{-1}) can be explained by the need for cells well adapted to the high salinity conditions of seawater to adapt to the new conditions, which increased the amount of Gibbs energy needed to produce a given amount of biomass. In addition, as previously mentioned in the section addressed to kinetic parameters, the high NaCl concentration in seawater-based media may have enhanced the photosynthetic CO_2 assimilation favoring biomass production [48]. However, it can be seen that, after the third day of cultivation, Y_{GX} decreased in the F/2 medium more quickly than in the alternative one, probably due to growth limitation resulting from depletion of some nutrient or microelement instead present in the formulation of the latter.

As is known, it is possible to describe cell metabolism through a set of reaction equations using stoichiometric coefficients for the formation of products from substrates [68]. In particular, knowing the compositions of substrates, products, and biomass, we can write the following overall material balance for the formation of 1 C-mol of *D. tertiolecta* biomass:

$$a\,HCO_3^- + b\,NO_3^- + c\,H_2O + d\,O_2 + e\,H^+ + CH_{1.944}O_{0.873}N_{0.249} = 0 \quad (13)$$

To this purpose, we used the elemental composition of biomass reported by Kim et al. [33]. Since Bezerra et al. [69] observed for *A. platensis* a negligible influence of a different elemental biomass composition (CHNOS) on the results of such a bioenergetic model, it was considered that it was not necessary to determine it at the end of each experiment, being sufficient, for the purposes of this study, the biomass composition data available in the literature.

The stoichiometric coefficients estimated for the growth of *D. tertiolecta* through material balances of carbon, nitrogen, oxygen, charge, and reduction degree of biomass (γ_X) are listed in Table 4 together with the Gibbs energies of formation under biological standard conditions of the compounds involved in growth.

Table 4. Gibbs energies of formation under biological standard conditions of the compounds involved in the growth of *Dunaliella tertiolecta* biomass, and stoichiometric coefficients estimated through material balances of carbon, nitrogen, oxygen, charge, and reduction degree of biomass (γ_X).

Compound	HCO_3^-	NO_3^-	H_2O	O_2	H^+	Biomass
$\Delta g_{f'_i}$ ($kJ \cdot mol^{-1}$)	−587.2	−111.4	−273.3	0	−39.9	−67.0
Stoichiometric Coefficients						
Symbol	*a*	*b*	*c*	*d*	*e*	γ_X
Value ($mol \cdot C\text{-}mol_{DM}^{-1}$)	−1	−0.24	1.0	1.52	−1.20	6.06

The reduction degree of *D. tertiolecta* biomass (γ_X = 6.06) was slightly higher than that reported for *A. platensis* (4.89–5.80) [69], because of a higher H content of its biomass compared to that of this cyanobacterium ($CH_{1.59}O_{0.50}N_{0.10}$), i.e., of a higher number of equivalents available electrons per C-mol of biomass.

Estimating the above stoichiometric coefficients by Equation (13), Y_{GX} by Equation (8) and using the standard Gibbs energies of formation of reactants, products and biomass listed in Table 4 [31], it was possible to calculate, by Equation (9), the moles of photons to sustain the autotrophic growth of 1 C-mol biomass (n_{Ph}), whose time course is illustrated in Figure 4. It can be observed that this bioenergetic parameter increased (in absolute value) throughout the cultivations, with the negative values indicating that photons were taken up by the system. In particular, at the beginning of cultures, when cell concentration was low, the n_{Ph} absolute values were very close to the theoretical one reported by Richmond [70] to support phototrophic growth under ideal conditions (8 moles of photons per C-mol of biomass). According to the widely accepted two-step model of photosynthesis, 8 mol quanta of light are in fact required to release 1 mol of O_2 [71]. A significant additional energy requirement with respect to such an ideal condition occurred during the stationary growth phase of batch cultures, which suggests that, under these hard environmental conditions, energy was preferentially used for maintenance of existing live cells rather than for growth. This effect was also described by Torre et al. [32] and Bezerra et al. [69] in their bioenergetics studies on *A. platensis*.

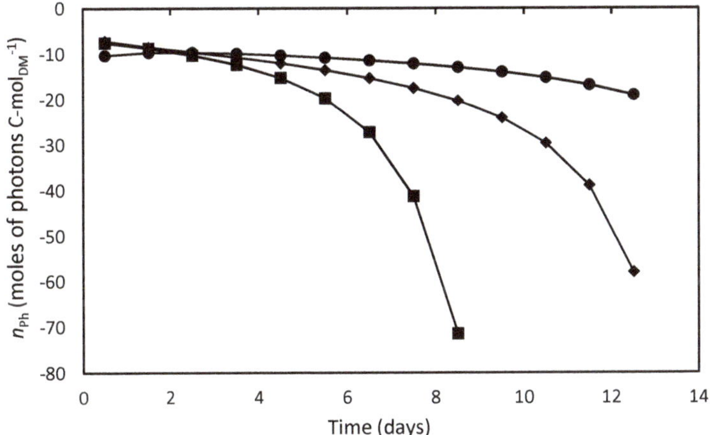

Figure 4. Moles of photons utilized to produce one C-mol biomass of *D. tertiolecta* in different culture media. F/2 medium (■), Bold's Basal medium (•), Alternative medium (◆).

Figure 4 also shows that the culture in F/2 medium needed higher n_{Ph} values to sustain autophototrophic growth compared to cultures in Bold's Basal and alternative media from the fifth day of cultivation onwards, likely due to the previously-supposed nutrient limitation. Consequently, cells reached lower concentration in the stationary growth phase, and most of energy was lost. It has been reported that nitrogen or phosphorus limitation affects the photosynthetic apparatus of *D. tertiolecta* cells by reducing their chlorophyll and carotenoids contents leading to chlorosis [34]. This effect is consistent with the pigment loss observed in *D. tertiolecta* cells cultivated in this medium (Figure 2). The balance between energy supplied to PSII by light harvesting and energy requested by photosynthesis and growth appears to be regulated by chlorophyll concentration [34,72]. Although less energy was needed to form 1 C-mol of biomass in the Bold's basal medium compared to the alternative one, the latter allowed achieving a higher cell concentration at the end of cultivation, which means that a greater fraction of absorbed energy was devoted to cell growth, thus proving a better medium for *D. tertiolecta* cultivation.

During photosynthesis, the light energy absorbed by the antenna pigments, mainly chlorophylls, xanthophylls and carotenes, is converted into redox energy, whose average value referred to 1 mol of photons (Δg_{Ph}) is 206.2 kJ·mol^{-1} at wavelength of 580 nm, that drives the formation of high-energy products such as ATP and NADPH [71]. According to the aforementioned photosynthesis model, 8 photons (*hv*) of light are needed to form 1 mol of O_2. Moreover, photorespiration was considered absent, as microalgae suppress photorespiration in the presence of light and inorganic carbon sources such as bicarbonate and CO_2, as occurs in C_4 plants [73]. Thus, the evolution of O_2 can be related to the flux of absorbed photons by the equation:

$$2\,H_2O + 8\,hv + 2\,NADP^+ \rightarrow 2\,NADPH_2 + O_2 \tag{14}$$

As show in Figure 5, the molar development of O_2, consumptions of H^+ and H_2O variation strictly followed cell growth. In particular, these activities were higher in the culture carried out on the alternative medium compared to the other media, consistently with its higher nutrient contents and higher photosynthesis rate, which resulted in higher cell concentrations.

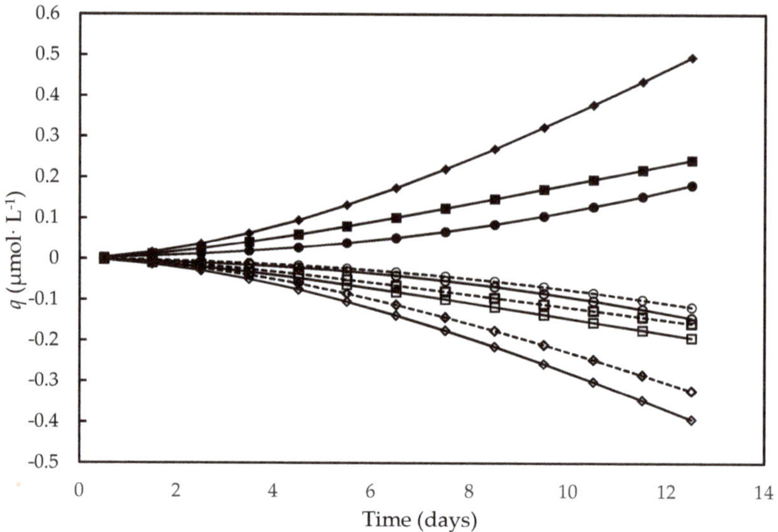

Figure 5. Development of O_2 (full symbols), consumption of H^+ (open symbols, continuous line) and H_2O variation (open symbols, dashed line) over time in different culture media. F/2 medium (□,■), Bold's Basal medium (○,•), Alternative medium (◇,◆).

The progressive consumption of H^+ described by the q_{H+} trend in this figure is consistent with the progressive rise in pH detected in *D. tertiolecta* autotrophic cultures, resulting from the OH^- release linked to photosynthetic fixation of CO_2 and NO_3^- uptake [36,74]. Indeed, such a pH increase appears to be a rule in photoautotrophic cultures, having been proposed that the uptake of two HCO_3^- moles leads to one mole of fixed CO_2 and one of CO_3^{2-} released in the medium [75].

As is well known, ATP is essential for several cell functions including the assembly of biopolymers as well as cell maintenance and division. For instance, in the autotrophic culture of *C. pyrenoidosa* up to 77% of total ATP produced by cell metabolism is due to the assimilation of CO_2 by the Calvin cycle [71]. Therefore, we assumed that a portion of molar Gibbs energy absorbed by the two photosystems (I and II) was used to convert ADP to ATP (ADP + Pi → ATP + H_2O, Δg^o = 30.5 kJ·mol^{-1}) by membrane-bound ATP synthase [76].

As shown in Figure 6, in all the experiments both the energy absorbed by the system (ΔG_a) and the fraction converted into ATP (ΔG_{ATP}) increased over time. The highest ΔG_a and ΔG_{ATP} values were associated to the low growth rates in the stationary growth phase, mainly in the F/2 medium, which can be ascribed to the additional amount of ATP needed to produce biomass from biopolymers, defined elsewhere as "growth-associated maintenance" [77,78]. As previously reported for the cyanobacterium *A. platensis* [69], unfavorable environmental conditions such as lack of nutrients or release of certain cell metabolites increase the energy needs of cells for transport, translocation, futile cycles of nutrients, and assembly of biopolymers into growing biomass. In addition, microalgae have a high degree of subcellular compartmentation of metabolism, whereby further transport reactions that consume large amounts of energy are involved in the metabolic reactions [71]. Moreover, Kliphuis et al. [79] reported that low cell growth rates are affected by the high energy maintenance requirements that result in low biomass yields.

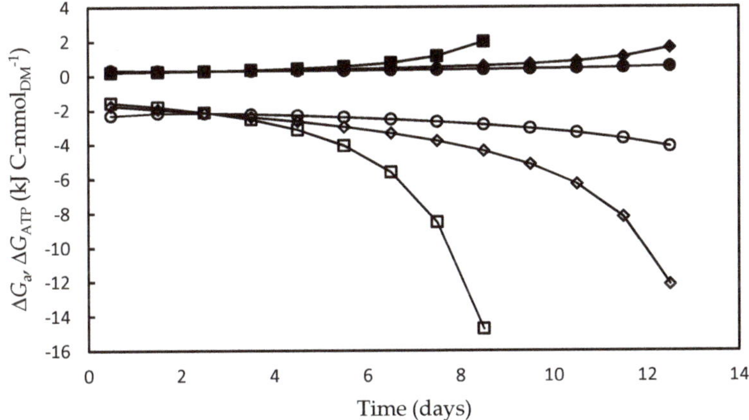

Figure 6. Total Gibbs energy absorbed by the photosynthetic apparatus (open symbols) (ΔG_a) and Gibbs energy transformed into ATP (full symbols) (ΔG_{ATP}) during *Dunaliella tertiolecta* cultivations carried out in different culture media: F/2 medium (□,■), Bold's Basal medium (○,•), Alternative medium (◇,◆).

Similarly to what was observed for ΔG_a (in absolute values) and ΔG_{ATP}, the enthalpy energy component (ΔH) and heat released (Q) also increased over time, and, as expected, the highest values of both occurred in the F/2 medium (Figure 7). This result suggests that, under the nutritional stress conditions occurred in this medium, a significant portion of excess light energy that entered the system (ΔG_a) was lost as heat, and, since growth was very poor, most of the energy fraction fixed by the photosynthesis (ΔH) was used for metabolic activities intended for cell maintenance. It is likely that most of the heat was dissipated by nonphotochemical reactions, i.e., by the so-called non-photochemical quenching (NPQ) photoprotection mechanism [73]. This mechanism is essential to remove the excess electrons in the antennae complexes, which would otherwise damage the photosynthetic apparatus by thermal dissipation in the xanthophyll cycle in *Dunaliella* cell [34,53,80].

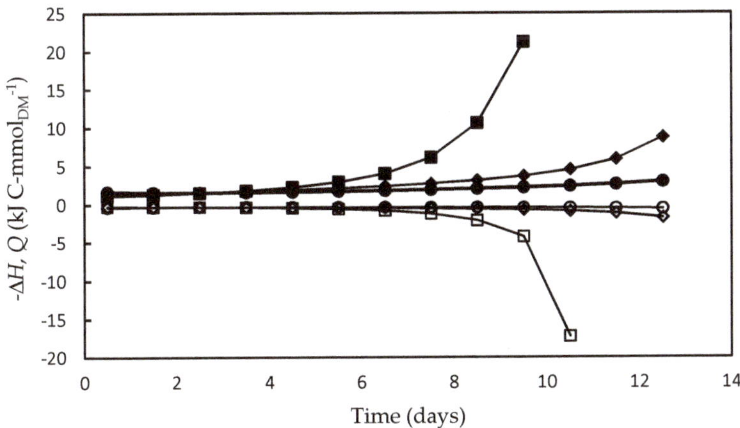

Figure 7. Energy fixed by the photosynthetic apparatus (open symbols) (ΔH) and energy released as heat (full symbols) (Q) during *Dunaliella tertiolecta* cultivations carried out in different culture media: F/2 medium (□,■), Bold's Basal medium (○,•), Alternative medium (◇,◆). The choice of illustrating -Q instead of Q was only due to the need to avoid overlapping with the ΔH curves.

Figure 8 shows the percentage distribution of the light energy absorbed versus time in the different culture media. In all the experiments, both the energy fraction stored in the phosphoanhydride bonds of ATP (η_{ATP}) and the enthalpic one fixed by the systems (η_F) decreased over time, whereas that released as heat (η_Q) increased. η_F and η_{ATP} values were higher at the beginning of cultivations in the media constituted by seawater (F/2 and alternative medium) and decreased more sharply when compared to Bold's Basal medium. The highest η_{ATP} values (9.6–14.2%) were close to that found for the autotrophic cultivation of *C. pyrenoidosa* (η_{ATP} = 10%) [71], and the slight decrease in η_F and η_{ATP} values in the Bold's Basal medium is consistent with the higher cell concentration observed compared to the F/2 medium. Furthermore, while in the F/2 medium the η_F values were practically null from the ninth day onwards, in the alternative one they remained positive, even if decreasing, during the entire course of the culture, suggesting that significant percentages of the energy fixed by the photosynthetic apparatus and of that transformed into ATP were directed to cell growth. It has been reported that the availability of nutrients, mainly nitrogen and phosphorus sources, is a main environmental factor capable of influencing the composition of the photosynthetic apparatus of *D. tertiolecta* [34]; therefore, the declining values of η_F could also have been the result of depletion of some nutrient especially at the end of cultivations. Bezerra et al. [69] also reported that high cell concentrations reduce the light availability to the cell through the so-called shading effect, thus reducing η_F more sharply. However, in the present work, the sharpest decrease in η_F occurred in the culture with the lowest cell concentration, indicating that light intensity was not the limiting factor.

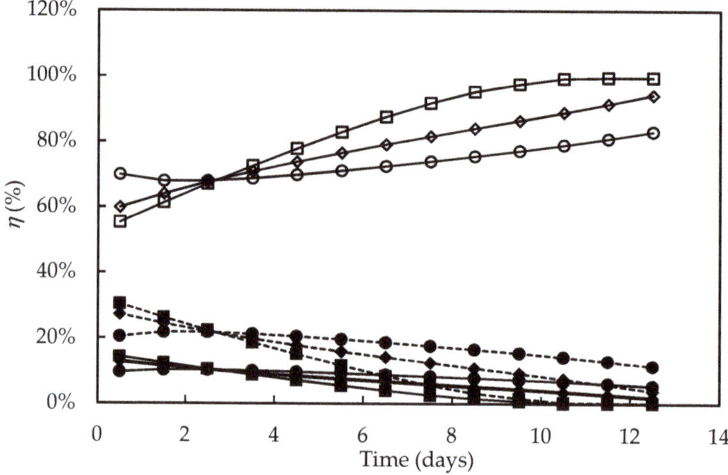

Figure 8. Percentage distribution of the light energy absorbed during *Dunaliella tertiolecta* cultivations in different culture media. Energy fixed by the photosystems (dashed line), energy transformed into ATP (continuous line, full symbols), energy released as heat (open symbols). F/2 medium (□,■), Bold's Basal medium (○,●), Alternative medium (◇,◆).

4. Conclusions

This work demonstrated that *Dunaliella tertiolecta* grown in Bold's Basal medium had high contents of carotenoids (52.0 ± 7.2 mg·g^{-1}) and chlorophyll (162.6 ± 15.7 mg·g^{-1}). However, to obtain high concentrations of biomass to be used for energy purposes, after the extraction and recovery of these precious components, an alternative medium based on seawater has proven to be an efficient alternative for the cultivation of this microalga, allowing a high cell productivity (102 mgDM·L^{-1}·d^{-1}) and photosynthesis rate. In fact, it was able to ensure, at the end of batch cultivation, more than twice the cell concentration obtained in the Bold's Basal medium. As regard the bioenergetic study, the optimal conditions for *D. tertiolecta* growth in the alternative medium were highlighted by high values of biomass yield on Gibbs energy dissipation (Y_{GX}), high molar development of O_2 and consumption of H^+ as well as a high energy fraction stored as ATP (η_{ATP}) during cultivation.

Author Contributions: Conceptualization: R.P.B. and D.d.A.V.-M.; methodology: R.P.B. and Y.A.S.M.; validation: all authors; formal analysis: Y.A.S.M., R.P.B., D.d.A.V.-M., and A.C.; investigation: Y.A.S.M., R.P.B., and D.d.A.V.-M.; resources: A.L.F.P.; data curation: Y.A.S.M., R.P.B., D.d.A.V.-M., and A.C.; writing—original draft preparation: Y.A.S.M., R.P.B., D.d.A.V.-M., and A.C.; writing—review and editing: A.C.; visualization, Y.A.S.M., R.P.B., and A.C.; supervision: A.C.; project administration: R.P.B., D.d.A.V.-M., and A.C.; funding acquisition: A.L.F.P. All authors have read and agreed to the published version of the manuscript.

Funding: This research was funded by the Foundation for Science and Technology of the State of Pernambuco, Brazil (FACEPE), grant number APQ-0252-5.07/14.

Acknowledgments: The authors acknowledge with thanks the Research Support Center (CENAPESQ, Recife, Brazil) and Laboratory of Technology of Bioactives (LABTECBIO, Recife, Brazil).

Conflicts of Interest: The authors declare no conflict of interest.

References

1. Chen, X.J.; Wu, M.J.; Jiang, Y.; Yang, Y.; Yan, Y.B. *Dunaliella salina* Hsp90 Is Halotolerant. *Int. J. Biol. Macromol.* **2015**, *75*, 418–425. [CrossRef]
2. Madkour, F.F.; Abdel-Daim, M.M. Hepatoprotective and Antioxidant Activity of *Dunaliella salina* in Paracetamol-Induced Acute Toxicity in Rats. *Indian J. Pharm. Sci.* **2013**, *75*, 642–648.
3. Srinivasan, R.; Chaitanyakumar, A.; Mageswari, A.; Gomathi, A.; Pavan Kumar, J.G.S.; Jayasindu, M.; Bharath, G.; Shravan, J.S.; Gothandam, K.M. Oral Administration of Lyophilized *Dunaliella salina*, a Carotenoid-Rich Marine Alga, Reduces Tumor Progression in Mammary Cancer Induced Rats. *Food Funct.* **2017**, *8*, 4517–4527. [CrossRef]
4. Caroprese, M.; Albenzio, M.; Ciliberti, M.G.; Francavilla, M.; Sevi, A. A Mixture of Phytosterols from *Dunaliella tertiolecta* Affects Proliferation of Peripheral Blood Mononuclear Cells and Cytokine Production in Sheep. *Vet. Immunol. Immunopathol.* **2012**, *150*, 27–35. [CrossRef]
5. Talebi, A.F.; Tohidfar, M.; Mousavi Derazmahalleh, S.M.; Sulaiman, A.; Baharuddin, A.S.; Tabatabaei, M. Biochemical Modulation of Lipid Pathway in Microalgae *Dunaliella* sp. for Biodiesel Production. *Biomed. Res. Int.* **2015**. [CrossRef]
6. Plaza, M.; Herrero, M.; Cifuentes, A.; Ibáñez, E. Innovative Natural Functional Ingredients from Microalgae. *J. Agric. Food Chem.* **2009**, *57*, 7159–7170. [CrossRef]
7. Khan, M.I.; Shin, J.H.; Kim, J.D. The Promising Future of Microalgae: Current Status, Challenges, and Optimization of a Sustainable and Renewable Industry for Biofuels, Feed, and Other Products. *Microb. Cell Fact.* **2018**, *17*, 36. [CrossRef] [PubMed]
8. Khozingoldberg, I.; Cohen, Z. The Effect of Phosphate Starvation on the Lipid and Fatty Acid Composition of the Fresh Water Eustigmatophyte *Monodus subterraneus*. *Phytochemistry* **2006**, *67*, 696–701. [CrossRef] [PubMed]
9. Hu, Q.; Sommerfeld, M.; Jarvis, E.; Ghirardi, M.; Posewitz, M.; Seibert, M.; Darzins, A. Microalgal Triacylglycerols as Feedstocks for Biofuel Production: Perspectives and Advances. *Plant J.* **2008**, *54*, 621–639. [CrossRef]
10. Prathima Devi, M.; Venkata Mohan, S. CO_2 Supplementation to Domestic Wastewater Enhances Microalgae Lipid Accumulation under Mixotrophic Microenvironment: Effect of Sparging Period and Interval. *Bioresour. Technol.* **2012**, *112*, 116–123. [CrossRef] [PubMed]
11. Ito, T.; Tanaka, M.; Shinkawa, H.; Nakada, T.; Ano, Y.; Kurano, N.; Soga, T.; Tomita, M. Metabolic and Morphological Changes of an Oil Accumulating Trebouxiophycean Alga in Nitrogen-Deficient Conditions. *Metabolomics* **2013**, *9*, 178–187. [CrossRef] [PubMed]
12. Rajesh, K.; Rohit, M.V.; Venkata Mohan, S. Microalgae-Based Carotenoids Production. *Algal Green Chem.* **2017**, 139–147. [CrossRef]
13. Khoo, K.S.; Chew, K.W.; Yew, G.Y.; Leong, W.H.; Chai, Y.H.; Show, P.L.; Chen, W.-H. Recent Advances in Downstream Processing of Microalgae Lipid Recovery for Biofuel Production. *Bioresour. Technol.* **2020**, *304*, 122996. [CrossRef] [PubMed]

14. Zabed, H.M.; Akter, S.; Yun, J.; Zhang, G.; Zhang, Y.; Qi, X. Biogas from Microalgae: Technologies, Challenges and Opportunities. *Renew. Sust. Energy Rev.* **2020**, *117*, 109503. [CrossRef]
15. Sun, J.; Xiong, X.; Wang, M.; Du, H.; Li, J.; Zhou, D.; Zuo, J. Microalgae Biodiesel Production in China: A Preliminary Economic Analysis. *Renew. Sust. Energy Rev.* **2019**, *104*, 296–306. [CrossRef]
16. Liang, M.-H.; Qv, X.-Y.; Chen, H.; Wang, Q.; Jiang, J.-G. Effects of Salt Concentrations and Nitrogen and Phosphorus Starvations on Neutral Lipid Contents in the Green Microalga *Dunaliella tertiolecta*. *J. Agric. Food Chem.* **2017**, *65*, 3190–3197. [CrossRef]
17. Rodríguez, M.B.R. Simulation of an Assisted Culture Medium for Production of *Dunaliella tertiolecta*. *Algal Res.* **2020**, *47*, 101838. [CrossRef]
18. García Morales, J.; López Elías, J.A.; Medina Félix, D.; García Lagunas, N.; Fimbres Olivarría, D. Efecto Del Estrés Por Nitrógeno y Salinidad En El Contenido de β-Caroteno de La Microalga *Dunaliella tertiolecta*//Effect of Nitrogen and Salinity Stress on the β-Carotene Content of the Microalgae *Dunaliella tertiolecta*. *Biotecnia* **2020**, *22*, 13–19. [CrossRef]
19. Lü, J.-M.; Lin, P.H.; Yao, Q.; Chen, C. Chemical and Molecular Mechanisms of Antioxidants: Experimental Approaches and Model Systems. *J. Cell. Mol. Med.* **2010**, *14*, 840–860. [CrossRef]
20. McWilliams, A. *The Global Market for Carotenoids*; BCC Research: Wellesley, MA, USA, 2018.
21. Guedes, A.C.; Amaro, H.M.; Sousa-Pinto, I.; Malcata, F.X. Algal Spent Biomass—A Pool of Applications. *Biofuels Algae* **2019**, 397–433. [CrossRef]
22. Sedjati, S.; Santosa, G.; Yudiati, E.; Supriyantini, E.; Ridlo, A.; Kimberly, F. Chlorophyll and Carotenoid Content of *Dunaliella salina* at Various Salinity Stress and Harvesting Time. *IOP Conf. Ser. Earth Environ. Sci.* **2019**, *246*, 012025. [CrossRef]
23. Tredici, M.R. Photobiology of Microalgae Mass Cultures: Understanding the Tools for the next Green Revolution. *Biofuels* **2010**, *1*, 143–162. [CrossRef]
24. Küçük, K.; Tevatia, R.; Sorgüven, E.; Demirel, Y.; Özilgen, M. Bioenergetics of Growth and Lipid Production in *Chlamydomonas reinhardtii*. *Energy* **2015**, *83*, 503–510. [CrossRef]
25. Demirel, Y.; Sieniutycz, S. *Nonequilibrium Thermodynamics: Transport and Rate Processes in Physical and Biological Systems*, 3rd ed.; Elsevier: Amsterdam, The Netherlands, 2014. [CrossRef]
26. Guillard, R.R.L.; Ryther, J.H. Studies on Marine Planktonic Diatoms I. *Cyclotella Nana* Hustedt and *Detonula Confervacea* (Cleve) Gran. *Can. J. Microbiol.* **1962**, *8*, 229–239. [CrossRef]
27. Bischoff, H.W.; Bold, H.C. Some Soil Algae from Enchanted Rock and Related Algal Species. *Phycol. Stud. IV* **1963**, *6318*, 1–95.
28. Jesus, S.S.; Maciel Filho, R. Modeling Growth of Microalgae *Dunaliella salina* under Different Nutritional Conditions. *Am. J. Biochem. Biotechnol.* **2010**, *6*, 279–283. [CrossRef]
29. Lichtenthaler, H. Chlorophylls and Carotenoids: Pigments of Photosynthetic Biomembranes. *Methods Enzym.* **1987**, *148*, 350–382.
30. Soletto, D.; Binaghi, L.; Ferrari, L.; Lodi, A.; Carvalho, J.C.M.; Zilli, M.; Converti, A. Effects of Carbon Dioxide Feeding Rate and Light Intensity on the Fed-Batch Pulse-Feeding Cultivation of *Spirulina platensis* in Helical Photobioreactor. *Biochem. Eng. J.* **2008**, *39*, 369–375. [CrossRef]
31. Heijnen, J.J. Stoichiometry and Kinetics of Microbial Growth from a Thermodynamic Perspective. In *Basic Biotechnology*; Cambridge University Press: Cambridge, UK, 2001; pp. 45–58. [CrossRef]
32. Torre, P.; Sassano, C.E.; Sato, S.; Converti, A.; Gioielli, L.A.; Carvalho, J.C. Fed-Batch Addition of Urea for *Spirulina platensis* Cultivation. *Enzym. Microb. Technol.* **2003**, *33*, 698–707. [CrossRef]
33. Kim, S.-S.; Ly, H.V.; Kim, J.; Lee, E.Y.; Woo, H.C. Pyrolysis of Microalgae Residual Biomass Derived from *Dunaliella tertiolecta* after Lipid Extraction and Carbohydrate Saccharification. *Chem. Eng. J.* **2015**, *263*, 194–199. [CrossRef]
34. Geider, R.; Macintyre; Graziano, L.; McKay, R.M. Responses of the Photosynthetic Apparatus of *Dunaliella tertiolecta* (Chlorophyceae) to Nitrogen and Phosphorus Limitation. *Eur. J. Phycol.* **1998**, *33*, 315–332. [CrossRef]
35. Wongsnansilp, T.; Juntawong, N.; Wu, Z. Effects of Phosphorus on the Growth and Chlorophyll Fluorescence of a *Dunaliella salina* Strain Isolated from Saline Soil under Nitrate Limitation. *J. Biol. Res. Boll. Soc. Ital. Biol. Sper.* **2016**, *89*. [CrossRef]
36. Kumar, A.; Guria, C.; Pathak, A.K. Optimal Cultivation towards Enhanced Algae-Biomass and Lipid Production Using *Dunaliella tertiolecta* for Biofuel Application and Potential CO_2 Bio-Fixation: Effect of Nitrogen Deficient Fertilizer, Light Intensity, Salinity and Carbon Supply Strategy. *Energy* **2018**. [CrossRef]

37. Polle, J.E.W.; Tran, D.; Ben-Amotz, A. History, Distribution, and Habitats of Algae of the Genus *Dunaliella* Teodoresco (Chlorophyceae). In *The Alga Dunaliella: Biodiversity, Physiology, Genomics and Biotechnology*; Ben-Amotz, A., Polle, J.E.W., Rao, D.V.S., Eds.; Science Publishers Inc.: Enfield, NH, USA, 2009; pp. 1–13.
38. Kulshreshtha, J.; Singh, G.P. Evaluation of Various Inorganic Media for Growth and Biopigment of *Dunaliella salina*. *Int. J. Pharm. Bio. Sci.* **2013**, *4*, 1083–1089.
39. Ohse, S.; Derner, R.B.; Ozório, R.Á.; Cunha, P.C.R.; Lamarca, C.P.; Santos, M.E.; Mender, L.B.B. Revisão: Sequestro de Carbono Realizado Por Microalgas e Florestas e a Capacidade de Produção de Lipídios Pelas Microalgas. *Insula* **2007**, *36*, 39–74.
40. Millero, F.J.; Feistel, R.; Wright, D.G.; McDougall, T.J. The Composition of Standard Seawater and the Definition of the Reference-Composition Salinity Scale. *Deep Sea Res. Part I Oceanogr. Res. Pap.* **2008**, *55*, 50–72. [CrossRef]
41. Jin, H.; Zhang, H.; Zhou, Z.; Li, K.; Hou, G.; Xu, Q.; Chuai, W.; Zhang, C.; Han, D.; Hu, Q. Ultrahigh-cell-density Heterotrophic Cultivation of the Unicellular Green Microalga *Scenedesmus acuminatus* and Application of the Cells to Photoautotrophic Culture Enhance Biomass and Lipid Production. *Biotechnol. Bioeng.* **2020**, *117*, 96–108. [CrossRef]
42. Xinyi, E.; Crofcheck, C.; Crocker, M. Application of Recycled Media and Algae-Based Anaerobic Digestate in *Scenedesmus* Cultivation. *J. Renew. Sust. Energy* **2016**, *8*, 1–14. [CrossRef]
43. Yeh, K.-L.; Chang, J.-S.; Chen, W. Effect of Light Supply and Carbon Source on Cell Growth and Cellular Composition of a Newly Isolated Microalga *Chlorella vulgaris* ESP-31. *Eng. Life Sci.* **2010**, *10*, 201–208. [CrossRef]
44. Chen, M.; Tang, H.; Ma, H.; Holland, T.C.; Ng, K.Y.S.; Salley, S.O. Effect of Nutrients on Growth and Lipid Accumulation in the Green Algae *Dunaliella tertiolecta*. *Bioresour. Technol.* **2011**, *102*, 1649–1655. [CrossRef]
45. Hexin, L.; Xianggan, C.; Zhilei, T.; Shiru, J. Analysis of Metabolic Responses of *Dunaliella salina* to Phosphorus Deprivation. *J. Appl. Phycol.* **2017**, *29*, 1251–1260. [CrossRef]
46. Tammam, A.A.; Fakhry, E.M.; El-Sheekh, M. Effect of Salt Stress on Antioxidant System and the Metabolism of the Reactive Oxygen Species in *Dunaliella salina* and *Dunaliella tertiolecta*. *Afr. J. Biotechnol.* **2011**, *10*, 3795–3808.
47. Katz, A.; Pick, U. Plasma Membrane Electron Transport Coupled to Na^+ Extrusion in the Halotolerant Alga *Dunaliella*. *Biochim. Biophys. Acta Bioenerg.* **2001**, *1504*, 423–431. [CrossRef]
48. Oren, A. A Hundred Years of *Dunaliella* Research: 1905–2005. *Saline Syst.* **2005**, *1*, 2. [CrossRef]
49. Chen, H.; Jiang, J.-G.; Wu, G.-H. Effects of Salinity Changes on the Growth of *Dunaliella salina* and Its Isozyme Activities of Glycerol-3-Phosphate Dehydrogenase. *J. Agric. Food Chem.* **2009**, *57*, 6178–6182. [CrossRef]
50. Jiang, Y.; Yoshida, T.; Quigg, A. Photosynthetic Performance, Lipid Production and Biomass Composition in Response to Nitrogen Limitation in Marine Microalgae. *Plant Physiol. Biochem.* **2012**, *54*, 70–77. [CrossRef]
51. Li, Y.; Horsman, M.; Wang, B.; Wu, N.; Lan, C.Q. Effects of Nitrogen Sources on Cell Growth and Lipid Accumulation of Green Alga *Neochloris oleoabundans*. *Appl. Microbiol. Biotechnol.* **2008**, *81*, 629–636. [CrossRef]
52. Wykoff, D.D.; Davies, J.P.; Melis, A.; Grossman, A.R. The Regulation of Photosynthetic Electron Transport during Nutrient Deprivation in *Chlamydomonas reinhardtii*. *Plant Physiol.* **1998**, *117*, 129–139. [CrossRef]
53. Zhang, L.; Happe, T.; Melis, A. Biochemical and Morphological Characterization of Sulfur-Deprived and H2-Producing *Chlamydomonas reinhardtii* (Green Alga). *Planta* **2002**, *214*, 552–561. [CrossRef]
54. Volgusheva, A.A.; Zagidullin, V.E.; Antal, T.K.; Korvatovsky, B.N.; Krendeleva, T.E.; Paschenko, V.Z.; Rubin, A.B. Examination of Chlorophyll Fluorescence Decay Kinetics in Sulfur Deprived Algae *Chlamydomonas reinhardtii*. *Biochim. Biophys. Acta Bioenerg.* **2007**, *1767*, 559–564. [CrossRef]
55. Srinivasan, R.; Mageswari, A.; Subramanian, P.; Suganthi, C.; Chaitanyakumar, A.; Aswini, V.; Gothandam, K.M. Bicarbonate Supplementation Enhances Growth and Biochemical Composition of *Dunaliella salina* V-101 by Reducing Oxidative Stress Induced during Macronutrient Deficit Conditions. *Sci. Rep.* **2018**, *8*, 6972. [CrossRef] [PubMed]
56. Gao, Y.; Yang, M.; Wang, C. Nutrient Deprivation Enhances Lipid Content in Marine Microalgae. *Bioresour. Technol.* **2013**, *147*, 484–491. [CrossRef] [PubMed]
57. Morita, M.; Watanabe, Y.; Saiki, H. Photosynthetic Productivity of Conical Helical Tubular Photobioreactor Incorporatin *Chlorella sorokiniana* under Field Conditions. *Biotechnol. Bioeng.* **2002**, *77*, 155–162. [CrossRef] [PubMed]
58. Lai, Y.-C.; Karam, A.L.; Sederoff, H.W.; Ducoste, J.J.; de los Reyes, F.L. Relating Nitrogen Concentration and Light Intensity to the Growth and Lipid Accumulation of *Dunaliella viridis* in a Photobioreactor. *J. Appl. Phycol.* **2019**, *31*, 3397–3409. [CrossRef]

59. Sheehan, J.; Dunahay, T.; Benemann, J.; Roessler, P.A. Look Back at the U.S. Department of Energy's Aquatic Species Program Biodiesel from Algae. *Nat. Renew. Energy Lab.* **1998**, *328*, 1–294.
60. Antal, T.K.; Volgusheva, A.A.; Kukarskikh, G.P.; Krendeleva, T.E.; Tusov, V.B.; Rubin, A.B. Examination of Chlorophyll Fluorescence in Sulfur-Deprived Cells of *Chlamydomonas reinhardtii*. *Biophysics* **2006**, *51*, 251–257. [CrossRef]
61. Shaker, S.; Morowvat, M.H.; Ghasemi, Y. Effects of Sulfur, Iron and Manganese Starvation on Growth, β-Carotene Production and Lipid Profile of *Dunaliella salina*. *J. Young Pharm.* **2017**, *9*, 43–46. [CrossRef]
62. Johnson, M.K.; Johnson, E.J.; Macelroy, R.D.; Speer, H.L.; Bruff, B.S. Effects of Salts on the Halophilic Alga *Dunaliella viridis*. *J. Bacteriol. Res.* **1968**, *95*, 1461–1468. [CrossRef]
63. Fazeli, M.R.; Tofighi, H.; Samadi, N.; Jamalifar, H. Carotenoids Accumulation by *Dunaliella tertiolecta* (Lake Urmia Isolate) and *Dunaliella salina* (CCAP 19/18 & Wt) under Stress Conditions. *DARU J. Pharm. Sci.* **2012**, *14*, 146–150.
64. Ye, Z.-W.; Jiang, J.-G.; Wu, G.-H. Biosynthesis and Regulation of Carotenoids in *Dunaliella*: Progresses and Prospects. *Biotechnol. Adv.* **2008**, *26*, 352–360. [CrossRef]
65. Ahmed, R.A.; He, M.; Aftab, R.A.; Zheng, S.; Nagi, M.; Bakri, R.; Wang, C. Bioenergy Application of *Dunaliella salina* SA 134 Grown at Various Salinity Levels for Lipid Production. *Sci. Rep.* **2017**, *7*, 8118. [CrossRef] [PubMed]
66. Silva, M.F.; Casazza, A.A.; Ferrari, P.F.; Perego, P.; Bezerra, R.P.; Converti, A.; Porto, A.L.F. A New Bioenergetic and Thermodynamic Approach to Batch Photoautotrophic Growth of *Arthrospira (Spirulina) Platensis* in Different Photobioreactors and under Different Light Conditions. *Bioresour. Technol.* **2016**, *207*, 220–228. [CrossRef] [PubMed]
67. Sassano, C.E.N.; Carvalho, J.C.M.; Gioielli, L.A.; Sato, S.; Torre, P.; Converti, A. Kinetics and Bioenergetics of *Spirulina platensis* Cultivation by Fed-Batch Addition of Urea as Nitrogen Source. *Appl. Biochem. Biotechnol.* **2004**, *112*, 143–150. [CrossRef]
68. Stephanopoulos, G.; Aristidou, A.A.; Nielsen, J. *Metabolic Engineering: Principles and Methodologies*, 1st ed.; Academic Press: San Diego, CA, USA, 1998.
69. Bezerra, R.P.; Matsudo, M.C.; Sato, S.; Perego, P.; Converti, A.; de Carvalho, J.C.M. Effects of Photobioreactor Configuration, Nitrogen Source and Light Intensity on the Fed-Batch Cultivation of *Arthrospira (Spirulina) Platensis*. Bioenergetic Aspects. *Biomass Bioenergy* **2012**, *37*, 309–317. [CrossRef]
70. Richmond, A. Phototrophic Microalgae. In *Biotechnology*; Rehm, H.J., Reed, G., Dellweg, H., Eds.; Verlag Chemie: Weinheim, Germany, 1983; Volume 3, pp. 109–143.
71. Yang, C.; Hua, Q.; Shimizu, K. Energetics and Carbon Metabolism during Growth of Microalgal Cells under Photoautotrophic, Mixotrophic and Cyclic Light-Autotrophic/Dark-Heterotrophic Conditions. *Biochem. Eng. J.* **2000**, *6*, 87–102. [CrossRef]
72. Escoubas, J.M.; Lomas, M.; LaRoche, J.; Falkowski, P.G. Light Intensity Regulation of Cab Gene Transcription Is Signaled by the Redox State of the Plastoquinone Pool. *Proc. Natl. Acad. Sci. USA* **1995**, *92*, 10237–10241. [CrossRef]
73. Mukhanov, V.S.; Kemp, R.B. Simultaneous Photocalorimetric and Oxygen Polarographic Measurements on *Dunaliella Maritima* Cells Reveal a Thermal Discrepancy That Could Be Due to Nonphotochemical Quenching. *Thermochim. Acta* **2006**, *446*, 11–19. [CrossRef]
74. Ben-Amotz, A.; Avron, M. The Wavelength Dependence Of Massive Carotene Synthesis In *Dunaliella Bardawil* (Chlorophyceae)1. *J. Phycol.* **1989**, *25*, 175–178. [CrossRef]
75. Binaghi, L.; Del Borghi, A.; Lodi, A.; Converti, A.; Del Borghi, M. Batch and Fed-Batch Uptake of Carbon Dioxide by *Spirulina platensis*. *Process Biochem.* **2003**, *38*, 1341–1346. [CrossRef]
76. Sakurai, H.; Masukawa, H.; Kitashima, M.; Inoue, K. Photobiological Hydrogen Production: Bioenergetics and Challenges for Its Practical Application. *J. Photochem. Photobiol. C Photochem. Rev.* **2013**, *17*, 1–25. [CrossRef]
77. Kayser, A.; Weber, J.; Hecht, V.; Rinas, U. Metabolic Flux Analysis of *Escherichia coli* in Glucose-Limited Continuous Culture. I. Growth-Rate-Dependent Metabolic Efficiency at Steady State. *Microbiology* **2005**, *151*, 693–706. [CrossRef] [PubMed]
78. Taymaz-Nikerel, H.; Borujeni, A.E.; Verheijen, P.J.T.; Heijnen, J.J.; van Gulik, W.M. Genome-Derived Minimal Metabolic Models for *Escherichia coli* MG1655 with Estimated in Vivo Respiratory ATP Stoichiometry. *Biotechnol. Bioeng.* **2010**, *107*, 369–381. [CrossRef] [PubMed]

79. Kliphuis, A.M.J.; Klok, A.J.; Martens, D.E.; Lamers, P.P.; Janssen, M.; Wijffels, R.H. Metabolic Modeling of *Chlamydomonas reinhardtii*: Energy Requirements for Photoautotrophic Growth and Maintenance. *J. Appl. Phycol.* **2012**, *24*, 253–266. [CrossRef] [PubMed]
80. Li, Y.; Li, L.; Liu, J.; Qin, R. Light Absorption and Growth Response of *Dunaliella* under Different Light Qualities. *J. Appl. Phycol.* **2020**, *32*, 1041–1052. [CrossRef]

Publisher's Note: MDPI stays neutral with regard to jurisdictional claims in published maps and institutional affiliations.

Article

Biorefinery-Based Approach to Exploit Mixed Cultures of *Lipomyces starkeyi* and *Chloroidium saccharophilum* for Single Cell Oil Production

Gaetano Zuccaro [1,*], Angelo del Mondo [2], Gabriele Pinto [3], Antonino Pollio [3] and Antonino De Natale [3]

1 Department of Chemical, Materials and Production Engineering, University of Naples Federico II, 80125 Napoli, Italy

2 Stazione Zoologica Anton Dohrn, Istituto Nazionale di Biologia, Ecologia e Biotecnologie Marine, Villa Comunale, 80121 Napoli, Italy; angelo.delmondo@unina.it

3 Dipartimento di Biologia, Universita' degli Studi di Napoli Federico II, Via Cinthia 26, 80126 Napoli, Italy; gabriele.pinto@unina.it (G.P.); antonino.pollio@unina.it (A.P.); denatale@unina.it (A.D.N.)

* Correspondence: gaetano.zuccaro@unina.it

Abstract: The mutualistic interactions between the oleaginous yeast *Lipomyces starkeyi* and the green microalga *Chloroidium saccharophilum* in mixed cultures were investigated to exploit possible synergistic effects. In fact, microalga could act as an oxygen generator for the yeast, while the yeast could provide carbon dioxide to microalga. The behavior of the two microorganisms alone and in mixed culture was studied in two synthetic media (YEG and BBM + G) before moving on to a real model represented by the hydrolysate of *Arundo donax*, used as low-cost feedstock, and previously subjected to steam explosion and enzymatic hydrolysis. The overall lipid content and lipid productivity obtained in the mixed culture of YEG, BBM + G and for the hydrolysate of *Arundo donax* were equal to 0.064, 0.064 and 0.081 $g_{lipid} \cdot g_{biomass}^{-1}$ and 30.14, 35.56 and 37.22 $mg_{lipid} \cdot L^{-1} \cdot day^{-1}$, respectively. The mixed cultures, in all cases, proved to be the most performing compared to the individual ones. In addition, this study provided new input for the integration of Single Cell Oil (SCO) production with agro-industrial feedstock, and the fatty acid distribution mainly consisting of stearic (C18:0) and oleic acid (C18:1) allows promising applications in biofuels, cosmetics, food additives and other products of industrial interest.

Keywords: mixed culture; *Lipomyces starkeyi*; *Chloroidium saccharophilum*; Single Cell Oils (SCOs); *Arundo donax*; biorefinery

Citation: Zuccaro, G.; del Mondo, A.; Pinto, G.; Pollio, A.; De Natale, A. Biorefinery-Based Approach to Exploit Mixed Cultures of *Lipomyces starkeyi* and *Chloroidium saccharophilum* for Single Cell Oil Production. *Energies* **2021**, *14*, 1340. https://doi.org/10.3390/en14051340

Academic Editor: José Carlos Magalhães Pires

Received: 3 January 2021
Accepted: 24 February 2021
Published: 1 March 2021

1. Introduction

A sustainable economic growth, devoted to the future generations, requires long-term available resources for industrial production, in terms of raw materials and energy. Thus, a biorefinery-based approach, aimed to the conversion of low-cost feedstocks into marketable chemicals, fuels and products, has to be preferred [1]. Thus far, several strategies have been explored to enhance the productivity and competitiveness of microbial-based processes and simultaneously improve the biorefinery efficiency [2]. Particular attention has been devoted to the Single Cell Oil (SCO) production or microbial oil synthesis used as supplier of functional oils and for biodiesel [3], but the high fermentation costs make this production still undeveloped industrially. Therefore, the identification of substrates that reduce the costs is considered a solution that positively affects the related industrial implementation as well as the identification of strategies able to contribute positively in the same direction. Among them, co-culturing oleaginous yeasts and microalgae has been studied in recent years for enhancing Single Cell Oil productivity by utilizing minimal resources in various fields such as wastewater treatment, biogas production, enzyme production and bioremediation [4]. While the benefits that can be derived from these systems are clear, the nature of mutualistic interactions between yeast and microalgae in co-culture systems are still largely unexplored [5]. The photoautotroph–heterotroph partnership has been defined as able to overcome the high oxygen accumulation that causes a significant problem for microalgal growth especially in closed systems, since it inhibits photosynthesis. Thus, the inclusion of a heterotroph partner able to consume the oxygen mitigates this problem and, at the same time, can contribute to increase microbial biomass and metabolite production [6,7]. In addition, microalgae can convert the dissolved CO_2 in the medium into bicarbonate. When it is consumed, releasing OH^- ions, it makes the medium alkaline. Conversely, yeast growth results in acidic medium, which can hinder microalgal growth. The combination of both can help to balance this phenomenon [4]. Additionally, the reduction of toxic reactive oxygen species (ROS) by the heterotroph partner has been shown capable of protecting the phototroph microorganisms from oxidative stress in these co-cultures systems [7]. To promote the microbial oil synthesis, it is necessary to provide sufficient organic carbon in culture medium or enhance the photosynthesis of microalgae. The research about the microalgae oil production has been focused mainly on the photoautotrophic growth mode, but there are significant drawbacks associated. In fact, it is difficult to find operating conditions for the simultaneous achievement of biomass accumulation and lipid synthesis during the microalgae life cycle [8]. In addition, light attenuation is unavoidable for photoautotrophic cultures from lab to pilot scale, leading to significant reductions in productivity [9]. For these reasons, other cultivation modes in which microalgae are also able to use the source of organic carbon have been explored to improve the productivity of microbial oils. Oleaginous yeasts can grow in the presence of different carbon sources, for example hexose and pentose sugars, with high growth rates [10]. Generally, they have the ability to accumulate Single Cell Oils (SCOs) at more than 20% of their total dry weight [11]. In culture medium with high C/N ratio, oleaginous microorganisms utilize the remaining carbon source for the synthesis of lipids, mainly triacylglycerols (TAGs). Thus far, yeasts, molds or microalgae have been used for lipid production more frequently than bacteria [12]. *Lipomyces starkeyi* [13] and *Chloroidium saccharophilum* (W. Krüger) [14] were chosen as species and their interactions in mixed cultures were investigated. *L. starkeyi* displays a greater capacity to accumulate lipids and a natural ability to assimilate several feedstocks. It also tolerates low pH [15,16] and can metabolize inhibitors present in cellulosic hydrolysates [17]. *C. saccharophilum* was chosen for its high capacity of CO_2 assimilation and high tolerance to acidic environments [18], as well as for its high capacity of lipid accumulation [19], which represents positive aspects with regard to the use of mixed cultures to increase the lipid yield but also for the potential use of mixed cultures for CO_2 mitigation that would make the process more economically feasible where CO_2-rich flue gases are available, i.e., in the vicinity of power plants. Moreover, *C. saccharophilum* is able to grow under heterotrophic conditions [20] that are not secondary in view of using different feedstocks. It is known that the costs of feedstock represent a bottleneck in the successful development of heterotrophic microbial cultures [21]. For this reason, we adopted a growth medium obtained by *Arundo donax* L. (Giant reed), a perennial grass largely diffused in the Mediterranean Region. *A. donax* is considered a promising crop for industrial applications, thanks to its high biomass productivity, its adaptability to different climatic and soil conditions (e.g., polluted or salinized soils) and for the efficient protection offered against the erosion of hilly soils [22]. *A. donax* hydrolysate has been successfully employed for growing oleaginous yeast strains [23], but, to the best of our knowledge, there are very few studies about mixed oleaginous yeast–microalgae cultures fed with lignocellulosic hydrolysates [4]. This study was performed to verify the presence of synergistic effects comparing the operating conditions and the dynamics of synthetic media inoculated by single and mixed cultures of *L. starkeyi* and *C. saccharophilum* strains with a real system represented by *Arundo donax* hydrolysate previously subjected to a steam explosion pre-treatment. The incidence of inhibitors and volatile organic acids was also evaluated.

2. Materials and Methods

2.1. Strains and Pre-Culture Media

The oleaginous yeast *Lipomyces starkeyi* (DBVPG 6193) was supplied by the Dipartimento di Biologia Vegetale di Perugia, Italy. The strains were maintained at 5 °C on a YPD agar slants and then transferred in YPD medium for the seed culture contained (per liter): yeast extract 10 g, peptone 20 g, D-glucose 20 g and agar 20 g, when required. Both media were sterilized at 121 °C for 20 min before use.

The green microalga *Chloroidium saccharophilum* strain 042 was supplied by the ACUF microalgae collection (http://www.acuf.net (accessed on 5 August 2019)) of the Department of Biology, at the University Federico II of Naples, Italy. Single colonies were picked up from the plates and suspended in BBM medium. Tubes and plates were grown at 25 °C under continuous light supply (100 μmol·photons·m^{-2}·s^{-1}). Once in the exponential phase, each pre-culture was previously diluted with physiological solution and then used as inoculum for the culture media to obtain an initial concentration of approximately 3.0×10^6 cells·mL^{-1}. The two microorganisms share a similar size range in the order of 5–10 micron.

2.2. Culture Media and Operating Conditions

Axenic cultures of *Lipomyces starkeyi* and *Chloroidium saccharophilum* were grown in 500 mL Erlenmeyer flasks with an initial volume of 150 mL which contained (g/L): KH_2PO_4 (1.0), $MgSO_4 \cdot 7H_2O$ (0.5), $(NH_4)_2SO_4$ (2.0), yeast extract (0.5), and glucose (10.0) (YEG). The culture flasks were inoculated, separately and simultaneously, to achieve the initial cell density of about 3.0×10^6 cells·mL^{-1} for both microorganisms. The pH was adjusted to 6–6.5, and, prior to inoculation, the culture medium was sterilized at 121 °C for 20 min. As seed medium, the Bold Basal Medium (BBM) supplemented with yeast extract and $(NH_4)_2SO_4$ was chosen to replace $NaNO_3$ as nitrogen source and glucose as carbon source (BBM + G). The medium was autoclaved for 20 min at 121 °C. The enriched Bold Basal Medium (BBM + G) medium contained the following components: $CaCl_2 \cdot 2H_2O$ (1.70×10^{-4} M), KH_2PO_4 (1.29×10^{-3} M), EDTA anhydrous (1.71×10^{-4} M), KOH (5.52×10^{-4} M), K_2HPO_4 (4.31×10^{-4} M), NaCl (4.28×10^{-4} M), $MgSO_4 \cdot 7H_2O$ (3.04×10^{-4} M), H_3BO_3 (1.85×10^{-4} M), $FeSO_4 \cdot 7H_2O$ (1.79×10^{-5} M), H_2SO_4 (1.79×10^{-5} M), $ZnSO_4 \cdot 7H_2O$ (3.07×10^{-5} M), $MnCl_2 \cdot 4H_2O$ (3.07×10^{-5} M), MoO_3 (4.93×10^{-6} M), $CuSO_4 \cdot 5H_2O$ ($6.29 \times 1\,0^{-6}$ M), $Co(NO_3)_2 \cdot 6H_2O$ (1.68×10^{-6} M), yeast extract (0.5 g·L^{-1}), $(NH_4)_2SO_4$ (2 g·L^{-1}) and glucose (10 g·L^{-1}). The initial pH was adjusted to 6–6.5. The flasks were incubated at 25 °C and under continuous and fluorescent cool white light intensity equal to 100 μmol·photons·m^{-2}·s^{-1} (36 WT12, Osram Germany). All chemicals were purchased from Sigma-Aldrich. All cultures were carried out in in 500 mL Erlenmeyer flasks with an initial volume of 150 mL under continuous agitation. The flasks were sealed with aluminum caps to avoid any contamination from the external environment. The culture flasks were inoculated, separately and simultaneously, to achieve the initial cell density of about 3×10^6 cells·mL^{-1} for both microorganisms. The synthetic media were compared to lignocellulosic hydrolysate from pretreated *Arundo donax*. Steam explosion was applied as pretreatment. *A. donax* (giant reed) was processed in a continuous pilot plant (mod. StakeTech System Digester) located at ENEA–Trisaia Research Centre (Rotondella, Matera, Italy). The biomass was treated processing 150–200 kg·h^{-1} of dry biomass, to which water was added to raise the intrinsic humidity up to 50%. The pretreatment was carried out at 210 °C for 4 min. The severity factor (SF) was determined to be 3.84 according to Equation (1) [24]:

$$SF = \log(t \times e^{\frac{T - 100}{14.75}}) \quad (1)$$

where t is the residence time in minutes, T is the temperature of pre-treatment, 100 is the reference temperature and 14.75 is an arbitrary constant.

Pretreated *Arundo donax* was mixed with distilled water (pH 5.2) to obtain a solution with 5% *w/v* solid content and was afterwards treated with commercial enzymes purchased from Sigma-Aldrich consisting of cellulase from *Trichoderma reesei* ATCC 26921 (15 FPU/g of cellulose) and β-glucosidase from *Aspergillus niger* (30 CBU/g of cellulose). Cellulase activity was measured following the NREL filter paper assay [25] and reported in filter-paper units (FPU) per milliliter of solution. β-glucosidase activity was measured using the method described by Wood and Bhat [26] and reported in cellobiase units (CBU). Enzymatic hydrolysis was carried out at 160 rpm and 50 °C for 48 h (Minitron, Infors HT, Switzerland). The initial pH was adjusted to 6–6.5. As the hydrolysate of *Arundo donax* (ADH), 8 mL of phosphate buffer (0.2 M) were added, and then it was inoculated with *Lipomyces starkeyi* and *Chloroidium saccharophilum* alone and mixed to achieve the initial cell density of 3.0×10^6 cells·mL^{-1} for both microorganisms. All tests were carried out in triplicate and the standard deviation was calculated on the biological triplicate.

2.3. Analytical Methods

Measurements of pH were made by an inoLab® Multi 740 Multimeters pH-meter (WTW). The optical density was monitored with a Shimadzu UV6100 spectrophotometer (Japan) and by measuring turbidity of liquid samples at 600 and 680 nm. Microbial biomass (g·L^{-1}) was determined by filtering 2–3 mL of culture over pre-weight PES filters (0.45 μm; Sartorius Biolab, Göttingen, Germany). The retained biomass on filters was washed, dried at 105 °C for 24 h and then stored in a desiccator before being weighed. The individual cell counts of yeast and microalga were determined with a hemocytometer, using the microscope, due to the different appearance of *L. starkeyi* and *C. saccharophilum* being different under the microscope. After centrifugation and filtration with 0.2 μm cut-off filters, the liquid samples were analyzed for residual substrate content (glucose) and soluble fermentation products (VFA, alcohols). Glucose and VFAs were analyzed by HPLC (LC2010, Shimadzu, Japan), equipped with a refractive index detector (RID-20A, Shimadzu, Japan). Samples were first centrifuged at 12,000 g for 15 min and then supernatants were filtered with 0.2 μm syringe filters. HPLC analysis were performed at a flow rate of 0.7 mL/min on an Aminex HPX-87H, 300 × 7.8 mm (Bio-Rad) column at a temperature of 35 °C. H_2SO_4 at 4 mM was used as the mobile phase. The total concentration of phenolic compounds was determined using Folin–Ciocalteu assay [27]. A simple method based on UV spectra was followed for the estimation of total furans (furfural and hydroxymethylfurfural) in the hydrolysates [28]. Lipids were extracted following a method adapted from Bligh and Dyer [29]. The samples were stirred in a $CHCl_3/CH_3OH$ mixture (2:1 *w/v*) over 24 h, and the oleaginous biomass was filtered off and washed with additional $CHCl_3$. This procedure was repeated three times. The solvent was then removed by evaporation under N_2 stream. The total lipid concentration was estimated by gravimetric method. To calculate the lipid concentration of the cells, they were dried to a constant weight in the oven at 80 °C. The lipids extracted were subjected to transesterification reaction in a stirred container at 60 °C for 10 min, using NaOH (1% *w/v*) as catalysts and methanol as reagent. The samples were dried by N_2 stream and subsequently 1 mL of heptane was added for the analysis. The content of fatty acid methyl esters was determined by gas chromatography. The GC (GC-MS 2010, Shimadzu, Japan) was equipped with a flame ionization detector and an Omegawax 250 (Supelco) column (30 m × 0.25 mm I.D., 0.25 μm). Helium was used as carrier gas (flow rate: 30 mL/min). The FAME samples were initially dissolved in 1 mL of heptane and 1 μL of this solution was loaded onto the column. The temperature of the column was kept at 50 °C for 2 min, then heated to 220 °C at a rate of 4 °C/min and finally kept constant for 2 min. Methyl decanoate was used as internal standard. The peak of each methyl ester was identified by comparing the retention time with the peak of the pure standard compound.

2.4. Parameter Analysis

The specific growth rate (μ_x) and yield factor ($Y_{x/s}$) were calculated according to Equations (2)–(4):

$$\frac{dX}{dt} = \mu X \tag{2}$$

$$\mu_x = [\ln(X_t/X_0)]/(t - t_0). \quad (3)$$

$$Y_{x/s} = \frac{X - X_0}{S_0 - S} \quad (4)$$

where X and X_0 are the concentrations of microbial biomass at time t and at initial time, respectively. μ_x is the specific growth rate of microbial biomass. S_0 and S are the values of substrate concentrations at initial time and during the cultivation time, respectively. Lipid content, lipid yield (Y_{lipid}) and lipid productivity were calculated according to Equations (5)–(7), respectively:

$$Lipid\ content\ [g_{lipid} \cdot g_{biomass}{}^{-1}] = \frac{m_{lipid}}{m_{microbial\ biomass}} \quad (5)$$

$$Y_{lipid}\ [g_{lipid} \cdot L^{-1}] = \frac{m_{lipid}}{V} \quad (6)$$

$$Lipid\ productivity\ [mg_{lipid} \cdot L^{-1} \cdot d^{-1}] = \frac{m_{lipid}}{\Delta t} \quad (7)$$

Lipid content is the total lipid amount divided by the total microbial biomass. The lipid yield (Y_{lipid}) is the lipid concentration, i.e., the mass of lipids on the (same) volume of culture. Lipid productivity is defined as the concentration obtained in relation to cultivation time.

3. Results and Discussion

3.1. Effects of Synthetic Media on C. saccharophilum and L. starkeyi Growth Performances in Mono and Mixed Culture

A preliminary experimental campaign was conducted to evaluate the behavior of the two pure strains selected for this study, *L. starkeyi* and *C. saccharophilum*, in two different synthetic media, BBM + G (Figure 1a–c) and YEG (Figure 2a–c), at a C/N ratio equal to 11, which was not very different from the Redfield one equal to about 7 [30], and thus avoiding initial N limitations. Tests were performed by inoculating each batch with single or mixed cultures. In this latter case, the inoculum ratio between the yeast and the microalga was 1:1. No evidence was observed in terms of microbial biomass and cell proliferation, even changing this ratio, as demonstrated in a previous study [31]. In addition, an increase in the microalga:yeast ratio would promote algal metabolic activity by inducing an increase in pH, which could inhibit yeast growth. In BBM + G, the growth of *C. saccharophilum* showed an initial trend correlated to the glucose consumption (Figure 1a). This could be justified by an initial photosynthetic activity lower than the metabolism of organic carbon, which could produce an endogenic source of CO_2 [31], reducing the CO_2-limiting effect and explaining the increase in microbial biomass once the predominantly heterotrophic metabolism has stopped in the latter part of the cultivation time (Figure 1a). In the meantime, the reduced cell proliferation—from 2.7×10^7 to 3.6×10^7 cells·mL^{-1}—would presuppose an increase in size more than in number, due to a phase of lipid accumulation. This hypothesis seemed to be confirmed by the trend of microbial biomass, which increased from 0.93 to 2.3 g·L^{-1} (Figure 1a). The biomass growth rate (μ_x), the related value of biomass productivity and the yield factor ($Y_{x/s}$), equal to 0.381 d^{-1}, 328.6 mg·L^{-1}·day^{-1}, 0.471 g·g^{-1}, respectively (Table 1), are comparable with the values obtained by Herrera-Valencia et al. and Tan and Johns [19,20], taking into account the different growth conditions. The single cultures of *L. starkeyi* did not show any growth (Figure 1b), despite the presence of yeast extract and glucose. The initial glucose consumption probably led to the production of ethanol and acetic acid due to a reduced activity of TCA-cycle, that, in turn, moved the culture to anaerobiosis, due to oxygen limitation in the medium, essential for glucose transport. The batch culture of *L. starkeyi* was exposed to transient fermentation inhibition (Custers effect), and the absence of an alternative oxygen-independent transport mechanism for glucose absorption led to the lack of microbial growth [32]. Figure 1c shows the performances of *C. saccharophilum* and *L. starkeyi* grown in mixed cultures. From the trends associated to the cell proliferation, it seemed evident the presence of a synergistic effect, as evidenced by the comparison of glucose consumption rate values. The microalga was able to provide the oxygen necessary for *L. starkeyi* to metabolize the glucose, which was partially assimilated by *C. saccharophilum* and this effect led to an increase of glucose consumption rate from 1.47 $g \cdot L^{-1} \cdot day^{-1}$ for *C. saccharophilum* alone to 2.15 $g \cdot L^{-} \cdot day^{-1}$ for mixed culture. The final constant values in terms of cell concentration (Figure 1c) are justified by the metabolic shift due to an imbalance in the C/N ratio, which induced lipid accumulation phase rather than cell duplication. Since the microbial biomass productivity is a result of gravimetric methodology, the relative value increased from 328.6 $mg \cdot L^{-1} \cdot day^{-1}$ for *C. saccharophilum* alone to 366.2 $mg \cdot L^{-1} \cdot day^{-1}$ for mixed culture (Table 1). An increase was also observed in the specific growth rate, but not in the yield factor ($Y_{x/s}$), which is likely affected by endogenous metabolism and a faster glucose consumption. In mixotrophic conditions, the microalga should be less dependent on the yeast for CO_2, whereas the latter still benefits from the alga for O_2 production, which is partially consumed by respiration of the alga, negatively affecting yeast growth. However, this aspect did not affect the metabolic activity of the yeast, which, on the contrary, was favored by the presence of alga (Figure 1b,c). Comparing the max values of $cells \cdot mL^{-1}$ for *C. saccarophilum* alone and in mixed culture (3.6×10^7 and 5.8×10^7 $cells \cdot mL^{-1}$, respectively), it emerges that the trend is similar; therefore, there was no real competition for N or P sources or an effective reduced production of O_2 by the microalga, necessary to metabolize glucose, which would otherwise have been the reason for a reduced growth of yeast as well. Moreover, the yeast proliferation has undoubtedly induced a progressive photo-inhibition effect which could be another reason for a slowdown of microalgal growth in the final part of the test. The latter effect was associated to a limiting O_2 concentration, which in turn did not allow the yeast to grow further and to produce CO_2 available to the microalga. Tests were carried out in an attempt to verify the behavior of *L. starkeyi* alone in a more suitable medium, classified as YEG since it was found to be unable to grow in a BBM + G medium, although enriched with yeast extract and glucose because of the Custers effect discussed above.

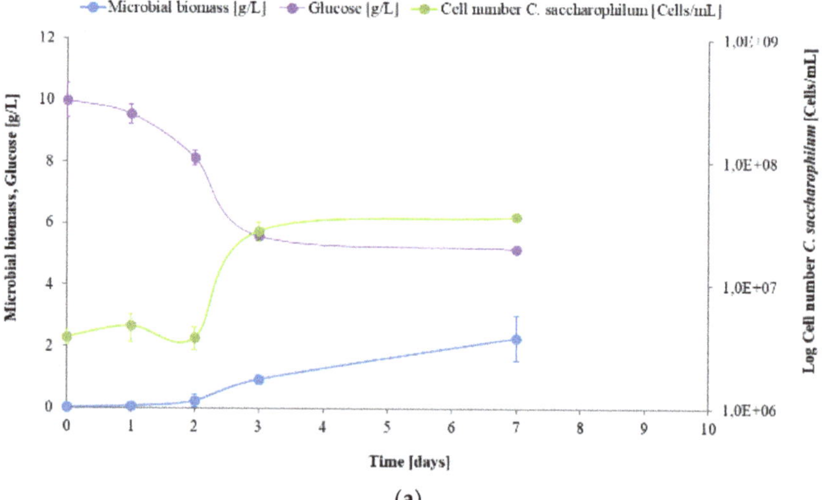

Figure 1. *Cont.*

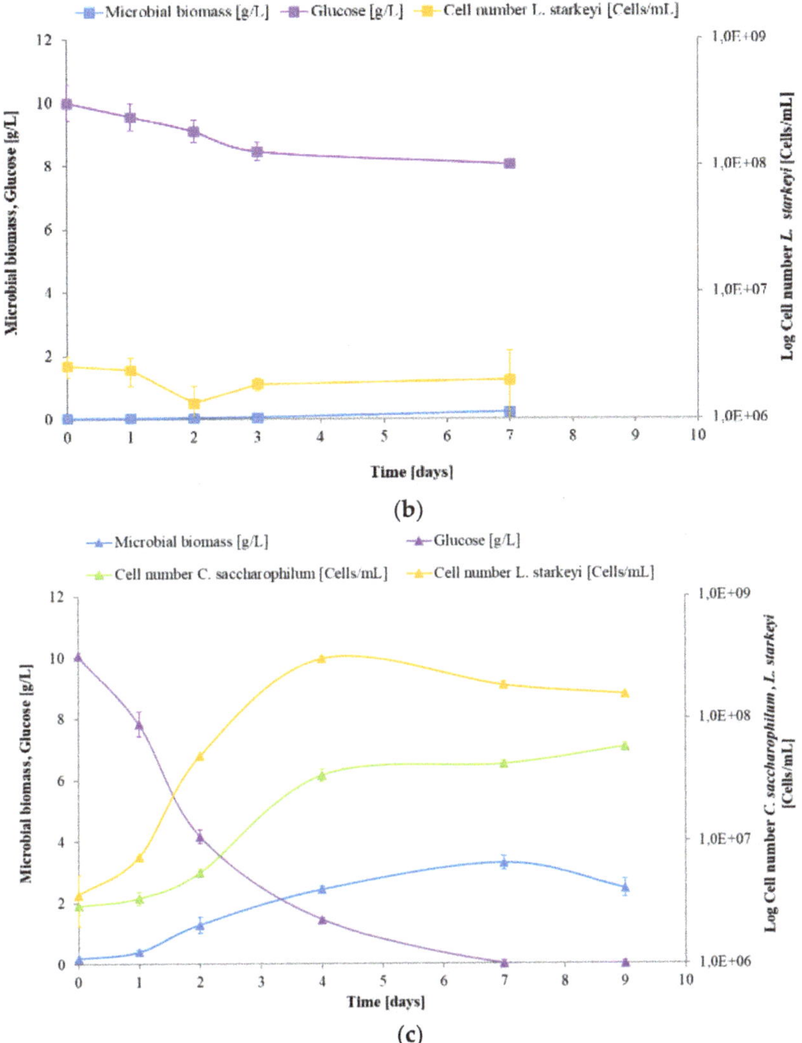

Figure 1. (**a**,**b**) Profiles of microbial biomass (g/L), glucose consumption (g/L) and cell number (cells/mL) in BBM + G medium supplemented with yeast extract and glucose inoculated by *C. saccharophilum* (**a**) and *L. starkeyi* (**b**). (**c**) Profiles of microbial biomass (g/L), glucose consumption (g/L) and cell number (cells/mL) in BBM + G medium supplemented with yeast extract and glucose inoculated by *C. saccharophilum* and *L. starkeyi* in mixed culture.

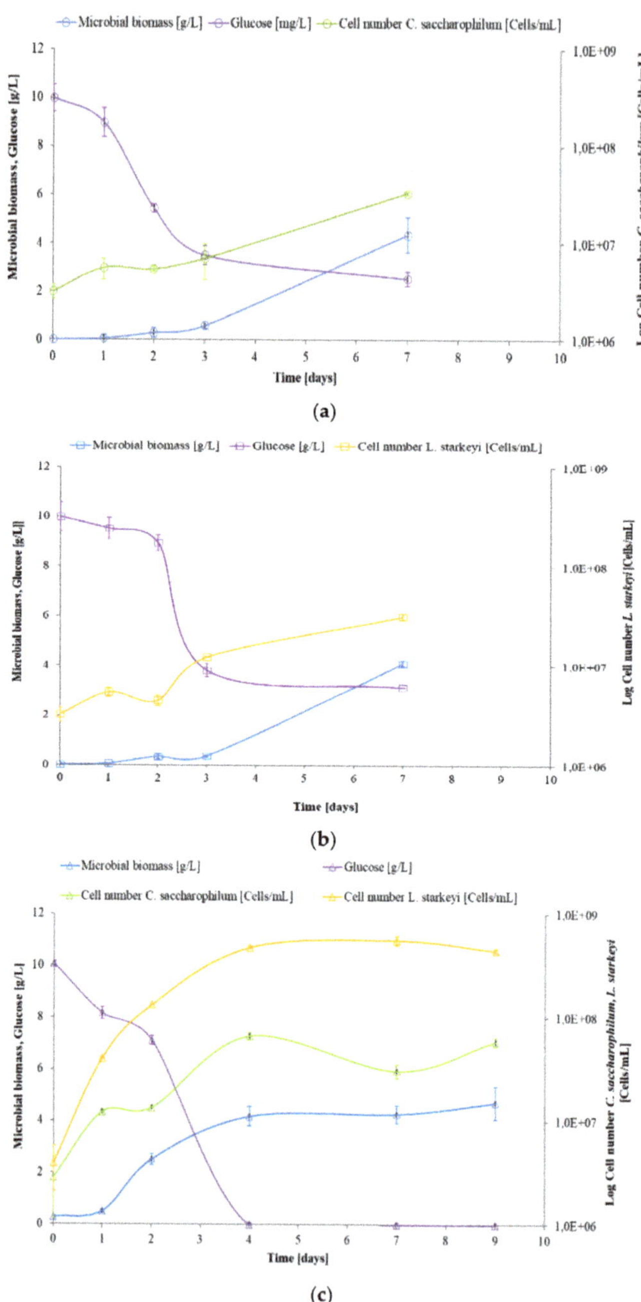

Figure 2. (**a**,**b**) Profiles of microbial biomass (g/L), glucose consumption (g/L) and cell number (cells/mL) in YEG medium inoculated by *C. saccharophilum* (**a**) and *L. starkeyi* (**b**). (**c**) Profiles of microbial biomass (g/L), glucose consumption (g/L) and cell number (cells/mL) in YEG medium inoculated by *C. saccharophilum* and *L. starkeyi* in mixed culture.

Table 1. Growth characteristics of the mixed cultures and the strains alone in BBM + G, YEG media and *Arundo donax* hydrolysates. Data shown as mean ± SD, $n = 3$.

Sample	μ_x [d^{-1}]	$Y_{x/s}$ [$g \cdot g^{-1}$]	μ_s [d^{-1}]	P_{CO2} [$mg \cdot L^{-1} \cdot d^{-1}$]	m Lipid [g]	Lipid Content [$g_{lipid} \cdot g_{biomass}^{-1}$]	Lipid Yield max $Y_{lipid\ max}$ [$g \cdot L^{-1}$]	Biomass Productivity [$mg \cdot L^{-1} \cdot d^{-1}$]	Lipid Productivity [$mg \cdot L^{-1} \cdot d^{-1}$]	$Cells_{max}$ *C. saccharophila* [10^6 cells·mL^{-1}]	$Cells_{max}$ *L. starkeyi* [10^6 cells·mL^{-1}]	Lipid Yield max/Cell max *C. saccharophila* [μg·$cells^{-1} \cdot 10^{-6}$]	Lipid Yield max/Cell max *L. starkeyi* [μg·$cells^{-1} \cdot 10^{-6}$]
BBM + G Chl	0.381	0.471	0.689	610.0	0.018	0.076 ± 0.03	0.175 ± 0.04	328.57	25.00	36.0 ± 1.76		4.9	
BBM + G Lip	0.037	0.095	0.279	49.9	0.002	0.093 ± 0.01	0.020 ± 0.04	30.65	2.87		2.0 ± 0.65		10.0
BBM + G Chl Lip	0.472	0.308	1.437	474.7	0.021	0.064 ± 0.02	0.211 ± 0.17	366.20	30.14	58.7 ± 1.93	309.3 ± 5.91	3.6	0.7
YEG Chl	0.944	0.583	1.065	1168.6	0.022	0.051 ± 0.01	0.221 ± 0.24	625.00	31.57	33.0 ± 1.27		1.5	
YEP Lip	0.803	0.596	0.978	1095.9	0.021	0.052 ± 0.03	0.212 ± 0.07	586.31	30.29		31.5 ± 0.77		6.7
YEG Chl Lip	1.166	0.440	1.118	925.4	0.032	0.068 ± 0.01	0.32 ± 0.28	611.90	35.56	66.7 ± 3.21	561.3 ± 16.64	1.0	0.1
ADH Chl	0.003	0.000 *	0.065	3.8	0.002	0.069 ± 0.03	0.016 ± 0.00	26.75	1.86	3.2 ± 0.37		5.0	
ADH Lip	0.614	0.422 *	0.883	837.2	0.032	0.074 ± 0.01	0.318 ± 0.05	552.38	35.33		32.45 ± 1.87		9.8
ADH Chl Lip	0.791	0.539 *	0.786	822.2	0.034	0.081 ± 0.04	0.335 ± 0.14	365.47	37.22	20.9 ± 1.23	110.0 ± 3.83	16.0	3.0

* Yield factor, $Y_{x/s}$, was calculated taking in account the contribute of all the monitored substrates.

Similar biomass profiles were observed in the presence of *C. saccharophilum* and *L. starkeyi* alone, when simultaneously no substrate inhibition was observed as shown by the corresponding glucose consumption profile (Figure 2a,b). The cessation of glucose consumption was rather attributed to the progressive consumption of CO_2 and O_2 for *C. saccharophilum* alone (Figure 2a) and *L. starkeyi* alone (Figure 2b), respectively. Moreover, this inhibition is due to a probable N source limitation. On the other hand, where the growth inhibition was observed (Figure 1b), the corresponding glucose consumption was lower. In the presence of co-cultures (Figures 1c, 2c and 3c), the virtuous loop CO_2-O_2 helped glucose consumption, but the progressive imbalance in the C/N ratio is also a form of inhibition that prevents a proportional increase in the concentration of microbial biomass. On the contrary, microalgae growth would be inhibited, due to excessive nutrient consumption caused by a high yeast:microalga ratio [33].

The individual cultures of *C. saccharophilum* (Figure 2a) and *L. starkeyi* (Figure 2b) show similar trends, as also confirmed by the values of specific growth rate (μ_x) equal to 0.94 and 0.8 d^{-1}, respectively, and the maximum value of cell concentration (cells·mL^{-1}) which for the alga was equal to 3.3×10^7 and for the yeast to 3.1×10^7. Mixotrophic activity of *C. saccharophilum* seemed to be confirmed by glucose consumption profile (Figure 2a). Instead, *L. starkeyi* alone showed an initial lag phase (Figure 2b). Therefore, although initially there were no limiting conditions with regard to oxygen concentration, the yeast needed an evident acclimation phase in terms of microbial biomass rather than the profile associated to the number of cells (cells·mL^{-1}), which increased with linear trend from the beginning (Figure 2b). Once the oxygen became limiting, i.e., after the third day, as observed also by the absence of further substrate consumption able to affect C/N ratio and consequently cell duplication, *L. starkeyi* showed a shift in metabolic activity directed on lipid accumulation. The effect of this metabolic turnover could be explained by the significant increase in microbial biomass, related to the occurrence of lipid bodies within the cell. Concerning *C. saccharophilum* alone, an initial phase of mixotrophic metabolism was followed by an autotrophic final phase, visible from the interruption of glucose assimilation. This was also done for the alga; therefore, during the final phase of the test, a lipid accumulation seems to be present, but it was related to a significant increase in the cell concentration due to the autotrophic metabolism. When *C. saccharophilum* and *L. starkeyi* were grown in mixed culture of YEG medium (Figure 2c), a synergistic effect was observed, confirming the results obtained with the enriched BBM + G medium. Therefore, the mixotrophic metabolism of the microalga and the yeast fermentation were not in competition for the organic C source or N and P, but rather they benefited from each other's metabolic activity, i.e., the consequent presence of a O_2-CO_2 virtuous loop. This seems evident looking at the initial glucose consumption rate in the case of *L. starkeyi* alone (Figure 2b) and *L. starkeyi* and *C. saccharophilum* mixed culture (Figure 2c). The presence of O_2 overproduction by the microalgal metabolism supported the fermentation and therefore the consumption of the organic substrate. If the specific growth rate (μ_x) (Table 1) was positively influenced by the mixed culture, different was the case related to the yield factor ($Y_{x/s}$), since the need of microbial biomass for organic carbon conversion increased as consequence of higher cell concentration in the inoculum. In fact, it changed from 0.583 g·g^{-1} for *C. saccharophilum* alone to 0.596 g·g^{-1} for *L. starkeyi* alone and 0.44 g·g^{-1} for mixed culture. The choice of a 1:1 ratio of microalga:yeast was made in relation to what was observed in a previous study [31] and according to what was highlighted, for example, by Ling et al. [34] and Li et al. [33]. An imbalance in the microalga:yeast inoculum ratio generally motivated by their different growth rates, besides not inducing a net increase in terms of lipid productivity, can sometimes lead to an accentuation for N and P competition, and consequently to an imbalance in pH. Figure 2c also shows that, once the carbon source was depleted, the consequence was, also in this case, the stopping of cell proliferation, as shown by the unmodified cell concentration.

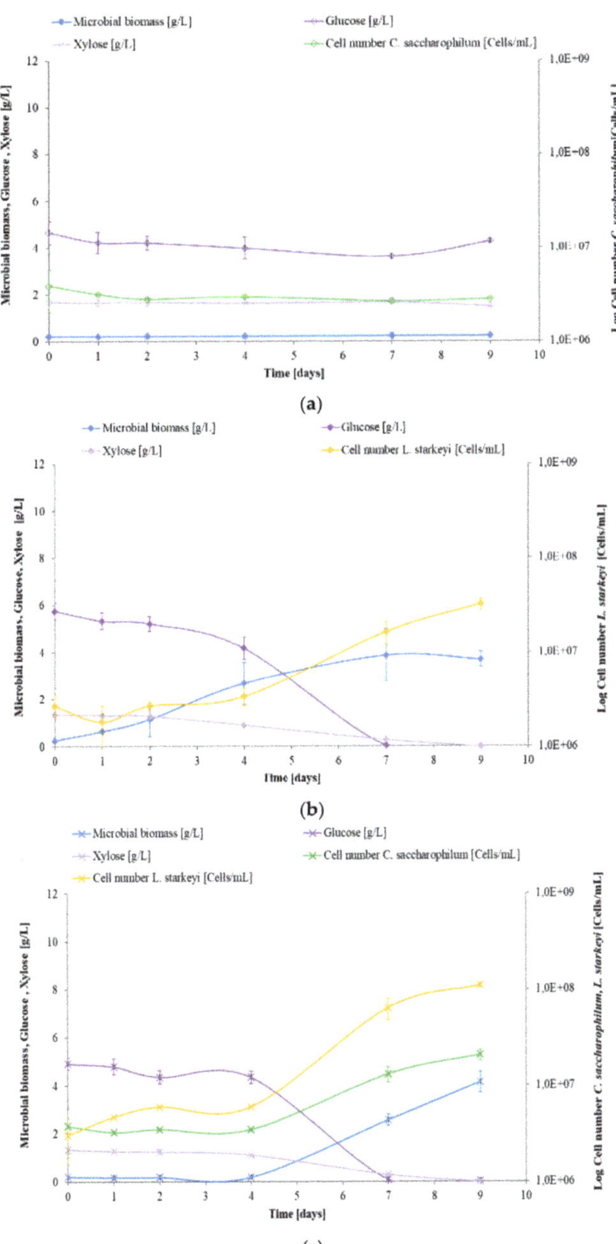

Figure 3. (**a**,**b**) Profiles of microbial biomass (g/L), glucose consumption (g/L), xylose consumption (g/L) and cell number (cells/mL) in *Arundo donax* hydrolysate (ADH) inoculated by *C. saccharophilum* (**a**) and *L. starkeyi* (**b**). (**c**) Profiles of microbial biomass (g/L), glucose consumption (g/L), xylose consumption (g/L) and cell number (cells/mL) in *Arundo donax* hydrolysate (ADH) inoculated by *C. saccharophilum* and *L. starkeyi* in mixed culture.

3.2. Study of *C. saccharophilum* and *L. starkeyi* Single and Mixed Cultures in *Arundo donax* Hydrolysate

The performances in synthetic media described in the previous paragraph were considered necessary before moving to a real system, represented by the hydrolysate, in which the performance of the alga and yeast was evaluated individually and then in a mixed culture. *Arundo donax* was previously subjected to steam explosion and enzymatic hydrolysis. These two preliminary processes were chosen considering evaluations supported by experimental campaigns previously performed and aimed to optimize the operating conditions, minimize the release of inhibiting compounds and maximize the concentration of fermentable sugars. The steam explosion represents a pre-treatment able to facilitate the access of the crystalline structure of lignocellulosic biomass to hydrolytic enzymes and to limit the negative effects just discussed. Enzymatic hydrolysis was chosen for its operating conditions, which remain less impacting compared to other hydrolytic processes, further reducing the risk to increase the production and the release of inhibitory compounds [35]. As observed in Figure 3a, *C. saccharophilum* alone did not present any form of microbial growth that could be associated to the presence of inhibitory compounds. The choice of *C. saccharophilum* was made also in view of its high tolerance to the presence, for example, of phenols, recognizing in these latter compounds also a positive effect on the regulation of enzymatic activity, cell membrane structure and macromolecule synthesis [36–38]. Furans could also be a reason for inhibition since they are able to cause long lag-phase [39]. Therefore, the permanence of this regime led to the exhaustion of CO_2 and subsequently of O_2 source, making the culture substantially anaerobic and therefore unable to grow. A further aspect preliminarily considered was the possible inhibition due to the turbidity of the culture medium that in other studies led to adopting a dilution factor to promote photosynthetic activity [40] involving, at the same time, a loss in organic substrates concentration. On the other hand, the inhibition could be attributed to acetate potentially present in the dissociated form at neutral pH, as in our case [41]. Russel [42] attributed the inhibitory effects of weak acids to two mechanisms: uncoupling and intracellular anion accumulation. For all these reasons, i.e., associated to the possible occurrence of limiting conditions in terms of CO_2 or O_2 and the potential inhibiting aspects such as turbidity or too high concentration in acetate, the study of mixed culture was found to be fundamental and potentially able to answer the doubts resulting from the single culture of *C. saccharophilum*. Another not secondary aspect that could have been the cause of algae inhibition is C/N ratio. In fact, the use of low-cost feedstock, which is motivated by the need to reduce the costs of the growth medium and the positive effect on the increase of lipid accumulation, presents, however, a high C/N ratio that exceeds the optimal values for the algae.

Before arriving at the mixed culture analysis, an additional control culture was monitored considering *L. starkeyi* alone as inoculum. Figures 3b and 4b show the relative trends in terms of metabolites (substrates and products) and microbial biomass. The lag phase associated to the first two days seemed caused by the presence of furans, which, although not in high concentration, represented a compound able to delay the growth phase. The hypothesis that the yeast inhibition is affected by the C/N ratio was ruled out, given its ability to tolerate values of even 100 [4]. The lag phase was therefore followed by a growth phase, more visible in terms of microbial biomass than in terms of cells$\cdot$mL^{-1}, which was the result of the simultaneous assimilation of glucose, xylose and acetate. This aspect confirmed what was observed by Anschau et al. [43], Gong et al. [44] and Yang et al. [45] related to the potential use of this microorganism in the presence of more complex feedstocks where co-metabolism is of great interest to solve inhibition problems due to volatile organic acids formed during the pretreatments and it should be valuable towards conversion of acetate lignocellulosic biomass materials into SCOs. Moreover, acetate assimilation by *L. starkeyi* represents a positive aspect with respect to pH adjustment, but it leads to an additional demand for dissolved oxygen.

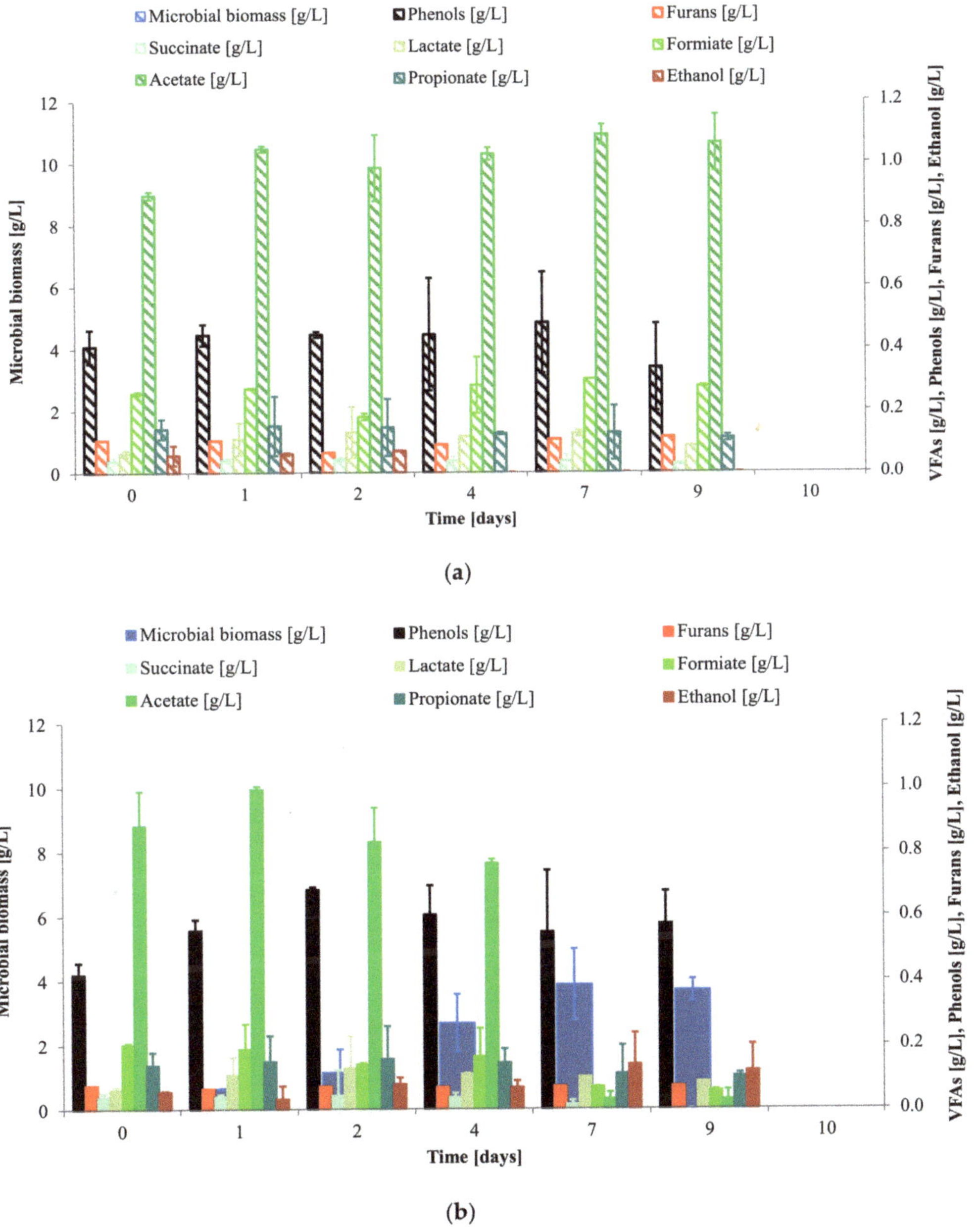

Figure 4. *Cont.*

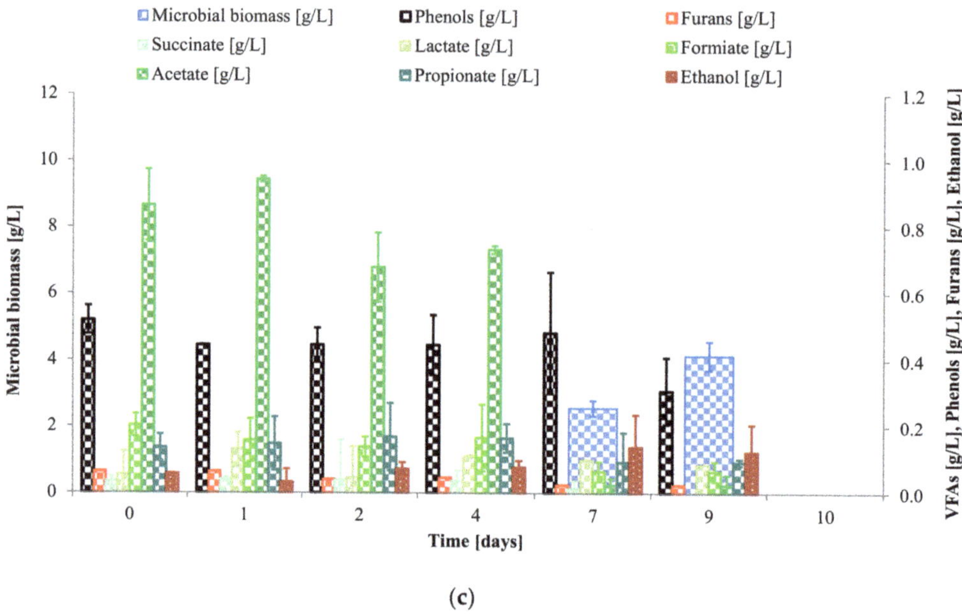

(c)

Figure 4. (**a**) Profiles of microbial biomass (g/L), VFAs, Phenols and Furans concentration (g/L) in *Arundo donax* hydrolysate (ADH) inoculated by *C. saccharophilum*. (**b**) Profiles of microbial biomass (g/L), VFAs, Phenols and Furans concentration (g/L) in *Arundo donax* hydrolysate (ADH) inoculated by *L. starkeyi*. (**c**) Profiles of microbial biomass (g/L), VFAs, Phenols and Furans concentration (g/L) in *Arundo donax* hydrolysate (ADH) inoculated by *C. saccharophilum* and *L. starkeyi*.

Figures 3b and 4b show how the consumption patterns of the substrates, i.e., mainly glucose, xylose and acetate, followed a diauxic mechanism. The consumption pattern of substrates is consistent with the metabolism becoming fermentative. The concentration and productivity of microbial biomass (Table 1) were 3.7 g·L^{-1} and 411 mg·L^{-1}·day^{-1}, respectively. These values can be considered in line with those obtained by Pirozzi et al. [46] considering that *Arundo donax* was submitted to different pre-treatment processes, but different by one order of magnitude when compared to hydrolysates of wheat corn, corn bran and corn stover [15,47,48].

Finally, *Arundo donax* hydrolysate was inoculated with a mixed culture of *L. starkeyi* and *C. saccharophilum* to observe the positive and less positive effects of the two-microorganism combination also in relation to what was previously studied with regard to synthetic media in which all aspects due to inhibitory effects resulting from the complexity of the hydrolysate were missing. Meanwhile, as shown in Figure 3c, the lag phase lasted for the first four days, apart from a slight increase in *L. starkeyi* cell number that passed from 3.3×10^6 to 5.9×10^6 cells·mL^{-1}, mainly as a result of glucose consumption rate, equal to 0.28 g·L^{-1}·day^{-1}. At this stage, inhibition due to the presence of inhibiting compounds, such as furans and phenols was confirmed, but once the lag phase was exceeded both microorganisms showed a growth capacity that, in the case of *C. saccharophilum*, was not observed in single culture. The max values of cells·mL^{-1} for *L. starkeyi* and *C. saccharophilum* in mixed culture were 1.1×10^8 and 2.1×10^7, respectively. These values were understandably lower than those found for mixed culture in YEG medium, respectively, equal to 5.6×10^8 and 6.7×10^7 cells·mL^{-1}, because of faster glucose consumption and absence of inhibition phenomena.

What seemed evident, confirming what has been observed in previous studies [31,32,49–51], was the presence of a synergistic effect that in our case was attributed to different processes. During the lag phase, *L. starkeyi* showed a latent growth phase

(Figure 3c) that induced a progressive endogenous source of CO_2 able to avoid limiting conditions for the microalga growth. As observed in the single culture (Figure 3a), the microalga did not grow according to a heterotrophic metabolism or mixotrophic because of the various forms of inhibition just discussed. This progressive accumulation of CO_2, not to the point of representing a form of inhibition for the yeast, could be the cause of a partial acidification of the culture medium, but both *C. saccharophilum* and *L. starkeyi* tolerate acid pH. This acidification was therefore counteracted by acetate consumption after the fourth day, since a pH increase is typically observed when a microorganism grows in a salt of an organic acid. The progressive consumption of the organic substrate, and thus the decrease in the C/N ratio that was crucial for the inhibition of single microalgae growth, together with the pH control due to acetate consumption to balance the potential acidification due to yeast respiration, were the positive aspects that favored the symbiotic growth of *C. saccharophilum* in mixed culture. In terms of the final microbial biomass, the differences were not evident when passing from 4.72 to 4.14 $g{\cdot}L^{-1}$ for mixed cultures in YEG medium and in *Arundo donax* hydrolysate (Figures 2c and 3c), this was different when evaluated in terms of microbial biomass productivity measured in the same time-range, equal, respectively, to 611.9 and 365.5 $mg{\cdot}L^{-1}/day^{-1}$, showing in the latter case the incidence of inhibiting phenomena that slowed down the growth phase. The microbial biomass growth rate (μ_x) and the yield factor ($Y_{x/s}$) in mixed culture were, respectively, 0.791 day^{-1} and 0.539 $g{\cdot}g^{-1}$ (Table 1). Once this synergistic effect stopped as a consequence of limiting conditions achievement in terms of O_2, yeast growth rate changed, switching to a fermentative metabolism and consequently to ethanol accumulation, negatively influencing also the alga, whose growth also suffered a slowdown (Figure 3c).

Studies about the tolerance of single microbe species to inhibitory compounds are quite limited. Several attempts have been made to promote the microbial growth by reducing toxic compounds before fermentation process with certain detoxification steps. However, the detoxification which directly leads to high cost will decrease the economy of the whole process [52]. The fact that depletions of hydrolysis degradation products were enhanced by the mixed culture mode should be regarded positively because the overall process of microbial lipid production from lignocellulosic hydrolysate can be simplified by omitting the need of a detoxification step [49].

3.3. Lipid Production and Fatty Acid Distribution

Different parameters were evaluated, as reported in Table 1, to compare the incidence of the culture conditions adopted onto the lipid content, lipid yield and lipid productivity. Nutrient limitation, notoriously an imbalance of C/N ratio, promotes lipogenesis. However, other factors besides nitrogen-limitation could induce lipogenesis in oleaginous yeasts, including the effect of phosphate and sulfur limitation. Indeed, soluble phosphates could be precipitated and removed by interaction with metal ions, such as Ca^{2+}, Mg^{2+} or Fe^{3+}, and the resulting hydrolysate, exhibiting high C/N and C/P ratio, could allow for even higher lipid accumulation [53]. The lipid content reached a maximum value of 0.081 $g_{lipid}{\cdot}g_{biomass}^{-1}$ for mixed culture in *Arundo donax* hydrolysate, which was higher than the mixed culture in BBM + G or YEG, where the same value was about 0.063–0.064 $g_{lipid}{\cdot}g_{biomass}^{-1}$. Therefore, these values were significantly lower to those obtained by Liu et al. [49], where the maximum lipid content was 0.53 $g_{lipid}{\cdot}g_{biomass}^{-1}$ for *Chlorella pyrenoidosa* and *Rhodotorula glutinis* consortium. The explanation could be found in the differences in substrate concentration, which, in the latter case, amounted from 30–60 $g{\cdot}L^{-1}$ promoting consequently an imbalance in the C/N ratio that could have favored microbial oil accumulation phase mainly by the yeast, whose metabolic activity is positively conditioned by the increase of this ratio [4]. The lipid yield values of mixed cultures in our study were 0.211 $g{\cdot}L^{-1}$ for BBM + G, 0.320 $g{\cdot}L^{-1}$ for YEG and 0.335 $g{\cdot}L^{-1}$ for *Arundo donax* hydrolysate. These values were very far from those measured by Liu et al. [49], where an average value of 7.73 $g{\cdot}L^{-1}$ was obtained; in line with those measured by Wang et al. [54] in the presence of *C. pyrenoidosa* and *R. glutinis* consortium, where the lipid yield was 0.75 g·L^{-1}; and much higher than those of Iasimone et al. [54,55], where the lipid yield was 0.05 g·L^{-1} in the presence of a mixed culture of *L. starkeyi* and algae consortium mainly represented by *Chlorella* sp. and *Scenedesmus* sp.

Cell number normalized lipid contents (Lipid Yield/Cell) were also calculated (Table 1), obtaining values for *C. saccharophilum* and *L. starkeyi* mixed culture respectively equal to 16 and 3 µg for 10^{-6} cells for *Arundo donax* hydrolysate. These values were comparable to those by Liu et al. [49] where for *C. pyrenoidosa* and *R. glutinis* were equal about to 16 and 14 µg for 10^{-6} cells, respectively, indicating that the C/N ratio of cassava bagasse hydrolysate probably enhanced the lipid accumulation metabolism of yeast rather than alga.

Fatty acid distribution was also monitored at the end of each culture, as shown in Figure 5a–c. Generally, the volumetric distribution in terms of C16:0, C16:1, C18:0, C18:1 and C18:2 for yeast monoculture was 7–20, 0.1–0.8, 3–12, 28–85 and 5–20, while, for alga monoculture, it was 12–40, 0.1–1, 2–32, 21–71 and 0–10. The effect of co-culture is a progressive increase in the concentration of saturated fatty acids (SFAs) and a reduction in polyunsaturated fatty acids (PUFAs) [4], which was previously observed by Zuccaro et al. [31] and partially confirmed in this study. Indeed, as described by Tkachenko et al. [56], at the intensification of aeration, the degree of lipid unsaturation and the relative amount of all groups of unsaturated acids increase. In fact, the compounds mainly present in our study were C18:0, C18:1 and C14:0, with an increasing concentration of C18:2 and C18:3 for single and mixed culture in *Arundo donax* hydrolysate. For the latter case, specifically, the distribution in terms of C14:0, C16:0, C18:0, C18:1, C18:2, C18:3 and C20:0 was the following: 7, 39, 30, 19, 2 and 0.1. In general, the lipids produced by microorganisms are converted into biodiesel via a process known as transesterification. The two most important properties of fatty acids that affect the fuel properties as listed above are: (a) the length of the carbon chain; and (b) the number of double bonds [57,58]. The degree of unsaturation in the fatty acids affects the oxidative stability of the biodiesel with SFAs being the most stable followed by MUFAs compared to the least stable PUFAs [59]. Additionally, C16:0 and C18:1 are used as food additives and cosmetics, indicating the SCOs from mixed culture using low-cost lignocellulosic feedstock such as *Arundo donax* could have several potential applications.

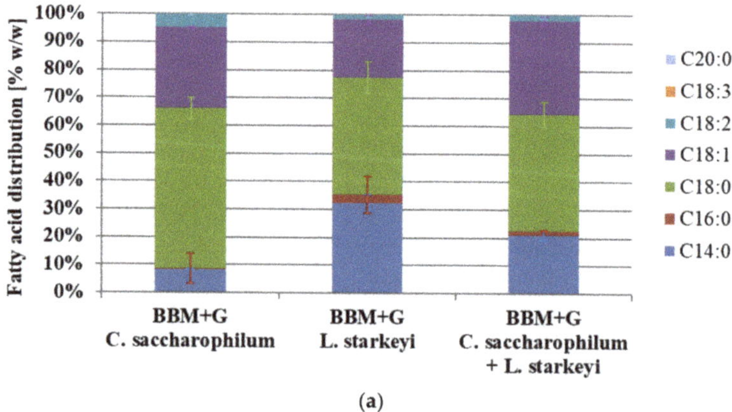

(a)

Figure 5. *Cont.*

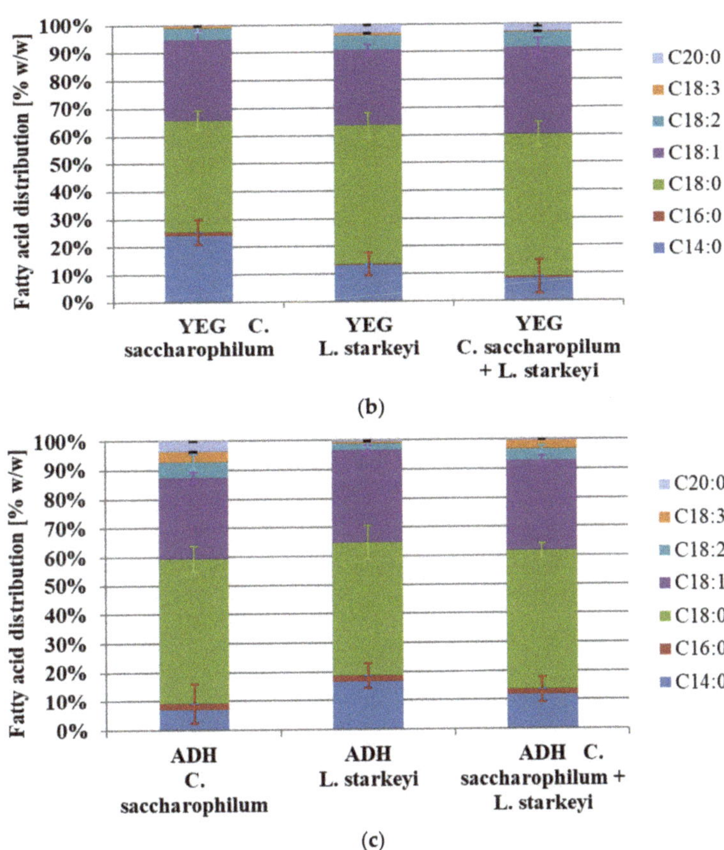

Figure 5. (**a**) Fatty acid distribution (% *w*/*w*) in microbial biomass (individual and mixed cultures) in BBM + G media. (**b**) Fatty acid distribution (% *w*/*w*) in microbial biomass (individual and mixed cultures) in YEG media. (**c**) Fatty acid distribution (% *w*/*w*) in microbial biomass (individual and mixed cultures) in ADH hydrolysates.

4. Conclusions

Lignocellulosic extracts have the potential to provide a complex substrate of fermentable sugars and volatile organic acids, mainly acetate, at low cost. In this study, it was shown that *C. saccharophilum* and *L. starkeyi* were able to grow according to a synergistic effect on complex substrate such as *Arundo donax* hydrolysate, while showing that this synergistic effect allowed overcoming the problems associated to inhibitory phenomena due to lignin or sugar degradation products and a non-optimal C/N ratio. The reason was mainly attributed to the virtuous exchange of O_2 and CO_2, as well as to the phenomenon of pH regulation. The promising results in terms of microbial growth and lipid accumulation were correlated with those of cultures in less complex synthetic media. The mixed cultures, in all cases, proved to be the best performing. The operational and economic impacts associated with the introduction of a detoxification phase in an attempt to overcome the inhibition effects of the above-mentioned products remain to be clarified and deepened. SCO production from lignocellulosic biomass offers a new direction for bio-refinery approach, and it will have a great future if the above-mentioned problems are properly handled. In fact, SCOs represent intermediates for biodiesel production, polymers and biosurfactants, and the control of unsaturation degree in their chain, for example by hydrogenation, could be critical to ensure selectivity and stability. The exploitation of SCOs related to the possibility of ensuring mono- or polyunsaturation could represent an alternative for the treatment of diseases such as atherosclerosis. Therefore, a decisive step in the direction of developing an economically sustainable method for the recovery of high purity SCOs is still awaited.

Author Contributions: A.d.M., G.P., A.P. and A.D.N.; formal analysis, G.Z. and A.d.M.; investigation, G.Z. and A.d.M.; resources, A.P. and A.D.N.; data curation, G.Z. and A.d.M.; writing—original draft preparation, G.Z.; writing—review and editing, G.Z., A.d.M., G.P., A.P. and A.D.N.; supervision, G.Z., A.P. and A.D.N.; project administration, A.D.N., funding acquisition, A.P. and A.D.N. All authors have read and agreed to the published version of the manuscript.

Funding: This research was funded by the project: FEAMP Campania 2014/2020-DRD n. 35 del 15.03.2019-Innovazione sviluppo sostenibilità settore pesca ed acquacoltura, WP4, misura 1.44.

Institutional Review Board Statement: Not applicable.

Informed Consent Statement: Not applicable.

Acknowledgments: G.Z. would like to thank ENEA–Trisaia Research Center (Rotondella, Matera, Italy). G.Z. gratefully acknowledges the Department of Chemical, Materials and Production Engineering (DICMaPI)—University of Naples Federico II for the support.

Conflicts of Interest: The authors declare no conflict of interest.

References

1. Pérez, A.T.E.; Camargo, M.; Rincón, P.C.N.; Marchant, M.A. Key challenges and requirements for sustainable and industrialized biorefinery supply chain design and management: A bibliographic analysis. *Renew. Sustain. Energy Rev.* **2017**, *69*, 350–359. [CrossRef]
2. Abghari, A.; Chen, S. *Yarrowia lipolytica* as an oleaginous cell factory platform for production of fatty acid-based biofuel and bioproducts. *Front. Energy Res.* **2014**, *2*, 1–21. [CrossRef]
3. Huang, C.; Chen, X.F.; Xiong, L.; Chen, X.D.; Ma, L.L.; Chen, Y. Single cell oil production from low-cost substrates: The possibility and potential of its industrialization. *Biotechnol. Adv.* **2013**, *31*, 129–139. [CrossRef] [PubMed]
4. Arora, N.; Patel, A.; Methani, J.; Pruthi, P.A.; Pruthi, V.; Poluri, K.M. Co-culturing of oleaginous microalgae and yeast: Paradigm shift towards enhanced lipid productivity. *Environ. Sci. Pollut. Res.* **2019**, *26*, 16952–16973. [CrossRef]
5. Naidoo, R.K.; Simpson, Z.F.; Oosthuizen, J.R.; Bauer, F.F. Nutrient exchange of carbon and nitrogen promotes the formation of stable mutualisms between *Chlorella sorokiniana* and *Saccharomyces cerevisiae* under engineered synthetic growth conditions. *Front. Microbiol.* **2019**, *10*, 609. [CrossRef]
6. Hays, S.G.; Yan, L.L.W.; Silver, P.A.; Ducat, D.C. Synthetic photosynthetic consortia define interactions leading to robustness and photoproduction. *J. Biol. Eng.* **2017**, *11*, 4. [CrossRef] [PubMed]
7. Li, T.; Li, C.T.; Butler, K.; Hays, S.G.; Guarnieri, M.T.; Oyler, G.A.; Betenbaugh, M.J. Mimicking lichens: Incorporation of yeasts strains together with sucrose secreting cyanobacteria improves survival, growth, ROS removal, and lipid production in a stable mutualistic co-culture production platform. *Biotechnol. Biofuels* **2017**, *10*, 55. [CrossRef]
8. Liu, J.; Huang, J.; Sun, Z.; Zhong, Y.; Jiang, Y.; Chen, F. Differential lipid and fatty acid profiles of photoautotrophic and heterotrophic *Chlorella zofingiensis*: Assessment of algal oils for biodiesel production. *Bioresour. Technol.* **2011**, *102*, 106–110. [CrossRef]
9. Wilhelm, C.; Jakob, T. From photons to biomass and biofuels: Evaluation of different strategies for the improvement of algal biotechnology based on comparative energy balances. *Appl. Microbiol. Biotechnol.* **2011**, *92*, 909–919. [CrossRef] [PubMed]
10. Chi, Z.; Zheng, Y.; Lucker, B.; Chen, S. Integrated System for Production of Biofuel Feedstock. U.S. Patent WO/2010/014797, 4 February 2010.
11. Beopoulos, A.; Nicaud, J.M. Yeast: A new oil producer? *OCL* **2012**, *19*, 22–28. [CrossRef]
12. Qin, L.; Liu, L.; Zeng, A.-P.; Wei, D. From low-cost substrates to Single Cell Oils synthesized by oleaginous yeasts. *Bioresour. Technol.* **2017**, *245*, 1507–1519. [CrossRef]
13. Lodder, J.; Kreger-van Rij, N.J.W. *The Yeast: A Taxonomic Study*, 1st ed.; North-Holland Publishing Company: Amsterdam, The Netherlands, 1952.
14. Darienko, T.; Gustavs, L.; Mudimu, O.; Menendez, C.R.; Schumann, R.; Karsten, U.; Friedl, T.; Proschold, T. *Chloroidium*, a common terrestrial coccoid green alga previously assigned to *Chlorella* (Trebouxiophyceae, Chlorophyta). *Eur. J. Phycol.* **2010**, *45*, 79–95. [CrossRef]
15. Calvey, C.H.; Su, Y.K.; Willis, L.B.; McGee, M.; Jeffries, T.W. Nitrogen limitation, oxygen limitation, and lipid accumulation in *Lipomyces Starkeyi*. *Bioresour. Technol.* **2016**, *200*, 780–788. [CrossRef]
16. McNeil, B.A.; Stuart, D.T. Optimization of C16 and C18 fatty alcohol production by an engineered strain of *Lipomyces starkeyi*. *J. Ind. Microbiol. Biotechnol.* **2018**, *45*, 1–14. [CrossRef]

17. Xavier, M.C.A.; Coradini, A.L.V.; Deckmann, A.C.; Franco, T.T. Lipid production from hemicellulose hydrolysate and acetic acid by *Lipomyces starkeyi* and the ability of yeast to metabolize inhibitors. *Biochem. Eng. J.* **2017**, *118*, 11–19. [CrossRef]
18. Beardall, J. CO_2 accumulation by *Chlorella saccharophilum* (Chlorophyceae) at low external pH: Evidence for active transport of inorganic carbon at the chloroplast envelope. *J. Phycol.* **1981**, *17*, 371–373. [CrossRef]
19. Herrera-Valencia, V.A.; Contreras-Pool, P.Y.; Lopez-Adrian, S.J.; Peraza-Echeverria, S.; Barahona-Perez, L.F. The green microalga *Chlorella saccharophilum* as a suitable source of oil for biodiesel production. *Curr. Microbiol.* **2011**, *63*, 151–157. [CrossRef]
20. Tan, C.K.; Johns, M.R. Fatty acid production by heterotrophic *Chlorella saccharophilum*. *Hydrobiologia* **1991**, *215*, 13–19. [CrossRef]
21. Zhang, T.-Y.; Hu, H.-Y.; Wu, Y.-H.; Zhuang, L.-L.; Xu, X.-Q.; Wang, X.-X.; Dao, G.-H. Promising solutions to solve te bottlenecks in the large-scale cultivation of microalgae for biomass/bioenergy production. *Renew. Sustain. Energy Rev.* **2016**, *60*, 1602–1614. [CrossRef]
22. Fagnano, M.; Impagliazzo, A.; Mori, M.; Fiorentino, N. Agronomic and Environmental Impacts of Giant Reed (*Arundo donax* L.): Results from a Long-Term Field Experiment in Hilly Areas Subject to Soil Erosion. *Bioenergy Res.* **2015**, *8*, 415–422. [CrossRef]
23. Zuccaro, G.; Travaglini, G.; Caputo, G.; Pirozzi, D. Enzymatic Hydrolysis and Oleaginous Fermentation of Steam-Exploded *Arundo donax* and *Lipomyces Starkeyi* in a Single Bioreactor for Microbial Oil Accumulation. *J. Multidiscip. Eng. Sci. Stud.* **2019**, *5*, 2577–2586.
24. Garrote, G.; Domínguez, H.; Parajó, J.C. Hydrothermal processing of lignocellulosic materials. *Holz Roh-Werkst* **1999**, *57*, 191–202. [CrossRef]
25. Adney, B.; Baker, J. *Measurement of Cellulase Activities*; NREL Analytical Procedure LAP-006; National Renewable Energy Laboratory: Golden, CO, USA, 1996.
26. Wood, T.M.; Bhat, K.M. Methods for measuring cellulase activities. *Methods Enzymol.* **1988**, *160*, 87–112.
27. Singleton, V.L.; Orhofer, R.; Lamuela-Raventos, R.M. Analysis of total phenols and other oxidation substrates and antioxidants by means of Folin-Ciocalteau reagent. *Methods Enzym.* **1999**, *299*, 152–178.
28. Martinez, A.; Rodriguez, M.E.; York, S.W.; Preston, J.F.; Ingram, L.O. Use of UV absorbance to monitor furans in dilute acid hydrolysates of biomass. *Biotechnol. Prog.* **2000**, *16*, 637–641. [CrossRef]
29. Bligh, E.G.; Dyer, W.J. A rapid method of total lipid extraction and purification. *Can. J. Biochem. Phys.* **1959**, *37*, 911–917. [CrossRef] [PubMed]
30. Redfield, A.C. On the Proportions of Organic Derivatives in Sea Water and Their Relation to the Composition of Plankton. In *James Johnstone Memorial Volume*; University Press of Liverpool: Liverpool, UK, 1934; pp. 176–192.
31. Zuccaro, G.; Steyer, J.-P.; van Lis, R. The algal trophic mode affects the interaction and oil production of a synergistic microalga-yeast consortium. *Bioresour. Technol.* **2019**, *273*, 608–617. [CrossRef] [PubMed]
32. Wijsman, M.R.; van Dijken, J.P.; van Kleeff, B.H.A.; Scheffers, W.A. Inhibition of fermentation and growth in batch cultures of the yeast *Brettanomyces intermedius* upon a shift from aerobic to anaerobic conditions (Custers effect). *Antonie Van Leeuwenhoek* **1984**, *50*, 183–192. [CrossRef]
33. Li, H.; Zhong, Y.; Lu, Q.; Zhang, X.; Wang, Q.; Liu, H.; Diao, Z.; Yao, C.; Liu, H. Co-cultivation of *Rhodotorula glutinis* and *Chlorella pyrenoidosa* to improve nutrient removal and protein content by their synergistic relationship. *RSC Adv.* **2019**, *9*, 14331–14342. [CrossRef]
34. Ling, J.; Nip, S.; Cheok, W.L.; Alves de Toledo, R.; Shim, H. Lipid production by a mixed culture of oleaginous yeast and microalga from distillery and domestic mixed wastewater. *Bioresour. Technol.* **2014**, *173*, 132–139. [CrossRef]
35. Zuccaro, G.; Pirozzi, D.; Yousuf, A. Lignocellulosic biomass to biodiesel. In *Lignocellulosic Biomass to Liquid Biofuels*; Academic Press: Cambridge, MA, USA; Elsevier: Amsterdam, The Netherlands, 2020; pp. 127–167.
36. Miazek, K.; Remacle, C.; Richel, A.; Goffin, D. Effect of lignocellulose related compounds on microalgae growth and product biosynthesis: A review. *Energies* **2014**, *7*, 4446–4481. [CrossRef]
37. Bajguz, A.; Czerpak, R.; Piotrowska, A.; Polecka, M. Effect of isomers of hydroxybenzoic acid on the growth and metabolism of *Chlorella vulgaris* Beijerinck (Chlorophyceae). *Acta Soc. Bot. Pol.* **2001**, *70*, 253–259. [CrossRef]
38. Larson, L.J. Effect of phenolic acids on growth of *Chlorella pyrenoidosa*. *Hydrobiologia* **1989**, *183*, 217–222. [CrossRef]
39. Larsson, S.; Palmqvist, E.; Hahn-Hägerdal, B.; Tengborg, C.; Stenberg, K.; Zacchi, G.; Nilvebrant, N.O. The generation of fermentation inhibitors during dilute acid hydrolysis of softwood. *Enz. Microb. Technol.* **1988**, *24*, 151–159. [CrossRef]
40. Cea Barcia, G.E.; Imperial Cervantes, R.A.; Torres Zuniga, I.; van Den Hende, S. Converting tequila vinasse diluted with tequila process water into microalgae-yeast flocs and dischargeable effluent. *Bioresour. Technol.* **2020**, *300*, 122644. [CrossRef]
41. Gong, Z.; Zhou, W.; Shen, H.; Yang, Z.; Wang, G.; Zuo, Z.; Huo, Y.; Zhao, Z.K. Co-fermentation of acetate and sugars facilitating microbial lipid production on acetate-rich biomass hydrolysates. *Bioresour. Technol.* **2016**, *207*, 102–108. [CrossRef] [PubMed]
42. Russell, J.B. Another explanation for the toxicity of fermentation acids at low pH: Anion accumulation versus uncoupling. *J. Appl. Bacteriol.* **1992**, *73*, 363–370. [CrossRef]
43. Anschau, A.; Xavier, M.C.; Hernalsteens, S.; Franco, T.T. Effect of feeding strategies on lipid production by *Lipomyces starkeyi*. *Bioresour. Technol* **2014**, *157*, 214–222. [CrossRef]
44. Gong, Z.; Wang, Q.; Shen, H.; Hu, C.; Jin, G.; Zhao, Z.K. Co-fermentation of cellobiose and xylose by *Lipomyces starkeyi* for lipid production. *Bioresour. Technol.* **2012**, *117*, 20–24. [CrossRef]
45. Yang, X.B.; Jin, G.J.; Gong, Z.W.; Shen, H.W.; Song, Y.H.; Bai, F.W.; Zhao, Z.B.K. Simultaneous utilization of glucose and mannose from spent yeast cell mass for lipid production by *Lipomyces starkeyi*. *Bioresour. Technol.* **2014**, *158*, 383–387. [CrossRef]

46. Pirozzi, D.; Ausiello, A.; Yousuf, A.; Zuccaro, G.; Toscano, G. Exploitation of oleaginous yeasts for the production of microbial oils from agricultural biomass. *Chem. Eng. Trans.* **2014**, *37*, 469–474.
47. Probst, K.V.; Vadlani, P.V. Production of single cell oil from *Lipomyces starkeyi* ATCC 56304 using biorefinery by-products. *Bioresour. Technol.* **2015**, *198*, 268–275. [CrossRef]
48. Matsakas, L.; Sterioti, A.A.; Rova, U.; Christakopoulos, P. Use of dried sweet sorghum for the efficient production of lipids by the yeast *Lipomyces starkeyi* CBS 1807. *Ind. Crop. Prod.* **2014**, *62*, 367–372. [CrossRef]
49. Liu, L.; Chen, J.; Lim, P.-E.; Wei, D. Enhanced single cell oil production by mixed culture of *Chlorella pyrenoidosa* and *Rhodotorula glutinis* using cassava bagasse hydrolysate as carbon source. *Bioresour. Technol.* **2018**, *255*, 140–148. [CrossRef] [PubMed]
50. Qin, L.; Wei, D.; Wang, Z.; Alam, M.A. Advantage assessment of mixed culture of *Chlorella vulgaris* and *Yarrowia lipolytica* for treatment of liquid digestate of year industry and cogeneration of biofuel feedstock. *Appl. Biochem. Biotechnol.* **2019**, *187*, 856–869. [CrossRef]
51. Zhang, Z.; Pang, Z.; Xu, S.; Wei, T.; Song, L.; Wang, G.; Zhang, J.; Yang, X. Improved carotenoid productivity and COD removal efficiency by co-culture of *Rhodotorula glutinis* and *Chlorella vulgaris* using starch wastewaters as raw material. *Appl. Biochem. Biotechnol.* **2019**, *189*, 193–205. [CrossRef]
52. Liu, Y.; Wang, Y.; Liu, H.; Zhang, J. Enhanced lipid production with undetoxified corncob hydrolysate by *Rhodotorula glutinis* using a high cell density culture strategy. *Bioresour. Technol.* **2015**, *180*, 32–39. [CrossRef]
53. Younes, S.; Bracharz, F.; Awad, D.; Qoura, F.; Mehlmer, N.; Brueck, T. Microbial lipid production by oleaginous yeasts grown on *Scenedesmus otusiusculus* microalgae biomass hydrolysate. *Bioprocess. Biosyst. Eng.* **2020**, *43*, 1629–1638. [CrossRef]
54. Wang, S.; Wu, Y.; Wang, X. Heterotrophic cultivation of *Chlorella pyrenoidosa* using sucrose as the sole carbon source by co-culture with *Rhodotorula glutinis*. *Bioresour. Technol.* **2016**, *220*, 615–620. [CrossRef] [PubMed]
55. Iasimone, F.; Zuccaro, G.; D'Oriano, V.; Franci, G.; Galdiero, M.; Pirozzi, D. Combined yeast and microalgal cultivation in a pilot-scale raceway pond for urban wastewater treatment and potential biodiesel production. *Water Sci. Technol.* **2018**, *77*, 1062–1071. [CrossRef]
56. Tkachenko, A.F.; Tigunova, S.M.; Shulga, S.M. Microbial lipids as a source of biofuel. *Cytol. Genet.* **2013**, *47*, 343–348. [CrossRef]
57. Stansell, G.R.; Gray, V.M.; Sym, S.D. Microalgal fatty acid composition: Implications for biodiesel quality. *J. Appl. Phycol.* **2011**, *24*, 791–801. [CrossRef]
58. Parsons, S.; Allen, M.; Chuck, C.J. Coproducts of algae and yeast-derived single cell oils: A critical review of their role in improving biorefinery sustainability. *Bioresour. Technol.* **2020**, *303*, 122862. [CrossRef]
59. Asraful, A.M.; Masjuki, H.H.; Kalam, M.A.; Rizwanul Fattah, I.M.; Imtenan, S.; Shahir, S.A.; Mobarak, H.M. Production and comparison of fuel properties, engine performance, and emission characteristics of biodiesel from various non-edible vegetable oils: A review. *Energy Convers. Manag.* **2014**, *80*, 202–228. [CrossRef]

Article

Consortium Growth of Filamentous Fungi and Microalgae: Evaluation of Different Cultivation Strategies to Optimize Cell Harvesting and Lipid Accumulation

Savienne M. F. E. Zorn [1,*], Cristiano E. R. Reis [1], Messias B. Silva [1], Bo Hu [2] and Heizir F. De Castro [1]

1 Department of Chemical Engineering, Engineering School of Lorena, University of São Paulo, Lorena, São Paulo 12602-810, Brazil; cristianoreis@usp.br (C.E.R.R.); messias.silva@usp.br (M.B.S.); heizir@usp.br (H.F.D.C.)
2 Department of Bioproducts and Biosystems Engineering, University of Minnesota, Saint Paul, MN 55108, USA; bhu@umn.edu
* Correspondence: savienne.elerbrock@usp.br; Tel.: +55-12-31595063

Received: 1 July 2020; Accepted: 13 July 2020; Published: 15 July 2020

Abstract: This study aims to evaluate the potential of consortium biomass formation between *Mucor circinelloides*, an oleaginous filamentous fungal species, and *Chlorella vulgaris*, in order to promote a straightforward approach to harvest microalgal cells and to evaluate the lipid production in the consortium system. A synthetic medium with glucose (2 g·L^{-1}) and mineral nutrients essential for both fungi and algae was selected. Four different inoculation strategies were assessed, considering the effect of simultaneous vs. separate development of fungal spores and algae cells, and the presence of a supporting matrix aiming at the higher recovery of algae cell rates. The results were evaluated in terms of consortium biomass composition, demonstrating that the strategy using a mature fungal mycelium with a higher algae count may provide biomass samples with up to 79% of their dry weight as algae, still promoting recovery rates greater than 97%. The findings demonstrate a synergistic effect on the lipid accumulation by the fungal strain, at around a fourfold increase when compared to the axenic control, with values in the range of 23% of dry biomass weight. Furthermore, the fatty acid profile from the samples presents a balance between saturated and unsaturated fatty acids that is likely to present an adequate balance for applications such as biodiesel production.

Keywords: fungi; algae; lichen; lipids; biofilm

1. Introduction

A bottleneck commonly reported in the viability of microalgal processes is found in the harvesting stage [1]. Although microalgae have the potential to become one of the main drivers of a new economic era due to the production of biofuels, the conditions in which microalgae are produced are still impaired by the energy intensity or the application of cost-limiting chemicals which can often cause environmental damage in the harvesting process [2]. Additional challenges are also related to the requirements of water supply to algae, in which, unlike multicellular organisms such as fungi, the cell densities obtained in conventional microalgae cultures usually present specific gravity values close to the culture medium [3]. Conventional algae harvesting and separation operations are, thus, dependent upon methods such as centrifugation, flocculation, flotation and energy-intensive filtration [2].

With the proper combination of fungi and microalgae, mimicking the natural concept of lichen, it has been demonstrated that it is possible to promote an attraction of fungal cells to microalgae, removing almost all the microalgal species in submerged growth [4]. While the growth in a consortium system occurs for reasons of nutrient exchange and structural support, resulting in complex structures such as lichens and symbiotic biofilms, several reports in the literature indicate the possibility of the combination of algae and fungi for industrial bioprocesses [5,6]. In this sense, considering that multiple wild strains are known to produce species-specific compounds [7], the possibility of utilizing consortium growth as a synergistic approach to optimize value-added metabolites by two or more strains is, thus, interesting to the development of a robust biorefinery. The utilization of a microbial-oil-based biorefinery to produce, for instance, biodiesel or polyunsaturated fatty acids with nutraceutical value is likely to be technically possible from a process perspective utilizing the combined growth of a photoautotrophic oleaginous microalgae strain with a supporting filamentous fungal mycelium. In this line, this article addresses some technical approaches in the development of a consortium growth strategy using a fungal strain known to accumulate considerable amounts of intracellular lipids, *Mucor circinelloides* University Recife Mycologia (URM) 4182, and *Chlorella vulgaris* Banco de Microrganismos Aidar & Kutner (BMAK) D07, a microalgae species that has been widely explored in the literature regarding its oil-bearing capacity.

2. Materials and Methods

2.1. Microorganisms: Maintenance, Culture Medium, and Growth

Mucor circinelloides f. griseo-cyanus URM 4182 was selected as the fungal component for the consortium development. The URM 4182 strain was acquired from the URM Bank (Federal University of Pernambuco, Recife, Brazil) and is maintained using a Potato Dextrose Agar (PDA) medium following laboratory routine practices [8]. The microalgae *Chlorella vulgaris* BMAK D07 was donated by the Aidar & Kutner Microbial Bank (Oceonagraphic Institute, University of São Paulo, São Paulo, Brazil). The BMAK D07 strain is maintained following similar procedures as those described by Loures et al. [9].

Cell cultivation assays were performed in 250 mL Erlenmeyer flasks containing 100 mL of culture medium A [10] composed of: glucose (2 $g \cdot L^{-1}$)—absent in the algal axenic assays, KNO_3 (1 $g \cdot L^{-1}$), KH_2PO_4 (0.075 $g \cdot L^{-1}$), K_2HPO_4 (0.1 $g \cdot L^{-1}$), $MgSO_4 \cdot 2H_2O$ (0.5 $g \cdot L^{-1}$), $Ca(NO_3)_2 \cdot 4H_2O$ (0.0625 $g \cdot L^{-1}$), $FeSO_4 \cdot 7H_2O$ (0.01 $g \cdot L^{-1}$), yeast extract (0.5 $g \cdot L^{-1}$) and metal solution (1 $mL \cdot L^{-1}$) composed of: H_3BO_3 (2.86 $g \cdot L^{-1}$), $Na_2MoO_4 \cdot 2H_2O$ (0.39 $g \cdot L^{-1}$), $ZnSO_4 \cdot 7H_2O$ (0.22 $g \cdot L^{-1}$), $MnCl_2 \cdot 4H_2O$ (1.81 $g \cdot L^{-1}$), $CuSO_4 \cdot 5H_2O$ (0.079 $g \cdot L^{-1}$) and $Co(NO_3)_2 \cdot 6H_2O$ (0.049 $g \cdot L^{-1}$), unless specified otherwise. Algae and consortia were grown under constant light intensity equivalent to 100 $\mu mol \cdot m^{-2} \cdot s^{-1}$, at 140 rpm and temperature of 26 °C. This study addresses four different development strategies of consortia, described as:

Strategy 1: 100 mL of culture medium was simultaneously inoculated with 2.55×10^8 microalgal cells and 8.5×10^5 fungal spores and incubated for 180 h.

Strategy 2: 100 mL of culture medium was simultaneously inoculated with 2.55×10^8 microalgal cells and 8.5×10^5 fungal spores and incubated for 180 h with a supporting cotton screen of 2.5 cm × 2.5 cm.

Strategy 3: 100 mL of culture medium was inoculated with 8.5×10^5 fungal spores and incubated for 72 h. After 72 h of fungal growth, 2.55×10^8 microalgae cells were added to the fungal culture, remaining in incubation until completing 180 h.

Strategy 4: 50 mL of culture medium was inoculated with 4.25×10^5 spores, and concomitantly, in another 250 mL Erlenmeyer flask, 50 mL of culture medium was inoculated with 1.275×10^8 microalgae cells. Both flasks were incubated for 72 h, followed by the transfer of the axenic cultures to a single 250 mL flask and continued incubation for 180 h.

2.2. Analytical Methods

Fungal and consortium biomass were harvested using vacuum filtration. Cell dry weight was determined via direct measurement of water loss using an infrared-coupled balance (MOC63u, Shimadzu). Glucose concentration was estimated using an adapted Dubois method [11].

The contribution of microalgal biomass in the consortium samples was determined indirectly by measuring chlorophyll-A (Chl-a) concentration present in the samples [10]. In this method, the chlorophylls present at a biomass sample with known weight and moisture content were extracted using 5 mL of aqueous methanol solution (90 vol.%) and glass beads, followed by stirring at 150 rpm at room temperature (25 °C), following by addition of 5 mL of distilled water. The liquid phase, containing the chlorophyll extract, was filtered using a 0.45 μm syringe filter. The Chl-a concentration was determined spectrophotometrically at 665 nm using a UV–VIS spectrophotometer (Varian Cary 5000), through a known correlation between algae dry weight and Chl-a absorbance at the wavelength at 665 nm. Similar measurements were made to the microalgae in suspension, i.e., to account for the microalgae cells that were not attached to the matrix in the consortium experiments or for the cell density cultures for the axenic algae assays. The fungal biomass in the consortium growth was estimated as being the difference between the consortium biomass and microalgae biomass.

Total lipids in the biomass samples were quantified by extraction performed in a digestion system with irradiation in the microwave region (CEM Discover DU-8081) using ethanol (vol. 96%) as extraction solvent, at temperature 60 °C, three cycles of 30 min [8]. The fatty acid composition was determined via methylation of lipids using a BF_3/methanol mixture following an adaptation to AOCS Ce 1-62 method. The fatty acid methyl esters (FAMEs) were identified by gas chromatography (GC) analysis using PerkinElmer®-Clarus 580 chromatograph, equipped with a flame ionization detector. A 30-m capillary column with a 0.25-mm internal diameter and 5% diphenyl 95% dimethylpolysiloxane stationary phase (non-polar) was employed during the GC analysis. Nitrogen was the carrier gas ($1\ mL{\cdot}min^{-1}$). The identification of methyl esters was carried out by comparing the retention times with MIX Supelco® Fatty Acid Methyl Acid (FAME) standard capric acid (C6: 0) to lignoceric acid (C24: 0) and quantification was performed by normalizing the calculated areas.

Attenuated Total Reflectance–Fourier Transform Infrared (ATR–FTIR) spectroscopy (Shimadzu FTIR spectrometer, model IRP PRESTIGE-21) was used to analyze the organic functions of the different consortium biomass samples. The ATR–FTIR spectra of the samples were obtained from the accumulation of a total of 32 scans at a range of 4000–400 cm^{-1} with a resolution of 4 cm^{-1}, and potassium bromide (KBr) used as a matrix.

2.3. Calculation of Biochemical Parameters

The assays were evaluated in terms of biomass and lipids, both in concentration and productivity, and in glucose consumption. The concentration of biomass, X, was determined as being the weight of the recovered biomass, which was then multiplied by the dry biomass factor, obtained via the measurement from the infrared-coupled balance, and divided by the volume used for such culture. The lipid content (% L) of each culture was determined using the Equation (1), in which m indicates the mass of either lipids or biomass (in mg), and MC is the moisture content.

$$\%\,L = \frac{m_{lipids}}{m_{biomass} \times (1 - MC)} \quad (1)$$

The productivity values of biomass (Q_X) and lipids (Q_L) were derived as being the ratio between the concentration of biomass and the cultivation time. The biomass productivity values are divided in some sections of the results as being the corresponding fungal or algal biomass. Considering that the microalgae in the flasks also can grow unattached to the consortium, this work defined the recovery efficiency of microalgae according to Equation (2), in which m indicates the dry biomass weight of the consortium in mg, w is the weight contribution, i.e., the weight of the microalgae over the total biomass weight, [microalgae] means the concentration of algae suspended in the medium at a given time in mg·L^{-1}, and V, the medium volume in L.

$$\text{Recovery Efficiency } (\%) = \frac{m_{consortium} \times w_{microalgae}}{m_{consortium} \times w_{microalgae} + [\text{microalgae}] \times V} \times 100\% \quad (2)$$

3. Results

3.1. Axenic and Combined Cell Growth

The culture of *C. vulgaris* on medium A in the presence and absence of glucose (Figure 1a.) presented a typical growth curve and demonstrated, in fact, a similar final concentration of cells at the period of 180 h, even though a slower growth is attained in the medium supplemented with 2 g·L^{-1} within the first days of growth. Medium A, supplemented with glucose at 2 g·L^{-1}, was also able to fully support the growth of the strain URM 4182 of the filamentous fungus *M. circinelloides*, providing a quick uptake of glucose within the first 24 hours of growth, which was able to be sustained for the whole period evaluated of 180 h, as depicted in Figure 1b.

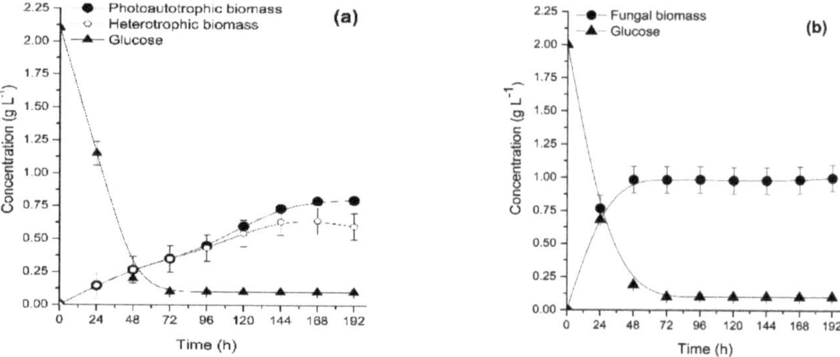

Figure 1. Growth curve of *C. vulgaris* (**a**) and *M. circinelloides* (**b**) on medium A.

The first approach considered for the recovery of *C. vulgaris* was to evaluate a simultaneous inoculation of algae cells and fungal spores, with a hypothesis that the combined growth would follow the profiles of the axenic behavior of each strain, which was named as strategy 1. The results demonstrate that the consortium system formed according this strategy was dispersed across the medium, without a clear formation of fungal pellets or other dense structures. Throughout the incubation period, a persistent and intense green color was observed in the culture medium, demonstrating that most of the microalgae cells were suspended or they did not adhere completely to the fungal biomass. Therefore, by simultaneous inoculation of cells and spores, there was poor adherence of the microalgal biomass to the fungal mycelium. The individual contributions of the consortium biomass were in average 25 ± 4% of microalgal cells and 75 ± 4% fungal mycelium. The microalgae recovery rate was within the range of 95%, demonstrating that approximately 5% of the total algal cell count was found in the liquid phase at the end of the total period analyzed. Figure 2a. illustrates in a flask the growth behavior of this strategy. Even though 95 ± 1% of algae recovery may seem like a result that would not require thorough optimization, the qualitative aspect of this given strategy demonstrates the persistence of the green color in the supernatant, which is an indication of suspended algae cells.

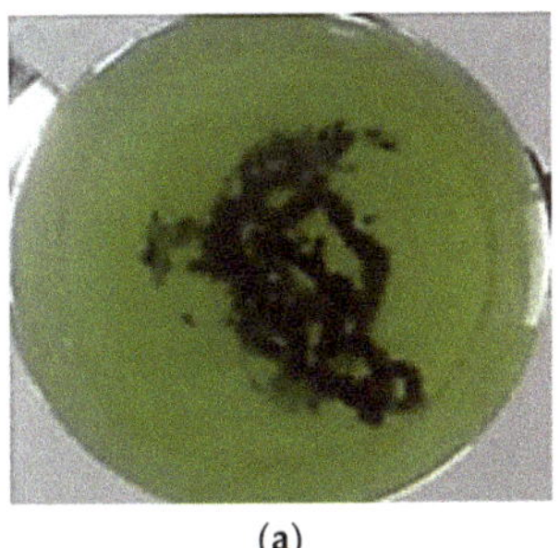

(a) (b) (c)

Figure 2. Culture flask with biomass employing strategy 1 (**a**) and employing strategy 2 (**b**). Axenic fungal mycelium in the presence of the supporting matrix used in strategy 2 (**c**).

An immobilization strategy was also taken into consideration in the evaluation of the performance of this cultivation condition, using a cotton lattice matrix as support, which was named strategy 2. The decision on testing the immobilization of the culture cells onto an external matrix was due to previously promising descriptions of algae recovery in systems assisted with fungal cells supported onto materials as cotton and polypropylene spun. The results evaluating the immobilized growth of the consortium biomass, as well as the axenic cultures of *C. vulgaris* and *M. circinelloides*, follow a similar description as described by Rajendran and Hu [10]. Both the axenic fungal culture and the consortium biomass were able to grow attached to the cotton lattice matrix, while the axenic *C. vulgaris* culture did not attach properly to the matrix. From the results it is observable a similar growth pattern as the ones without the supporting matrix in terms of biomass accumulation and glucose consumption for *M. circinelloides*.

The growth performance of strategy 2 was similar to the system without the addition of the matrix, i.e., strategy 1. The fungal mycelium was completely adhered to the cotton structure (Figure 2b) in a similar fashion to the axenic growth of Figure 2c. The composition of the consortium biomass was greater in terms of algae when compared to the non-immobilized growth (1224 mg × 1076 mg), however, the weight contribution of algae to the total biomass was still lower than 50 ± 1%, meaning that the majority of the consortium biomass was composed of fungal mycelium in terms of dry cell weight. Interestingly, on the other hand, was that the consortium biomass was greater than the axenic growth of fungal cells in the immobilized growth throughout all the cultivation process. The algal cell recovery process was also increased from the average of 95 ± 1% in the non-immobilized medium to values within the range of 99.7 ± 0.4% in the system with the cotton matrix, indicating a clear improvement in terms of algae harvesting. The microalgae in the liquid were apparently present in a smaller amount, since the color of the medium showed a much clearer aspect compared to the first approach. Due to both systems, immobilized and non-immobilized, having been inoculated simultaneously, the glucose uptake cannot be distinguished in terms of being utilized by the fungal or the algal cells. In this sense, still considering the algal cell recovery objective of this work, glucose utilization by *C. vulgaris* is unwanted, due to the fact that fungal growth, which is limited in these conditions by the sugar availability in the medium, can be partially limited by the sugar consumption by the algae cells, indicated by the lower fungal contribution to the biomass in the consortium, at approximately 560 mg·L^{-1}, corresponding to approximately 52 ± 1% of the consortium biomass, when compared to its axenic behavior in the same medium, which achieved biomass concentrations in the range of 880 mg·L^{-1}. In this sense, the results obtained by strategy 2 seem not to fulfill the objectives set by this work, but may represent an indication for applications in which the algal cell contribution to the total biomass is not an important to be considered, for example, those related to bioremediation of algae-contaminated systems. The adherence of the cell system to the cotton matrix could be considered as an irreversible process from a practical point of view. Therefore, the stability of the consortia immobilized onto the matrix could be beneficial to applications in which such asset is desired, as those related to the removal of algae cells from aqueous systems, e.g., algae blooms.

3.2. Evaluation of Different Strategies for Higher Algae Recovery Efficiencies

Two additional strategies were tested in order to assay higher yields of the expected outcomes, as depicted in Figure 3. Strategy 3 was comprised of an axenic culture of *M. circinelloides*, which grew for 72 h, to which the same number of algal cells was added, leading to an overall consortium growth time of 108 h. The evolution of the results involved in this growth were based on: an initial dispersion of the algae cells onto the medium, providing a green liquid phase, followed by the fact that, approximately 24 h after the addition of *C. vulgaris* cells, the medium became clear and transparent. After 24 h, the algae cells were close to full adherence to the fungal biomass. Following the initial 24 h of combined growth, the consortium biomass presented similar total dry weight. However, the highest recovery of microalgae biomass was attained at the conditions of this consortium biomass formation, at over 99.9%, while the contribution of the algae cells to overall dry weight of the consortium biomass was within the range of 11.9 ± 1.1%. The consortium biomass, which was composed of a great majority of fungal mycelia, was greater in terms of dry weight if compared to the previous strategies, achieving values close to 2 $g{\cdot}L^{-1}$. Despite the fact that lower relative concentrations of algae were observed in this particular growth, it can be observed that the absolute concentration values of algae were similar to the strategies described in strategies 1 and 2, demonstrating that the early maturation of the fungal mycelia with the subsequent inoculation of algae at a lower cell count may provide a positive effect for the fungal cells, but neutral or negative growth for the algae.

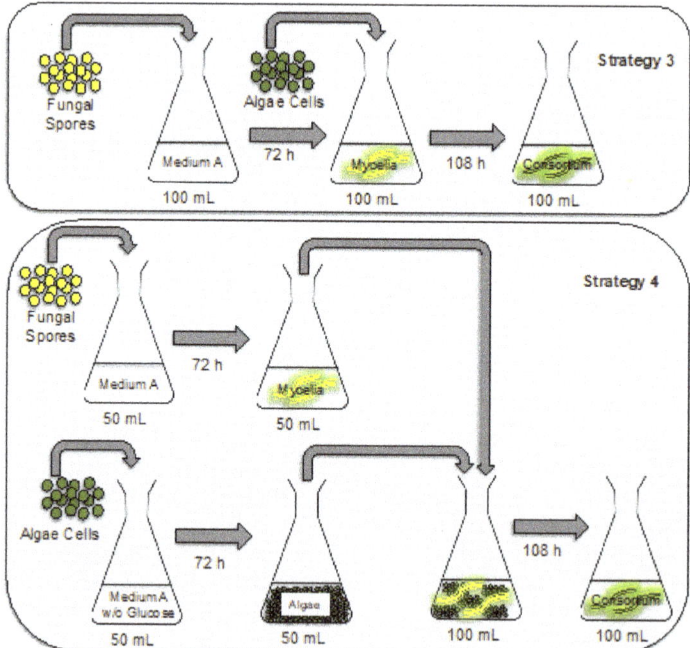

Figure 3. Schematic illustration of strategies 3 and 4.

The application of strategy 4, with a higher initial algae cell count and a mature fungal mycelium, was then tested. Microalgae cells and fungal spores were led to growth in axenic conditions for 72 h each, which were then combined, following again the overall process time of 180 h, leading to a consortium maturation time of 108 h. In this strategy, not only the fungal spores consumed the glucose prior to the development of the consortium biomass, but the algae were able to develop in a photoautotrophic growth, achieving a concentration of approximately 350 mg·L^{-1} ± 4.1 (Figure 1a). Figure 4 summarizes the results of the strategies.

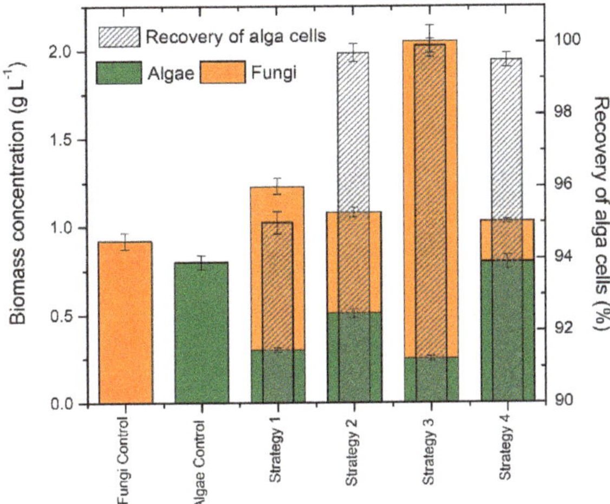

Figure 4. Summary of results according to biomass accumulation and distribution fungi (orange), algae (green) and % recovery of algae cells (grey).

3.3. Effects on Lipid Content and Fatty Acid Distribution

Table 1 summarizes the quantification of lipids across the strategies employed herein. The results demonstrate that, not only the total lipids are increased in all the consortium strategies when compared to the axenic cultures, but the lipid productivity of all strategies are at least 20% greater when compared to the axenic photoautotrophic growth of *C. vulgaris* under the same conditions and could be increased more than fourfold when compared to the axenic *M. circinelloides* growth on medium A.

Table 1. Summary of lipid accumulation data from different culture strategies.

Culture Condition	Total Lipids (% Dry Biomass Weight)	Total Lipids (mg·L^{-1})	Q_L (mg·L^{-1}·day^{-1})
Axenic photoautotrophic *C. vulgaris*	23.8 ± 1.0	190.4 ± 1.0	25.4 ± 1.0
Axenic *M. circinelloides*	7.1 ± 0.2	63.9 ± 0.8	8.5 ± 0.8
Strategy 1	22.7 ± 0.4	272.4 ± 0.5	36.3 ± 0.5
Strategy 2	23.5 ± 0.8	253.8 ± 1.0	33.8 ± 0.5
Strategy 3	19.0 ± 0.5	190.0 ± 0.5	22.0 ± 0.5
Strategy 4	22.2 ± 0.4	230.8 ± 0.5	30.8 ± 0.5

Even though the lipid productivity obtained by the fourth strategy was lower than counterparts 1 and 2, the consortium growth using strategy 4 was selected as the most appropriate for the objectives of cell recovery and microalgae cell contribution to the total biomass weight, as discussed in the previous sections. In this sense, a screening of process time was carried in order to evaluate the composition stability over the period of up to 384 h. The factor of a prolonged cultivation time was evaluated to assess the individualized effects on the total biomass growth, on the distribution of algae and fungi to the total biomass weight, on the microalgae cell recovery, and on the lipids, which were characterized in terms of total lipids and on fatty acid distribution. A few interesting points can be seen on the data presented in Table 2. Even though small fluctuations in the consortium dry weight were observed during the assay, the total biomass was approximately constant over the whole period.

Table 2. Summary of strategy 4 lipid accumulation at different incubation times.

Parameter	Culture Time (Days)				
	8	10	12	14	16
Consortium Biomass (mg·L^{-1})	1037 ± 7	806 ± 8	932 ± 7	1025 ± 7	1150 ± 6
Microalgae Recovery (%)	99.4 ± 0.2	98.8 ± 0.2	98.5 ± 0.2	98.3 ± 0.2	98.0 ± 0.2
Total Lipids (% of Dry Biomass)	31.1 ± 0.5	32.8 ± 0.5	31.6 ± 0.5	30.1 ± 0.5	30.5 ± 0.5
Fatty Acids	**Weight contribution to the total Fatty Acids (%)**				
C 12:0	1.25	1.12	1.27	1.11	1.01
C 14:0	0.91	0.87	0.90	0.76	0.71
C 15:0	6.79	6.67	6.78	7.27	7.06
C 16:0	26.58	26.79	25.56	23.81	26.25
C 17:0	1.70	1.62	1.28	2.68	2.22
C 18:0	1.31	1.29	1.63	1.79	2.03
C 16:1	0.91	0.92	0.86	0.87	0.97
C 18:1	32.98	33.06	35.0	31.34	30.07
C 18:2	11.81	11.72	12.14	11.74	11.55
C 18:3	15.76	15.93	14.57	18.64	18.13

There was no significant change in the total lipid content. The algal biomass contribution to the total consortium weight was of 79 ± 0.4%, which suggests an efficient method to obtain algae-rich biomass in coculture systems. Regarding the lipid portion of the consortium, it can be observed that the lipid content of the consortium biomass remains approximately constant throughout the evaluation period of 16 days. The fatty acid profile, as also demonstrated in Table 2, presents, at all growth times, a predominance of palmitic acid (C16: 0), ranging from 23.81 up to 26.79% in regard to the total fatty acids.

4. Discussion

4.1. Evaluation of the Culture Medium and Simultaneous Inoculation of Algae Cells and Fungal Spores

The medium selected for this study, medium A, has been described in the literature to fully support the growth of different *C. vulgaris* strains, such as UTEX 2714 [4]. The mixotrophic behavior of *C. vulgaris*, which has been widely demonstrated in multiple studies [12,13] is confirmed in this study for the strain BMAK D07. However, the growth of *C. vulgaris* in the medium supplemented with glucose was turbid and the cells were likely to be decanted, as also described by Zhang and Hu [5]. Considering that the two strains were able to grow in the same medium, the selection of the nutritional characteristics of medium A were appropriate for the evaluation of different parameters involving the interactions between the two strains.

The effect of different inoculation ratios between algae and fungi, pH, and the initial concentration of glucose, among other factors, were evaluated by Gultom et al. [4], on the interaction between *Aspergillus niger* and *C. vulgaris*, and the optimum conditions suggested by the authors were adopted for the assays involving the growth of *M. circinelloides* and *C. vulgaris* herein. Therefore, the first approach considered for the recovery of *C. vulgaris* was to evaluate a simultaneous inoculation of algae cells and fungal spores, with a hypothesis that the combined growth would follow the profiles of the axenic behavior of each strain, which was named strategy 1.

The results observed for strategy 1 indicate a potential for consortium development between the two strains, even though a higher proportion of fungal cells were observed. While there may be applications in which the composition of the consortium biomass is composed in its major proportion by fungal cells, such observation diverts from the objectives set by the current study, which are to promote recoverable biomass rich in algae cells, rather than the counterpart fungal component of the biomass.

An immobilization strategy was also taken into consideration in the evaluation of the performance of this cultivation condition, using a cotton lattice matrix as support, which was named strategy 2.

Rajendran and Hu [10] evaluated different matrices for the formation of lichen-type biofilm consisting of microalgae and filamentous fungi, resulting in a recommendation of using a polypropylene-spun lattice matrix as an efficient support for the immobilized growth of the consortium biomass. The immobilization of filamentous fungal mycelia and consortium biomass is likely to be derived from the intrinsic need for the cells to produce adhering structures to the rough cotton surface, which, despite being a complex phenomenon, is linked to the capacity of the cells to produce cellulolytic enzymes that are able to promote an initial slow degradation of the cotton matrix, from which, protein–carbohydrate structures are formed, linking the cellulose and cellular structures [14]. The results evaluating the immobilized growth of the consortium biomass, as well as the axenic cultures of *C. vulgaris* and *M. circinelloides*, follow a similar description as that described by Rajendran and Hu [10].

4.2. Other Strategies Involving the Use of Mature Fungal Mycelium

The cultivation strategy aimed for optimum results in this study should be comprised of the highest concentration of algae biomass within the consortium dry weight. Considering the results given by strategies 1 and 2, it can be observed that simultaneous inoculation may provide subpar results to the objectives set by the current study. In this sense, two further strategies were developed in order to address such issues based on the axenic separation of the fungal mycelia, with a direct consequence of depletion of glucose, which was then combined with the algae cells in order to evaluate the performance of the biofilm formation.

Strategy 3 comprised of an axenic culture of *M. circinelloides*, which grew for 72 h, to which the same number of algal cells was added, leading to an overall consortium growth time of 108 h. The evolution of the results involved in this growth were based on: an initial dispersion of the algae cells onto the medium, providing a green liquid phase, followed by the fact that 24 h approximately after the addition of *C. vulgaris* cells, the medium became clear and transparent. After 24 h, the algae cells were close to full adherence to the fungal biomass. Following the initial 24 h of combined growth, the consortium biomass presented similar total dry weight. However, the highest recovery of microalgae biomass was attained at the conditions of this consortium biomass formation, at over 99.9%, while the contribution of the algae cells to overall dry weight of the consortium biomass was within the range of 11.9%. The consortium biomass, which was composed of a great majority of fungal mycelia, was greater in terms of dry weight if compared to the previous strategies, achieving values close to 2 $g \cdot L^{-1}$. Despite the fact that lower relative concentrations of algae were observed in this particular growth, it can be observed that the absolute concentration values of algae were similar to the strategies described previously, in axenic and combined cell growth, demonstrating that the early maturation of the fungal mycelia with the subsequent inoculation of algae at a lower cell count may provide a positive effect for the fungal cells, but neutral or negative growth to the algae.

The variations in biomass contribution were evaluated according to the infrared spectrum of the different consortium samples obtained throughout the study based on the simultaneous inoculation of both fungal spores and algae cells without the influence of the cotton matrix to avoid interferences from the material used in the cell immobilization process (Strategies 1, 3 and 4). Interestingly, the ATR–FTIR spectra (Figure 5), which evaluated the functional groups in the consortium samples, presented similar profiles, demonstrating a possible similarity to the mechanism of formation, regardless of the strategy employed.

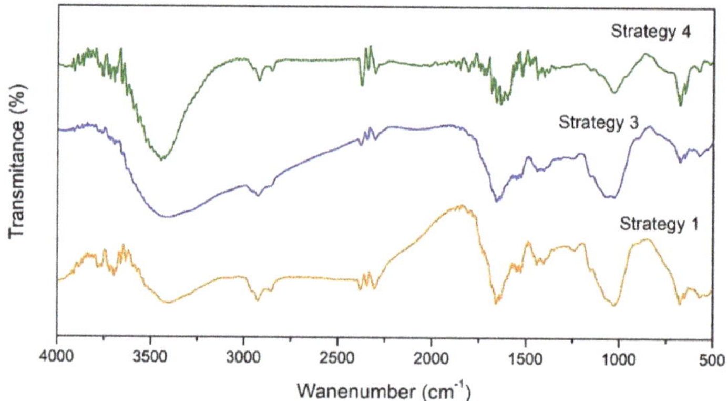

Figure 5. FTIR spectra of the consortium biomass samples from strategies 1, 3, and 4.

A thorough reading on the ATR–FTIR spectra show that, for all samples, a strong band between 3402 and 3444 cm^{-1}, the presence of -OH, likely due to the residual water activity of the samples [15]. No particular difference was either noticed within the other particular band regions of amide I and amide II, between 1480 and 1550 cm^{-1}, characterized by a particular amide C=O elongation, nor between 1444 and 1395 cm^{-1}, characteristic of vibrations of CH_2/CH_3, C=O, C-N, N-H, which constitute major protein structures (significant part of cell walls), and the C-O-C elongation vibrations between the wave numbers of 1047 to 1026 cm^{-1}, which could represent some structural difference in the polysaccharides constituting the biomass [16]. Lipids, which can be identified in similar samples by the bands related to C-H stretching vibrations [17] (=C-H stretch in the region of 3010 cm^{-1}, C-H stretching in -CH_3 and -CH_2 at 2855 and 2920 cm^{-1}) are also found in the three samples. Between 2927 and 2842 cm^{-1}, there is a characteristic fatty acid stretch with symmetrical CH_2 elongation and asymmetrical CH_2 elongation [17]. Other characteristic groups related to the presence of lipid components are the ester groups: C=O stretching at the 1740 cm^{-1} region and the C-O-C stretching (1070–1250 cm^{-1}). The C=O stretch vibrations are also characteristic to fatty acid ester bonds, which are found in all samples between 1629 and 1657 cm^{-1}.

The data presented herein are in consonance with some reports in the literature. Table 3 summarizes the strategies utilized for some recent works, demonstrating that, for the objectives of algae cell harvesting, as described herein, the utilization of a mature fungal biomass instead of simultaneous fungal spore germination and growth with the algae cells may provide better results.

Considering the multiple objectives described in the literature with the data presented herein, diverse authors describe the harvest efficiency and the composition of the consortium biomass as a determining factor for their purposes. For instance, Barnharst et al. [18] verified the application of artificial lichens using *C. vulgaris* and *Mucor indicus* in intensive aquaculture bioremediation process, observed a reduction in phosphate and total ammonia to undetectable limits, while axenic cultures were suitable for target removal of solely nitrogen or phosphorus. Rajendran et al. [22] verified the effectiveness of lichens between *C. vulgaris* and *M. circinelloides* by employing a polypropylene and cotton yarn in the ethanol co-products industry, in which the high concentration of P (818 mg·L^{-1}) and N (924 mg·L^{-1}) nutrients in the samples were recovered in the microalgae biomass by 55.7% and 74%, respectively, with a COD reduction in up to 65.6%. Yang et al. [20] also proposed ammonia and total solids recovery, by lichens of *Chorella* sp. and *Aspergillus* sp. in molasses wastewater, as well as achieving a microalgae recovery efficiency of over 97%.

Table 3. Comparison of microalgal cell recovery, consortium biomass and weight distribution with literature data.

Fungal strain	Microalgae Strain	Culture Medium	Immobilization Matrix	Microalgae Recovery (%)	Consortium Biomass (mg·L^{-1})	Microalgae: Fungal Weight Contribution to the Consortium	Reference
Mortierella isabellina, *Fusarium equiseti*, *F. lacertarum*, *Nigrospora oryzae*, *Altermaria alternate*, *F.equiseti*, *M.hiemalis*, and *M. circinelloides*	*C. vulgaris* UTEX 2714	Medium A	Polypropylene spun and tape yarn	34.85 ± 8.1–99.94 ± 0.02 (strain specific)	611.6 ± 11.9–1243.0 ± 37 (strain specific)	4.9: 95.1–48.9:51:0	Rajendran and Hu [10]
M. indicus ATCC 24905	*C. vulgaris* UTEX 2714	Medium A supplemented with ammonium concentrations ranging from 0 to 100 mg L^{-1}	Polypropylene spun and tape yarn	61.26–97.70 ([NH_4^+]-specific)	657.2–1125.0 ([NH_4^+]-specific)	19.72:80.28–80:99-19.01 ([NH_4^+]-specific)	Barnharst et al. [18]
A.fumigatus	*C. vulgaris*, *Chlamydomonas reinhardtii*, *Pseudokirchneriella subcapitata*, *Scenedesmus quadricauda*, *Thraustochytrid sp*, *Dunaliella tertiolecta*, *D.salina*, *Nannochloropsis oculata*, *Tetraselmis chuii*, *and Pyrocystis lunula*	Fungal growth broth with glucose or acid-pre-treated wheat straw and various dilutions of anaerobically digested swine wastewater	Self-fungal pelletization	≅ 25 -> 95 (strain and medium specific)	110–1060 (strain and medium specific)	Not specified	Wrede et al. [19]
Aspergillus sp.	*C. vulgaris*	Pre-treated molasses	Not Disclosed	>95	Up to 4125 (condition specific)	26.3:73.7–95.2:4.8 (condition specific)	Yang et al. [20]
M.elongata AG 77	*N.oceanica* CCMP 1779	Guillard F/2 medium	Not Disclosed	>60	Approximately 1000	Not specified	Du et al. [21]
M. circinelloides URM 4182	*C. vulgaris* BMAK D07	Medium A	Self-fungal pelletization	99.5 ± 0.2	1023 ± 27	79 ± 0.4: 21 ± 0.4	This work

4.3. The Effect of Different Strategies on the Lipid Accumulation and Properties

Both the microalgae and the fungal strain evaluated in this study have been reported in the literature as oil-accumulating microorganisms [23,24]. This is of interest to some biotechnological applications, including the production of biodiesel and other modifications of microbial lipids for nutraceutical applications, for example. Medium A is not an ideal growth medium intended for oil accumulation, due to an inadequate balance of carbon to nitrogen in the system, as well as a low concentration of glucose, which impairs high cell concentrations of fungi. Therefore, considering the growth of *M. circinelloides*, while the conversion of glucose to biomass was moderately high, as demonstrated in Table 1, it can be easily seen that the metabolic routes leading to oil accumulation are insufficient, attaining lipid contents in the range of 7.1 ± 0.2% in regard to the dry biomass weight in the axenic culture, whereas the same fungal strain has been reported to accumulate lipids in concentrations greater than 40% in other culture media.

The axenic photoautotrophic growth of *C. vulgaris*, however, is not dependent on sugars available in the medium [13], and, despite presenting lower biomass accumulation when compared to the fungi, the accumulation of lipids in regard to the dry biomass weight was within the range of 20% as also seen in Table 1, close to values previously reported for the same species in other photoautotrophic media [13].

The factor of a prolonged cultivation time was evaluated to assess the individualized effects on the total biomass growth, on the distribution of algae and fungi to the total biomass weight, on the microalgae cell recovery, and on the lipids, which were characterized in terms of total lipids and on fatty acid distribution. Even though small fluctuations in the consortium dry weight were observed during the assay, the total biomass was approximately constant over the whole period.

There was no significant change in the total lipid content as well. The algal biomass contribution to the total consortium weight was of 79 ± 0.4%, which suggests an efficient method to obtain algae-rich biomass in co-culture systems. The observations demonstrate a few important points, namely that the harvesting the fungal–algal system at an early time may be preferable to avoid fungal proliferation and dominance to the consortium [25,26], and the consortium system is possibly providing a cellular stress to both strains, due to the increased lipid content when compared to the axenic cultures, following a similar effect as those reported by Reis et al. [27].

Regarding the lipid portion of the consortium, it can be observed that the lipid content of the consortium biomass remains approximately constant throughout the evaluation period of 16 days. The fatty acid profile, as also demonstrated in Table 2, presents, at all growth times, a profile of fatty acids ranging from C12 up to C18, with a major concentration within the C16: 0 and C18: 1 acids (≅ 60%). It is known that the degree of saturation of fatty acids had a direct impact on the properties of the saponifiable lipids and their consequent modified products, as biodiesel. For instance, high degrees of saturation are usually linked to the greater viscosity and density of biodiesel products [28]. The abundance of unsaturated fatty acids is also known to decrease the oxidative stability of biodiesel. In this sense, it is expected that the distribution of fatty acids, which is close to some of the values also reported by Talebi et al. [28] will promote a good equilibrium between properties of flow and viscosity, which are enhanced due to the presence of unsaturated fatty acids, and oxidative stability, due to the presence of saturated fatty acids.

The distribution of fatty acids illustrated in Figure 6 demonstrates that the consortium biomass, using strategy 4 at eight days of cultivation, presents a lower contribution of saturated fatty acids when compared to the axenic controls of fungal and algal biomass using the same medium. While these results cannot distinguish between the individual contributions of fungal and algal cells and, therefore, the contributions of each species towards the characterization of the total fatty acids, the fatty acid profile obtained for the consortium presents some typical characteristics when compared to other references involving the growth of *M. circinelloides*, demonstrated by a balance between saturated and unsaturated fatty acids, as seen in Figure 6.

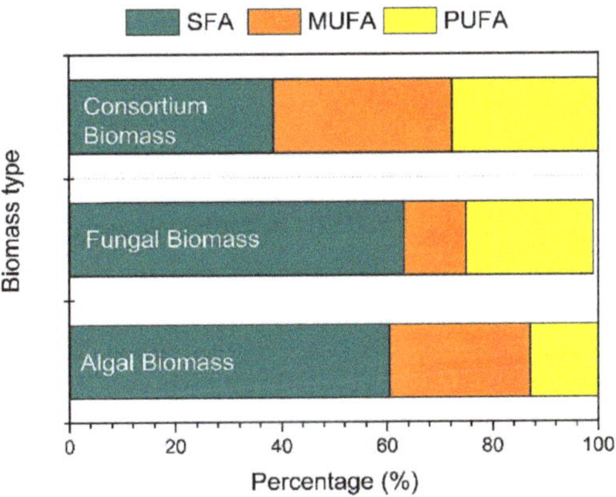

Figure 6. Distribution of fatty acids according to their category on the consortium biomass (strategy 4 at eight days of process). SFA: Saturated Fatty Acids, MUFA: Monounsaturated Fatty Acids, PUFA: Polyunsaturated Fatty Acids.

There is a wide range of factors that affect the fatty acid distribution of *M. circinelloides*. *M. circinelloides* is known to produce lipids with characteristics that are comparable to vegetable oils, thus, having been considered as a potential candidate for the establishment of microbial-biodiesel production units [6]. In fact, the stress factor related to the carbon to nitrogen ratio available in the medium regulates the content and distribution of fatty acids by *M. circinelloides*, as described by Wynn et al. [29]. As demonstrated herein, the accumulation of lipids in the consortium system has as a regulating factor, given the experimental conditions, the presence of the algae, which impairs the nitrogen availability to the fungal cells, thus promoting a higher effective carbon to nitrogen ratio, i.e., while the concentration of nitrogen is the same in both axenic and consortium systems, the consumption of nitrogen is greater in a system with two species competing for the same source of nitrogen, which induces higher accumulation of lipids by the fungi. It is important to stress that the medium selected herein has not been defined as a medium with characteristics that induce high lipid accumulation by *M. circinelloides*; in fact, this observation can be concluded by both the low lipid productivities and the rather abnormal characterization of *M. circinelloides* fatty acids when compared to the several applications in the literature [6], which often is described with relatively higher concentrations of both mono- and polyunsaturated fatty acids. The same is true for the composition of the fatty acids by *C. vulgaris*, which is often presented as a biomass rich in polyunsaturated fatty acids [30]. Therefore, the consortium biomass is likely to introduce some stress factors that induce lower accumulation of saturated fatty acids by both species and simultaneously, higher lipid productivities, possibly by decreasing the nitrogen availability to both strains involved in the culture. While there are multiple factors that affect the accumulation of lipids and the distribution of fatty acids, it can be foreseen that the application of consortium growth could induce the required characteristics for results that are likely to provide lipids with a distribution of fatty acids, as seen in Figure 6, closer to one would expect for applications in the biodiesel-production chain. In this sense, given the opportunities presented in this study, there is a potential for implementing consortium systems of algae and fungi with a possible control of the fatty acid distribution, given the stress conditions each strain is subjected to.

5. Conclusions

The article addresses an effort to recover microalgae cells using a straightforward approach based on co-cultivation with a filamentous fungus. From an initial screening of four different strategies to perform the consortium system, based on the effect of simultaneous vs. separate development of fungal spores and algae cells, and the presence of a supporting matrix, the results demonstrate that all of those attained promising initial results in the recovery of algal cells from the culture broth.

The results indicate high recovery rates, all of which are over 95%, and a synergistic effect on the lipid accumulation of both species, especially considering the fourfold increase observed on the lipid content by the axenic fungal growth when compared to the consortium system. The fatty acid distribution of the consortium system presents contributions of saturated, mono- and polyunsaturated fatty acids that could be applied for the production of biodiesel.

Author Contributions: S.M.F.E.Z., C.E.R.R. and H.F.D.C. conceived and designed research. S.M.F.E.Z. conducted experiments. M.B.S., B.H. and H.F.D.C. analyzed data. S.M.F.E.Z. and C.E.R.R. wrote the manuscript. All authors have read and agreed to the published version of the manuscript.

Funding: This study was funded by Fundação de Amparo à Pesquisa do Estado de São Paulo (FAPESP, grants #16/10636-8, #17/12908-8, and #18/01386-3), Coordenação de Aperfeiçoamento de Pessoal de Nível Superior—Brazil (CAPES, finance code 001), and Conselho Nacional de Desenvolvimento Científico e Tecnológico—CNPq (process number 433248/2018-1).

Conflicts of Interest: The authors declare no conflict of interest.

References

1. Henderson, R.K.; Parsons, S.A.; Jefferson, B. Successful removal of algae through the control of zeta potential. *Sep. Sci. Technol.* **2008**, *43*, 1653–1666. [CrossRef]
2. Mata, T.M.; Martins, A.A.; Caetano, N.S. Microalgae for biodiesel production and other applications: A review. *Renew. Sustain. Energy Rev.* **2010**, *14*, 217–232. [CrossRef]
3. Pragya, N.; Pandey, K.K.; Sahoo, P.K. A review on harvesting, oil extraction and biofuels production technologies from microalgae. *Renew. Sustain. Energy Rev.* **2013**, *24*, 159–171. [CrossRef]
4. Gultom, S.; Hu, B. Review of microalgae harvesting via co-pelletization with filamentous fungus. *Energies* **2013**, *6*, 5921–5939. [CrossRef]
5. Zhang, J.; Hu, B. A novel method to harvest microalgae via co-culture of filamentous fungi to form cell pellets. *Bioresour. Technol.* **2012**, *114*, 529–535. [CrossRef]
6. Reis, C.E.R.; Bento, H.B.S.; Carvalho, A.K.F.; Rajendran, A.; Hu, B.; De Castro, H.F. Critical applications of *Mucor circinelloides* within a biorefinery context. *Crit. Rev. Biotechnol.* **2019**, *39*, 555–570. [CrossRef] [PubMed]
7. Olson, D.G.; Mc Bride, J.E.; Shaw, A.J.; Lynd, L.R. Recent progress in consolidated bioprocessing. *Curr. Opin. Biotechnol.* **2012**, *23*, 396–405. [CrossRef] [PubMed]
8. Carvalho, A.K.F.; Rivaldi, J.D.; Barbosa, J.C.; De Castro, H.F. Biosynthesis, characterization and enzymatic transesterification of single cell oil of *Mucor circinelloides*—A sustainable pathway for biofuel production. *Bioresour. Technol.* **2015**, *181*, 47–53. [CrossRef] [PubMed]
9. Loures, C.C.A.; Amaral, M.S.; Da Rós, P.C.M.; Zorn, S.M.F.E.; De Castro, H.F.; Silva, M.B. Simultaneous esterification and transesterification of microbial oil from *Chlorella minutissima* by acid catalysis route: A comparison between homogeneous and heterogeneous catalysts. *Fuel* **2018**, *211*, 261–268. [CrossRef]
10. Rajendran, A.; Hu, B. Mycoalgae biofilm: Development of a novel platform technology using algae and fungal cultures. *Biotechnol. Biofuels* **2016**, *9*, 112. [CrossRef] [PubMed]
11. Dubois, M.; Gilles, A.K.; Hamilton, K.J.; Rebers, A.P.; Smith, F. Colorimetric method for determination of sugars and related substances. *Anal. Chem.* **1956**, *28*, 350–356. [CrossRef]
12. Liang, F.; Jin, F.; Liu, H.; Wang, Y.; Chang, F. The molecular function of the yeast polo-like kinase Cdc5 in Cdc14 release during early anaphase. *Mol. Biol. Cell* **2009**, *20*, 3671–3679. [CrossRef]
13. Heredia-Arroyo, T.; Wei, W.; Ruan, R.; Hu, B. Mixotrophic cultivation of *Chlorella vulgaris* and its potential application for the oil accumulation from non-sugar materials. *Biomass Bioenergy* **2011**, *35*, 2245–2253. [CrossRef]
14. Shong, J.; Diaz, M.R.J.; Collins, C.H. Towards synthetic microbial consortia for bioprocessing. *Curr. Opin. Biotechnol.* **2012**, *23*, 798–802. [CrossRef]
15. Movasaghi, Z.; Rehman, S.; ur Rehman, D.I. Fourier transform infrared (FTIR) spectroscopy of biological tissues. *Appl. Spectrosc. Rev.* **2008**, *43*, 134–179. [CrossRef]
16. Szeghalmi, A.; Kaminskyj, S.; Gough, K.M. A synchrotron FTIR microspectroscopy investigation of fungal hyphae grown under optimal and stressed conditions. *Anal. Bioanal. Chem.* **2007**, *387*, 1779–1789. [CrossRef]

17. Dean, A.P.; Sigee, D.C.; Estrada, B.; Pittman, J.K. Using FTIR spectroscopy for rapid determination of lipid accumulation in response to nitrogen limitation in freshwater microalgae. *Bioresour. Technol.* **2010**, *101*, 4499–4507. [CrossRef]
18. Barnharst, T.; Rajendran, A.; Hu, B. Bioremediation of synthetic intensive aquaculture wastewater by a novel feed-grade composite biofilm. *Int. Biodeterior. Biodegrad.* **2018**, *126*, 131–142. [CrossRef]
19. Wrede, D.; Taha, M.; Miranda, A.F.; Kadali, K.; Stevenson, T.; Ball, A.S.; Mouradov, A. Co-cultivation of fungal and microalgal cells as an efficient system for harvesting microalgal cells, lipid production and wastewater treatment. *PLoS ONE* **2014**, *9*, 1–22. [CrossRef]
20. Yang, L.; Li, H.; Wang, Q. A novel one-step method for oil-rich biomass production and harvesting by co-cultivating microalgae with filamentous fungi in molasses wastewater. *Bioresour. Technol.* **2019**, *275*, 35–43. [CrossRef]
21. Du, Y.Z.; Alvaro, J.; Hyden, B.; Zienkiewicz, K.; Benning, N.; Zienkiewicz, A.; Bonito, G.; Benning, C. Enhancing oil production and harvest by combining the marine alga *Nannochloropsis oceanica* and the oleaginous fungus *Mortierella elongata*. *Biotechnol. Biofuels* **2018**, *11*, 2–16. [CrossRef] [PubMed]
22. Rajendran, A.; Fox, T.; Hu, B. Nutrient recovery from ethanol co-products by a novel mycoalgae biofilm: Attached cultures of symbiotic fungi and algae. *J. Chem. Technol. Biotechnol* **2017**, *92*, 1766–1776. [CrossRef]
23. Xia, C.; Zhang, J.; Zhang, W.; Hu, B. A new cultivation method for microbial oil production: Cell pelletization and lipid accumulation by *Mucor circinelloides*. *Biotechnol. Biofuels* **2011**, *4*, 15. [CrossRef] [PubMed]
24. Tran, D.T.; Yeh, K.L.; Chen, C.L.; Chang, J.S. Enzymatic transesterification of microalgal oil from *Chlorella vulgaris* ESP-31 for biodiesel synthesis using immobilized *Burkholderia* lipase. *Bioresour. Technol.* **2012**, *108*, 119–127. [CrossRef]
25. Zamalloa, C.; Gultom, S.O.; Rajendran, A.; Hu, B. Ionic effects on microalgae harvest via microalgae-fungi co-pelletization. *Biocatal. Agric. Biotechnol.* **2017**, *9*, 145–155. [CrossRef]
26. Rajendran, A.; Fox, T.; Reis, C.R.; Wilson, B.; Hu, B. Deposition of manure nutrients in a novel mycoalgae biofilm for Nutrient management. *Biocatal. Agric. Biotechnol.* **2018**, *14*, 120–128. [CrossRef]
27. Reis, C.E.R.; Rajendran, A.; Silva, M.B.; Hu, B.; De Castro, H.F. The application of microbial consortia in a biorefinery context: Understanding the importance of artificial lichens. In *Sustainable Biotechnology-Enzymatic Resources of Renewable Energy*; Singh, O., Chandel, A., Eds.; Springer: Cham, Switzerland, 2018; pp. 423–437. [CrossRef]
28. Talebi, A.F.; Mohtashami, S.K.; Tabatabaei, M.; Tohidfar, M.; Bagheri, A.; Zeinalabedini, M.; Mirzaei, H.H.; Mirzajanzadeh, M.; Shafaroudi, S.M.; Bakhtiari, S. Fatty acids profiling: A selective criterion for screening microalgae strains for biodiesel production. *Algal Res.* **2013**, *2*, 258–267. [CrossRef]
29. Wynn, J.P.; Hamid, A.A.; Li, Y.; Ratledge, C. Biochemical events leading to the diversion of carbon into storage lipids in the oleaginous fungi *Mucor circinelloides* and *Mortierella alpina*. *Microbiology* **2001**, *147*, 2857–2864. [CrossRef]
30. Hultberg, M.; Jönsson, H.L.; Bergstrand, K.J.; Carlsson, A.S. Impact of light quality on biomass production and fatty acid content in the microalga *Chlorella vulgaris*. *Bioresour. Technol.* **2014**, *159*, 465–467. [CrossRef]

Article

Improving the Energy Balance of Hydrocarbon Production Using an Inclined Solid–Liquid Separator with a Wedge-Wire Screen and Easy Hydrocarbon Recovery from *Botryococcus braunii*

Kenichi Furuhashi [1,*], Fumio Hasegawa [1], Manabu Yamauchi [2], Yutaka Kaizu [1] and Kenji Imou [1]

1 Department of Biological and Environmental Engineering, The University of Tokyo, 1-1-1, Yayoi, Bunkyo-ku, Tokyo 113-8657, Japan; pxl05260@nifty.ne.jp (F.H.); kaizu@g.ecc.u-tokyo.ac.jp (Y.K.); k-imou@g.ecc.u-tokyo.ac.jp (K.I.)

2 Toyo Screen Kogyo Co., Ltd. 2-10-6, Kozen, Ikaruga-cho, Ikoma-gun, Nara 636-0103, Japan; yamauchi@toyoscreen.co.jp

* Correspondence: kfuruhashi@g.ecc.u-tokyo.ac.jp; Tel.: +81-3-5841-5360

Received: 27 July 2020; Accepted: 6 August 2020; Published: 10 August 2020

Abstract: The green colonial microalga *Botryococcus braunii* produces large amounts of hydrocarbons and has attracted attention as a potential source of biofuel. When this freshwater microalga is cultured in a brackish medium, the hydrocarbon recovery rate increases; furthermore, the colony size becomes large. In this study, the effects of such changes on the energy balance of harvesting and hydrocarbon recovery were studied via filtrate experiments on an inclined separator and extraction from a concentrated slurry. The inclined separator was effective for harvesting large-colony-forming algae. The water content on the wire screen of slit sizes larger than 150 μm was <80% and a separation rate of >85% could be achieved. The input energy of the harvesting using the brackish medium with this separator was ≈44% of that using the freshwater medium with vacuum filtration, while the input energy of the hydrocarbon recovery using the brackish medium was ≈88% of that using the freshwater medium with pre-heating before *n*-hexane extraction. Furthermore, the energy profit ratio of the process in the brackish medium was 2.92, which was ≈1.2 times higher than that in the freshwater medium. This study demonstrated that filtration techniques and hydrocarbon recovery from *B. braunii* with a low energy input through culture in a brackish medium are viable.

Keywords: microalgae; inclined solid–liquid separator; hydrocarbon recovery; biofuel; energy balance; harvesting

---

1. Introduction

Botryococcus braunii (*B. braunii*) is an autotrophic alga that can potentially be employed as a biofuel resource because this alga produces hydrocarbons during cell division and secretes them in an oleophilic extracellular biopolymer. First, *B. braunii* produces and stores hydrocarbons up to 86% of its dry weight [1,2]. Although hydrocarbons are produced in species of all alga pyhla, their contents are low and the lipids that most microalgae can store with a high content rate are triglycerides or fatty acids [2,3]. Meanwhile, *B. braunii* produces hydrocarbons that can be classified based on their type [4]; for example, the hydrocarbons accumulated by the A, B, and L races of *B. braunii* are *n*-alkadienes and *n*-alkatrienes, mainly C_{30} to C_{37} triterpenoids, and tetraterpenoids, respectively. Among the specified races, race B has been widely studied for liquid biofuel production because it stores hydrocarbons at relatively high volumes and can be easily decomposed into suitable fuels for internal combustion engines via hydrocracking [5]. Second, *B. braunii* produces hydrocarbons during cell division [6]. Most microalgae start to produce lipids under environmental stress conditions, such as nutrient depletion. This feature allows for easier culture management under continuous conditions on a large scale and a high algal density can be maintained to prevent contamination; however, the growth rate of *B. braunii* is slower than other oil-producing microalgae. This may be because not only saccharides or proteins but also high energy hydrocarbons similar to fossil fuel are produced with cell division. Third, *B. braunii* secretes hydrocarbons in the oleophilic extracellular biopolymer that connects cells undergoing cell division [7,8]. Other algae store lipids in the cell interior and hence extra processes involving degradation of the cell wall are required to effectively recover them [9,10]. Therefore, biofuel production from *B. braunii* can potentially save extra input energy and production costs during the culture, lipid-recovery, and fuel-conversion processes.

Various methods have been reported for hydrocarbon recovery from *B. braunii*. Efficient hydrocarbon recovery may be achieved using an amphiphilic solvent, such as a methanol/chloroform mixture or dimethyl ether [11,12]. Hydrothermal liquefaction or supercritical carbon dioxide extraction, in which organic solvents are not required, is also reportedly effective for hydrocarbon recovery from *B. braunii* [13,14]. These methods require high-temperature or high-pressure conditions; otherwise, the recovery of some polar solvents from the water phase is difficult. Meanwhile, non-polar solvent extraction, for example, *n*-hexane extraction of soybean oil, can also be applied to *B. braunii* because this alga stores hydrocarbons in the oleophilic extracellular biopolymer. However, the direct recovery rate from wet slurries is low, which is a limitation of this methodology [15]. Cells and thin layers containing hydrocarbons are entirely enclosed by fibrillar structures, i.e., the colony sheath, on the retaining wall [16,17]. This sheath mainly consists of saccharide components that prevent the entry of non-polar solvents into the colony interior [18]. Various pretreatments, such as drying [19], homogenization [20], and pre-heating [21], have been reported to improve hydrocarbon recovery using *n*-hexane, but these methods also require thermal or electrical energy as an input for treating algal slurries. Previously, we reported that the hydrocarbon recovery rate could be improved from *B. braunii* cultured in a brackish medium (BM) with a salinity of 3 g L^{-1} without growth inhibition [22]; this improvement was attributed to a shortening in the colony sheath and low fibrilliform density via a shift in algal metabolism from the biosynthesis of the colony sheath consisting of polysaccharides that surround the algal colonies to produce osmolytes of disaccharides and obtain salt tolerance [18]. When the salinity in the medium rises above 3 g L^{-1}, the hydrocarbon recovery rate is improved more but the growth is inhibited. It is considered that this salinity is the limit of salt tolerance for the culture of freshwater microalga *B. braunii*. This is different from the approach in which a high osmotic shock is used to disrupt cell walls to extract lipids [23]. Saga et al. evaluated the energy profit ratio (EPR) of the hydrocarbon recovery process with a pre-heating treatment by changing the ratio of *n*-hexane to the algal slurry or the water content of the slurry [24]. EPR is the index used to evaluate the energy investment efficiency of energy production equipment and is obtained by dividing the production energy by the fossil energy input. Culturing in the BM can decrease the input energy required for hydrocarbon recovery compared to a pre-heating process.

Moreover, algal culture in a BM increased the colony size, with the median particle diameter of colonies increasing from ≈60 μm in a freshwater medium (FM) to over 200 μm in a BM [22]. Increases in the colony size of *B. braunii* are achieved using a high light intensity [25] or in the presence of 40 mM glucose [26]. Our light condition (100 μmol m^{-2} s^{-1}) is classified into low density compared with Zhang and Kojima [25]. Tanoi et al. reported that it was suggested that the high osmotic pressure from glucose in a culture medium affects the colony size [26]. They inferred the colony size was related to the change in polysaccharides or other matrix components, such as algaenan, via the culture conditions created by the high light intensity or osmotic pressure. The BM in this study contained 30 mM NaCl of the main components and an ion concentration of NaCl such that the osmotic pressure was doubled. The osmotic pressure was not significantly different from the literature. The brackish medium caused a shift in the algal metabolism of saccharides, as mentioned above. We consider that the changes in saccharide metabolism via the high osmotic pressure in the BM affected the colony size. Although *B. braunii* is a colonial and large alga compared to other unicellular algae, it is difficult to harvest concentrated algal slurries before hydrocarbon recovery. Centrifugation, flocculation, or filtration via cloth is generally used for harvesting algae [27]. In this context, inclined solid–liquid separators with wedge-wire screens are often used for sewage treatment. In this apparatus, wastewater comes into contact with an inclined screen with slits in the tangential direction; the filtrate passes through the screen and the concentrated slurry or solid particles are continuously recovered on the screen. This separator requires only pump power for harvesting and ensures easy maintenance. No case has yet been made for introducing the inclined separator for harvesting *B. braunii*. The harvesting of large-colony-forming *B. braunii* cultured in the BM via this separator would lead to obtaining under 80% water content of slurry-like sewage treatment and decreasing the input energy required for harvesting.

In this study, the effect of the hydrocarbon recovery rate and colony size of *B. braunii* cultured in a BM on the EPR was investigated. We first evaluated the harvesting process based on filtrate experiments using an inclined solid–liquid separator with a varying slit size for the different colony sizes of *B. braunii* cultured in different media. Second, hydrocarbon recovery was evaluated based on solvent extraction experiments on a concentrated algal slurry because we previously performed the solvent extraction only for low algal concentrations in a culture fluid. Lastly, the EPR was analyzed based on the data from these experiments and the literature.

2. Materials and Methods

2.1. Microalgae Cultivation

B. braunii race B (Showa strain) was cultured in a 10 L culture bag aerated with 1% CO_2 at 25 °C under illumination (100 μmol m^{-2} s^{-1}) with a 12 h light–dark cycle for 30 days. Two types of culture media were used: a modified Chu13 medium as the freshwater medium (FM) and a brackish medium (BM) with a salinity of 3 g L^{-1}, which was prepared by diluting commercial artificial seawater (Daigo's Artificial Seawater SP for Marine Microalgae Medium, Wako Pure Chemical Industries, Japan; total salinity, 36 g L^{-1}) [22]. In our previous study, we found that this salinity did not inhibit algal growth or the hydrocarbon production rate (0.031 g-hydrocarbons L^{-1} d^{-1} in both the FM and the BM) [22]. Nutrients and trace metals, such as KNO_3, $K_2HPO_4{\cdot}3H_2O$, and FeNaEDTA, were already present in the Chu13 medium, but not in the artificial seawater. Therefore, they were added to the brackish medium at levels similar to those in the Chu13 medium.

2.2. Filtration Using an Inclined Solid–Liquid Separator with a Wedge-Wire Screen

In this test, culture liquids with an algal density of 1.0–1.5 g L^{-1} were used. The algal density was measured after the algal culture liquid was filtered through a glass fiber filter (GF/A 110 mm diameter, Whatman, Germany), which was rinsed with deionized water to wash the salt in the culture media and dried at 105 °C for 24 h. The concentrated algal slurry on the screen was recovered and the water content was also measured by drying at 105 °C for 24 h. Algal fluid (20 L) was allowed to flow into the first screen of the separator at a flow rate of 3 L min^{-1} from a tank positioned at a height of 350 mm above the separator. Figure 1 shows the side view of the inclined solid–liquid separator. The filtrate was passed through the wire screen and collected. The wedge wires were triangles pointing toward the filtrate and were lined vertically relative to the flow direction. The shape of the triangular cross-section led to minimizing the contact area and clogging of the screen when solids passed through the screen. The separator consisted of screens positioned at two different angles, where the inclination angles of the first and second screens were 66° and 51°, respectively. The concentrated slurry could be recovered after the culture liquid passed through the second screen. The size distribution of the algal colonies in this study was measured using a laser diffraction particle-size-distribution analyzer (MT−3300EXII, Nikkiso, Japan). Based on the measured colony size, the slit size of the wedge-wire screen was determined to be 10, 20, 30, 50, and 75 μm for samples cultured in the FM, and 50, 100, 150,

200, and 300 μm for samples cultured in the BM. The dry-matter separation rate (R_s) of the separator was calculated as follows:

$$R_s = M_s TS_s \ / \ M_c TS_c \quad (1)$$

where M_s and M_c (kg) are the masses of the recovered algal slurry and algal culture liquid, respectively, and TS_s and TS_c (kg kg^{-1}) are total algal solids in the recovered algal slurry and algal culture liquid, respectively.

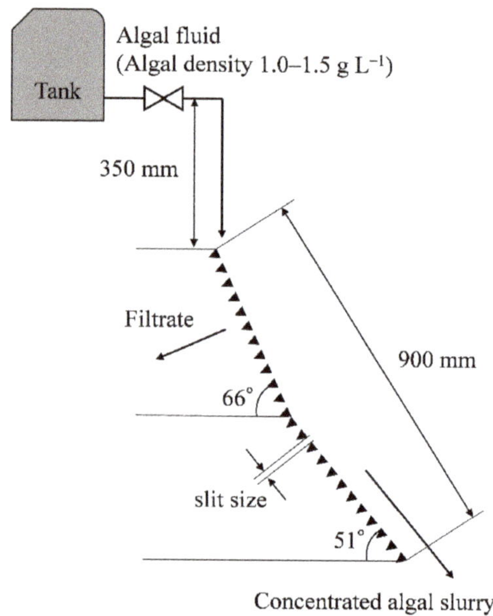

Figure 1. Side view of the inclined solid–liquid separator consisting of two wedge-wire screens at different angles.

2.3. Hydrocarbon Recovery from Concentrated Algal Slurries

The alga was concentrated using vacuum filtration through a 20 μm nylon mesh and the obtained slurry was rinsed with deionized water. The water content and total solids (TS) content in the concentrate were adjusted by the degree of the vacuum and measured using a moisture analyzer (MX-50, A&D, Japan). Solvent extraction from 10 g of the wet algal slurry was conducted in a 250 mL Teflon vessel. In this process, the algal slurry was dispersed in *n*-hexane (weight of *n*-hexane/dry weight of algal slurry = 6/1) and the mixture was stirred at 800 rpm. The sediment was rinsed using the same amount of *n*-hexane twice to reflect the actual counter-current multistage (three-stage) extraction. The hydrocarbon mass dissolved in the recovered solvent was measured using programmed-temperature gas chromatography with flame ionization detection (GC-FID) by using a capillary column (GC-2014 with Rtx-1 capillary column, Shimadzu, Japan). The temperature program was as follows: the column was maintained at 50 °C for 1 min, heated from 50 to 220 °C at 10 °C min^{-1}, maintained at 220 °C for 3 min, heated from 220 to 260 °C at 2 °C min^{-1}, and equilibrated at 260 °C for 3 min [28]. The hydrocarbon mass was calculated from the sum of the peak areas corresponding to hydrocarbons relative to the peak areas for known amounts of standard hydrocarbons extracted and purified from freeze-dried algal samples; this process is described below.

The hydrocarbons contained in the alga were extracted from dry algal samples. The extraction from dried *B. braunii* using *n*-hexane or *n*-heptane was carried out to measure the content of external hydrocarbons [15] rather than the total lipids in the alga [29,30]. Freeze-dried algal samples were soaked in *n*-hexane. The extraction process was repeated until the yellow pigments that co-existed with the hydrocarbons in the extracellular matrices were lost. Subsequently, the extracts were combined and subjected to silica-gel column chromatography. The fraction of pure hydrocarbons obtained after the chromatography was weighed and a standard curve was constructed for GC-FID [22]. GC-electron impact mass spectrometry analysis (GCMS-QP2010, Shimadzu, Japan) with a capillary column (InertCap 1MS, GL Science, Japan) was carried out under the same temperature program as the GC-FID analysis. The major eight components that are typical of methylated squalene (C30–34 botryococcene) were identified in accordance with Atobe et al. [28]. The number of peaks and hydrocarbon elution patterns were similar between the GC-FID and GC-MS. The hydrocarbon recovery rate was expressed as the ratio of the amounts of hydrocarbons recovered from the wet algal slurry to the hydrocarbon content.

2.4. Energy Profit Ratio Calculation

Figure 2 compares the hydrocarbon production from the BM and the FM (with pre-heating before the hydrocarbon extraction). Alga cultured in the BM entered the hydrocarbon recovery process directly after the inclined solid–liquid separator or vacuum filtration. In contrast, alga cultured in the FM was introduced into the hydrocarbon recovery process through vacuum filtration and pre-heating the algal slurry containing 92% water at 95 °C. The thermal pretreatment dissolved all amphiphilic fibrils of the colony sheath into the hot water. The slurry was rinsed twice after pre-heating and subjected to vacuum filtration because the released fibrils formed emulsions with *n*-hexane and prevented hydrocarbons extraction [24]. Although the thermal pretreatment was effective for *B. braunii* cultured in BM, the pretreatment with BM was not considered in order to show the energy profit ratio and hydrocarbon recovery rate from the concentrated slurry that had a different colony surface structure. Under these conditions, it was assumed that the water content reduced from 92% to 70%. The input energy can be categorized into thermal energy and electrical energy for mechanical operations (e.g., for the pump). In this study, the receiving and efficiency in Japan were set at 36.9%. Assuming a rotary vacuum belt filter for the vacuum filtration (TSK belt filter, TSK, Japan), the filtration time was set at 20 s based on the dipping rate and rotational frequency of the drum (Table 1). The algal filtration throughput over 20 s was determined at a gauge pressure of 40 kPa using the vacuum leaf test (Leaf Tester, Miyamoto, Japan); in this case, a 20 µm nylon mesh was used as the filter cloth. The electrical energies of the pump (2) and belt filter (3) were calculated using the following equations:

$$E_s = \rho V g H / \Phi \tag{2}$$

$$E_b = P_g V / \Phi \tag{3}$$

where E_s and E_b (W) represent the electric power required for transferring the algal culture liquid (1.0 g L^{-1}) to the filtration unit and vacuum pump of the belt filter, respectively; ρ (kg L^{-1}) is the culture density (1.0 kg L^{-1}); V (L s^{-1}) is the throughput flow rate; g (m s^{-2}) represents acceleration due to gravity; H (m) is the pump head; P_g (Pa) is the gauge pressure; and Φ (-) indicates the pump efficiency (0.7). The EPR was calculated using the following relationship on the basis of 1 kg of dry algae (4):

$$\text{EPR} = \text{HHV of recovered hydrocarbons/Input energy} \tag{4}$$

Here, HHV (MJ) represents the higher heating value. It was calculated from the elemental composition using Dulong's formula, as shown in Equation (5) [31]:

$$Q = 0.3383C + 1.442(H - O/8) \tag{5}$$

where Q (MJ) is the heating value. The elemental compositions of the *B. braunii* and the purified hydrocarbons were analyzed using two elemental analyzers (NCH-22F, Sumika Chemical Analysis Service, Ltd., Japan and EMGA-920, Horiba, Ltd., Japan). The mixing power of the slurry was measured using a mixing torque meter (ST-300II, Satake Chemical Equipment, Japan) and the slurry was prepared by mixing *B. braunii* with a 70% water content and *n*-hexane (*n*-hexane/dry weight of algal slurry = 6/1).

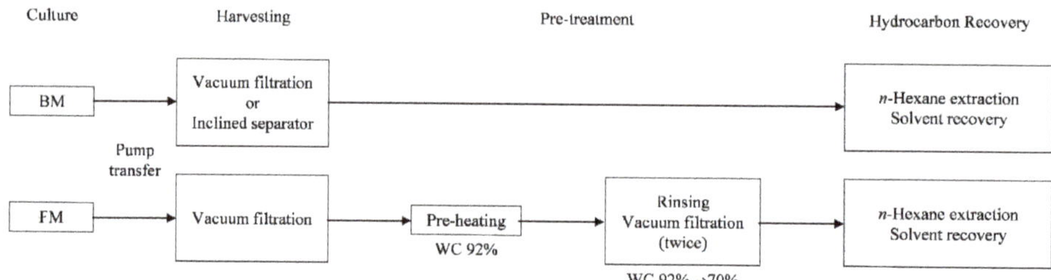

Figure 2. Hydrocarbon production from algal cultures in brackish medium (BM) or freshwater medium (FM) with pre-heating before the hydrocarbon extraction. WC: Water content.

Table 1. Parameter values used to calculate the energy profit ratio.

Item		Units	Value	Ref.
Rotary vacuum belt filter	Drum diameter	m	4.2	TSK
	Drum length	m	7.6	
	Rotational speed	rpm	1.0	
	Dipping rate	-	0.33	
	Motor output	kW	10	
	Gauge pressure	kPa	40	
Pump head		m	10	Set in this study
Algal throughput by one belt filter unit	FM	kg s^{-1}	0.23	Leaf test
	BM	kg s^{-1}	0.58	Leaf test
Elemental composition of *B. braunii*	C	%	71.4	Elemental analyzer
	H	%	10.3	
	N	%	1.21	
	O	%	13.6	
Elemental composition of the hydrocarbons	C	%	87.4	Elemental analyzer
	H	%	12.1	
Higher heating value	*B. braunii*	MJ kg^{-1}	36.6	Dulong's formula
	Hydrocarbons	MJ kg^{-1}	47.0	
	n-hexane	MJ kg^{-1}	46.2	
Hydrocarbon content	FM	kg kg^{-1}	0.35	[22]
	BM	kg kg^{-1}	0.35	[22]
Hydrocarbon recovery rate	Pre-heating	%	95.0	[24]
	BM		95.0	From this study
Pinch temperature for pre-heating		°C	20	[19]
Specific heat	*B. braunii*	kJ kg^{-1} K^{-1}	1.9	[19]
	Water	kJ kg^{-1} K^{-1}	4.18	
Latent heat	*n*-hexane	MJ kg^{-1}	0.34	[19]
	Water	MJ kg^{-1}	2.25	
n-hexane recovery rate			99.9	[19]
Evaporated water loss/residual hexane		kg kg^{-1}	154	[19]
Stirring power on extraction from slurry		W kg^{-1}	15	Torque meter
Receiving and efficiency	, Japan	%	36.9	

Table 1 shows the values of the parameters used in this study to calculate the EPR. The hydrocarbon contents of the alga cultured in both media were set at 35.0% [22]. In this previous report, the brackish medium led to a tendency toward a high hydrocarbon content with no significant difference rather than no effect. However, a slight difference in hydrocarbon contents has a huge effect on the energy profit ratio. Hence, the hydrocarbon contents were set at the same ratio in this study in order to show the difference of input energy in each process. The elemental compositions of the *B. braunii* cultured in the BM were supposed to be equal to the alga in the FM regarding hydrocarbon contents, although the saccharide metabolites were particularly changed by the BM.

3. Results and Discussions

3.1. Harvesting Using the Inclined Solid–Liquid Separator with Wedge-Wire Screens

Figure 3 shows the size distributions of the algal colonies used in this study. Table 2 shows the results of harvesting using the inclined solid–liquid separator with wedge-wire screens of different slit sizes selected based on the measured colony size (Figure 3). Figure 4 shows the apparatus used in this experiment and *B. braunii* recovered from the BM culture on the screen. High separation rates of >90% were achieved with a slit size of less than 30 μm in the FM, but the water content in the slurry recovered on the screen was >99.0%. This was because the water did not pass through the wire screen due to its high surface tension. When the slit size was larger than 50 μm, the separation rate decreased sharply and the water content was found to be >99.0%. At a slit size of 75 μm, almost all the alga passed through the screen and could not be recovered on the screen. In conclusion, the alga in the FM could not be concentrated by the inclined separator due to its small colony size. Pressure injection from the nozzle might be needed to recover small colonies and a low slurry content from the wedge-wire screen with a slit size of less than 30 μm. In contrast, this separator was effective for separating the algae cultured in the BM. Although the water content of the slurry on the screen with a slit size of 50 μm was 96.6%, the water content at slit sizes larger than 150 μm was below 80%. Moreover, unlike in the case of the FM culture, there was no sharp decrease in the separation rate from the BM. A separation rate of >85.0% was achieved with a slit size of less than 150 μm and the water content of the slurry on the screen did not change; this rate decreased only when the slit size was larger than 150 μm. The water content after centrifugation was about 99% and after the vacuum belt filter, it was about 75% [32], which was similar to the inclined separator in this study. This separator had the same effect as the vacuum belt filter without using any electrical energy input when the *B. braunii* cultured in the BM was used.

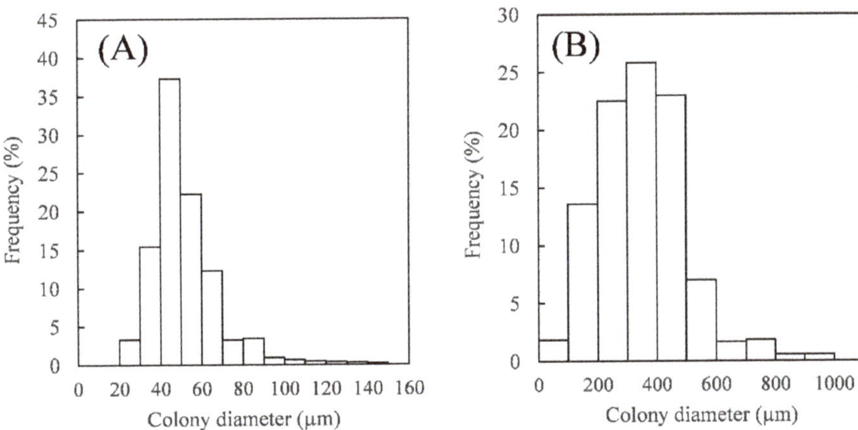

Figure 3. Size distributions of algal colonies cultured in (**A**) the freshwater medium and (**B**) the brackish water medium.

Table 2. Effect of the slit size of the wedge-wire screen on the separation rate and water content of the algal slurries recovered from different media. FM: Freshwater medium, BM: Brackish medium. At a slit size of 75 μm, almost all the algae cultured in the FM passed through the screen, and hence the water content and the separation ratio could not be determined (ND).

Medium	Slit Size (μm)	Aperture Ratio (%)	Separation Rate (%)	Water Content (%)
FM	10	2.0	96.3	99.9
	20	3.8	93.0	99.7
	30	5.7	91.3	99.3
	50	4.8	51.3	99.1
	75	7.0	ND	ND
BM	50	4.8	96.0	96.6
	100	9.1	92.6	83.0
	150	13.0	86.7	78.9
	200	16.7	77.8	77.4
	300	16.7	66.2	78.8

Figure 4. Harvesting of *B. braunii* cultured in the brackish medium (BM) using an inclined solid–liquid separator with a wedge-wire screen (150 μm slit size).

Moreover, we considered that the colony increase was caused by the changes in a factor related to saccharide metabolism that accelerated or inhibited the colony division via high osmotic pressure in the BM; however, the obvious reason, as mentioned in the introduction, was not found to be the case. The colony organization of *B. braunii* is like a flower bouquet in which the cell apex is directed toward the surface of the colony and matrix that stores hydrocarbons fills the space between the cells [18]. Cells in the center of the large formed colony clearly stored the hydrocarbons in cells and matrices, as seen using fluorescent microscopy and freeze-substitution electron microscopy. *B. braunii* produces hydrocarbons in cell division. This was not directly evidenced by the cell viability in the inner colony; however, the growth and hydrocarbon contents in the BM with 3 g L^{-1} was equal to that from the

FM [22]. This also shows that the colony increase was not caused by the simple accumulation of lipids. As another possibility, the improvement of light transmittance via colony enlargement may have affected the photosynthetic efficiency of cells outside the colony, although cells in the inner colony were dead. Further analysis of the colony structure and metabolome is required in the next challenge in order to confirm whether the colony enlargement was the cause.

Figure 5 shows the input electrical energy required for each harvesting process. The inclined separator with the FM (Figure 2) was not considered because the alga cultured in the FM could not be recovered by the filtration process described earlier. The total input energies of the BM harvesting with the inclined separator, the BM with the vacuum filtration, and the FM with the vacuum filtration were 0.14, 0.21, and 0.32 MJ kg^{-1}, respectively. Using the BM, the input energy could be reduced by 56% when compared to that using the FM. Two points must be particularly mentioned here. The first is that the pump transfer consumed a large proportion of the input energy. In this study, the pump head was set at 10 m on the basis of a general centrifugal pump. When the algal slurry was transferred from the edge of a wide pond to the harvesting apparatus, the pump head exceeded the selected value in some cases. However, the inclined separator did not require electrical energy and was portable, unlike the vacuum-filtration setup. This means that the algae could be harvested at the edge of the pond. The second point is that in this study, it was assumed that during the vacuum filtration, no cracks were formed in the residue on the filter cloth. The occurrence of cracks increased the input energy of the vacuum pump due to air suction [33]. Therefore, the BM–inclined separator combination required lower input energy than the FM-based approaches. The operating costs would reduce as the input energy reduces compared with vacuum filtration. The introduction motive of the inclined separator is the low initial and management costs if the target of the filtration can be recovered. This is because the wire screen is easy to wash due to being made of stainless steel and requires no motor. The cost calculation should be considered by obtaining the maximum throughput via filtrating further amounts of algae in the next step.

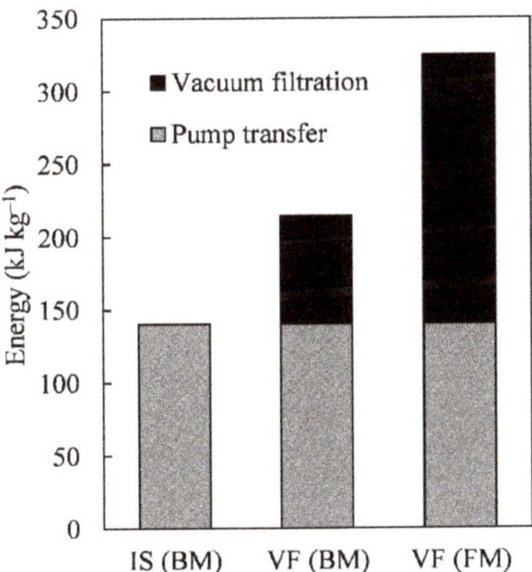

Figure 5. Input electrical energy per unit of dry algae in the algal harvesting from cultures in different media using an inclined solid–liquid separator (IS) or using vacuum filtration (VF). FM: Freshwater medium, BM: Brackish medium.

3.2. Hydrocarbon Recovery Rate from the Concentrated Algal Slurry

Figure 6 shows the effects of the water content in the algal slurry and culture medium on the hydrocarbon recovery rate. A hydrocarbon content of >95% could be recovered in 4 h when the water content in the slurry cultured in the BM was 70%. Furthermore, the hydrocarbon recovery rate increased with a decrease in the water content. In the case of the FM, the recovery rates were only 20% or 47% within 9 and 12 h, respectively. Saga et al. reported that a hydrocarbon content of >95% could be recovered from a pre-heated *B. braunii* slurry (same strain as that used in this study) within 2 h under similar solvent conditions [24]. Furthermore, it took only 45 min to recover 90% of the hydrocarbon content from 60%-water-content algal pellets. Pre-heating was found to exert a more significant effect on the hydrocarbon recovery than the BM culture. The extraction time for a 95% hydrocarbon recovery was almost 4 h in the BM system but only 2 h was required with the FM and pre-heating.

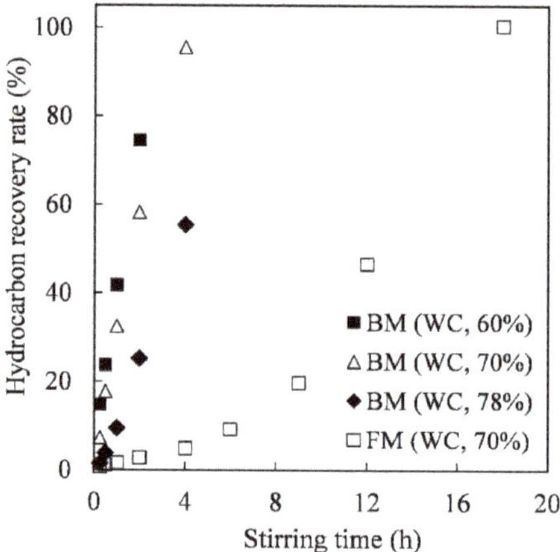

Figure 6. Hydrocarbon recovery rates from concentrated algal slurries of different water contents cultured in different media. FM: Freshwater medium, BM: Brackish medium, WC: Water content.

Figure 7A shows the input energy in the hydrocarbon recovery process and the recovered hydrocarbon energy from dry algae per unit mass. The HHV of the recovered hydrocarbon per unit weight of dry algae was 15.6 MJ kg^{-1} and the input energy recovery from the FM and BM cultures were 5.69 and 4.98 MJ kg^{-1}, respectively. This difference was attributed to the exclusion of pre-heating in the BM process and a low stirring power. Hexane loss contributed to the HHV of the *n*-hexane in the residue without being evaporated and recovered. In addition, extraction for long periods had little effect on the EPR. However, if the extraction time for a 95% hydrocarbon recovery was >16 h in the FM without pre-heating, as shown in Figure 6, the input energy for the stirrer was >2.3 MJ kg^{-1} and larger than that for pre-heating. In this context, one point is to be particularly observed. The evaporation of water from the residue was required to recover 99.9% of the *n*-hexane content using liquid–liquid extraction. The proportion of the input energy for the evaporation of water was the largest. Therefore, the balance between the amount of the evaporated water and residual *n*-hexane or using solvents that evaporate more easily than *n*-hexane should be considered in the future for liquid–liquid extraction from microalgae.

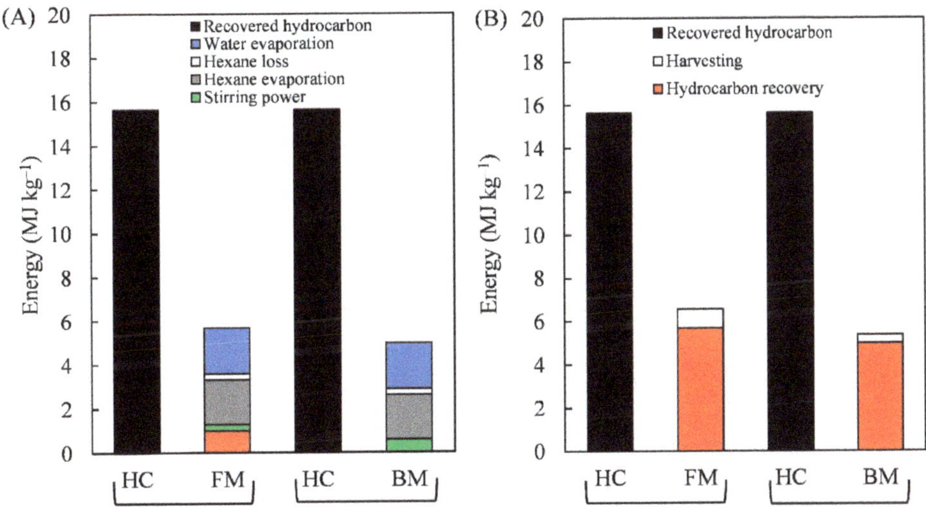

Figure 7. Input energy per unit of dry algae of hydrocarbon recovery (**A**) and the process from harvesting to hydrocarbon recovery (**B**) from algae cultured in different media. HC: Hydrocarbon, FM: Freshwater medium, BM: Brackish medium.

3.3. The Energy Profit Ratio from Harvesting to Hydrocarbon Recovery

Figure 7B shows the total input energy from harvesting to hydrocarbon recovery and the HHVs of the recovered hydrocarbons. In this experiment, the inclined separator was used for harvesting algae cultured in the BM. The total input energies in the FM and BM processes was 6.57 and 5.36 MJ kg^{-1}, respectively. Meanwhile, the EPR of the BM process was 2.92, which was ≈1.2 times higher than that of the FM process (2.38). The input energy for harvesting was much lower than that for the hydrocarbon recovery with both the FM and BM. When reviewing the energy balances in the biofuel production from microalgae, in most models, it was found that harvesting and drying accounted for the largest proportion of the energy consumption [34]. This was thought to be due to the incorporation of the drying process into the harvesting process. *B. braunii* grows in large colonies, has good filterability, secretes hydrocarbons in the oleophilic extracellular biopolymer, and does not require drying or cell disruption. These unique characteristics lead to a decrease in the ratio of the energy of the harvesting process to the total input energy. Finally, the input energy of the culture process was omitted in this study. This was because the input energy of the culture process in open ponds had various estimations in reports and we observed that there was no difference in the growth rate and hydrocarbon productivity between the BM and FM in our previous study [22]. In this study, the utilization of the extraction residue was not considered. The lipid-extracted microalgae could be utilized as a mediator and substrate in microbial fuel cells [35], and the quorum-sensing molecules extracted from sludge in this cell enhanced the productivity of the algal culture [36]. Thus, a further improvement of the energy balance is required via the energy transformation of the extraction residue. Moreover, in the future, culture development by taking advantage of the large algal colony size in the BM should be considered.

4. Conclusions

In this study, the effects of a *B. braunii* culture in a BM on the EPR of hydrocarbon recovery and harvesting were analyzed via the extraction from concentrated slurries and filtration using an inclined solid–liquid separator, respectively. The inclined solid–liquid separator with a wedge-wire screen, which was used for filtration, did not require electrical energy and was effective for large-colony-forming algae cultured in the BM. The water content in the slurry on the wire screen at slit sizes larger than

150 μm was <80% and a separation rate of >85.0% could be achieved at slit sizes less than 150 μm. In contrast, in the case of the FM culture, the water content in the slurry was found to be >99.0% at any slit size due to its small colony size and high surface tension. The input electrical energy of the harvesting process using the BM with the inclined separator was 0.14 MJ kg^{-1}, which was ≈0.44 times higher than that observed using the FM with vacuum filtration (0.32 MJ kg^{-1}). The EPR of the process in the BM was 2.92, which was ≈1.2 times higher than that in the FM with pre-heating before the liquid–liquid extraction (2.38). The brackish medium of this study can be prepared in the filed scale because only seawater was added into the culture medium. This study showed that the filtration techniques and hydrocarbon recovery with a low energy input were viable. However, we consider that the main bottlenecks of biofuel production from microalgae are the stable culture technology and culture costs. In the next step, culture technology will be developed by taking advantage of the large algal colony size through the elucidation of the mechanism of colony enlargement.

Author Contributions: Conception and design of the study: K.F., F.H., M.Y., Y.K., and K.I. Collection and analysis of the data: K.F., F.H., and M.Y. Data interpretation, manuscript drafting, and final approval of the article: K.F., F.H., M.Y., Y.K., and K.I. All authors have read and approved the final version of the manuscript.

Funding: This work was supported by JSPS (Japan Society for the Promotion of Science) KAKENHI (grant number JP 19K15939).

Conflicts of Interest: The authors declare no conflict of interest.

References

1. Brown, A.C.; Knights, B.A.; Conway, E. Hydrocarbon content and its relationship to physiological state in the green alga *Botryococcus braunii*. *Phytochemistry* **1969**, *8*, 543–547. [CrossRef]
2. Qin, J.G. Hydrocarbons from algae. In *Handbook of Hydrocarbon and Lipid Microbiology*; Timmis, K.N., Ed.; Springer: Heidelberg/Berlin, Germany, 2010; pp. 2817–2826.
3. Sharma, K.K.; Schuhmann, H.; Schenk, P.M. High Lipid Induction in Microalgae for Biodiesel Production. *Energies* **2012**, *5*, 1532–1553. [CrossRef]
4. Metzger, P.; Metzger, P.; Largeau, C.; Largeau, C. *Botryococcus braunii*: A rich source for hydrocarbons and related ether lipids. *Appl. Microbiol. Biotechnol.* **2005**, *66*, 486–496. [CrossRef] [PubMed]
5. Tran, N.H.; Bartlett, J.R.; Kannangara, G.S.K.; Milev, A.S.; Volk, H.; Wilson, M.A. Catalytic upgrading of biorefinery oil from micro-algae. *Fuel* **2010**, *89*, 265–274. [CrossRef]
6. Okada, S.; Murakami, M.; Yamaguchi, K. Hydrocarbon production by the yayoi, a new strain of the green microalga *Botryococcus braunii*. *Appl. Biochem. Biotechnol.* **1997**, *67*, 79–86. [CrossRef]
7. Largeau, C.; Casadevall, E.; Berkaloff, C.; Dhamelincourt, P. Sites of accumulation and composition of hydrocarbons in *Botryococcus braunii*. *Phytochemistry* **1980**, *19*, 1043–1051. [CrossRef]
8. Suzuki, R.; Ito, N.; Uno, Y.; Nishii, I.; Kagiwada, S.; Okada, S.; Noguchi, T. Transformation of Lipid Bodies Related to Hydrocarbon Accumulation in a Green Alga, *Botryococcus braunii* (Race B). *PLoS ONE* **2013**, *8*, e81626. [CrossRef]
9. Lee, A.K.; Lewis, D.M.; Ashman, P.J. Disruption of microalgal cells for the extraction of lipids for biofuels: Processes and specific energy requirements. *Biomass Bioenrgy* **2012**, *46*, 89–101. [CrossRef]
10. Zhang, S.; Hou, Y.; Liu, Z.; Ji, X.; Wu, D.; Wang, W.; Zhang, D.; Wang, W.; Chen, S.; Chen, F. Electro-fenton based technique to enhance cell harvest and lipid extraction from microalgae. *Energies* **2020**, *13*, 3813. [CrossRef]
11. Kanda, H.; Li, P.; Yoshimura, T.; Okada, S. Wet extraction of hydrocarbons from *Botryococcus braunii* by dimethyl ether as compared with dry extraction by hexane. *Fuel* **2013**, *105*, 535–539. [CrossRef]
12. Hidalgo, P.; Ciudad, G.; Navia, R. Evaluation of different solvent mixtures in esterifiable lipids extraction from microalgae Botryococcus braunii for biodiesel production. *Bioresour. Technol.* **2016**, *201*, 360–364. [CrossRef] [PubMed]

13. Mendes, R.L.; Nobre, B.P.; Cardoso, M.T.; Pereira, A.P.; Palavra, A.F. Supercritical carbon dioxide extraction of compounds with pharmaceutical importance from microalgae. *Inorg. Chim. Acta* **2003**, *356*, 328–334. [CrossRef]
14. Dote, Y.; Sawayama, S.; Inoue, S.; Minowa, T.; Yokoyama, S. Recovery of liquid fuel from hydrocarbon-rich microalgae by thermochemical liquefaction. *Fuel* **1994**, *73*, 1855–1857. [CrossRef]
15. Frenz, J.; Largeau, C.; Casadevall, E.; Kollerup, F.; Daugulis, A.J. Hydrocarbon recovery and biocompatibility of solvents for extraction from cultures of *Botryococcus braunii*. *Biotechnol. Bioeng.* **1989**, *34*, 755–762. [CrossRef]
16. Weiss, T.L.; Roth, R.; Goodson, C.; Vitha, S.; Black, I.; Azadi, P.; Rusch, J.; Holzenburg, A.; Devarenne, T.P.; Goodenough, U. Colony organization in the green alga *Botryococcus braunii* (Race B) is specified by a complex extracellular matrix. Eukaryot. *Cell* **2012**, *11*, 1424–1440.
17. Uno, Y.; Nishii, I.; Kagiwada, S.; Noguchi, T. Colony sheath formation is accompanied by shell formation and release in the green alga *Botryococcus braunii* (race B). *Algal Res.* **2015**, *8*, 214–223. [CrossRef]
18. Furuhashi, K.; Noguchi, T.; Okada, S.; Hasegawa, F.; Kaizu, Y.; Imou, K. The surface structure of *Botryococcus braunii* colony prevents the entry of extraction solvents into the colony interior. *Algal Res.* **2016**, *16*, 160–166. [CrossRef]
19. Saga, K.; Hasegawa, F.; Miyagi, S.; Atobe, S.; Okada, S.; Imou, K.; Osaka, N.; Yamagishi, T. Comparative evaluation of wet and dry processes for recovering hydrocarbon from *Botryococcus Braunii*. *Appl. Energy* **2015**, *141*, 90–95. [CrossRef]
20. Lee, J.; Yoo, C.; Jun, S.; Ahn, C.; Oh, H. Comparison of several methods for effective lipid extraction from microalgae. *Bioresour. Technol.* **2010**, *101*, S75–S77. [CrossRef]
21. Kita, K.; Okada, S.; Sekino, H.; Imou, K.; Yokoyama, S.; Amano, T. Thermal pre-treatment of wet microalgae harvest for efficient hydrocarbon recovery. *Appl. Energy* **2010**, *87*, 2420–2423. [CrossRef]
22. Furuhashi, K.; Hasegawa, F.; Saga, K.; Kudou, S.; Okada, S.; Kaizu, Y.; Imou, K. Effects of culture medium salinity on the hydrocarbon extractability, growth and morphology of *Botryococcus braunii*. *Biomass Bioenergy* **2016**, *91*, 83–90. [CrossRef]
23. Yoo, G.; Park, W.; Kim, C.W.; Choi, Y.; Yang, J. Direct lipid extraction from wet *Chlamydomonas reinhardtii* biomass using osmotic shock. *Bioresour. Technol.* **2012**, *123*, 717–722. [CrossRef] [PubMed]
24. Saga, K.; Magota, A.; Atobe, S.; Okada, S.; Imou, K.; Osaka, N.; Matsui, T. Hydrocarbon Recovery from Concentrated Algae Slurry via Thermal Pretreatment. *J. Jpn. Inst. Energy* **2013**, *92*, 1212–1217. [CrossRef]
25. Zhang, K.; Kojima, E. Effect of light intensity on colony size of microalga *Botryococcus braunii* in bubble column photobioreactors. *J. Ferment. Bioeng.* **1998**, *86*, 573–576. [CrossRef]
26. Tanoi, T.; Kawachi, M.; Watanabe, M.M. Iron and glucose effects on the morphology of *Botryococcus braunii* with assumption on the colony formation variability. *J. Appl. Phycol.* **2014**, *26*, 1–8. [CrossRef]
27. Ishizaki, R.; Noguchi, R.; Putra, A.S.; Ichikawa, S.; Ahamed, T.; Watanabe, M.M. Reduction in energy requirement and CO_2 emission for microalgae oil production using wastewater. *Energies* **2020**, *13*, 1641. [CrossRef]
28. Atobe, S.; Saga, K.; Hasegawa, F.; Magota, A.; Furuhashi, K.; Okada, S.; Suzuki, T.; Imou, K. The effect of the water-soluble polymer released from *Botryococcus braunii* Showa strain on solvent extraction of hydrocarbon. *J. Appl. Phycol.* **2015**, *27*, 755–761. [CrossRef]
29. Folch, J.; Lees, M.; Sloane-Stanley, G.H. A simple method for the isolation and purification of total lipids from animal tissues. *J. Biol. Chem.* **1957**, *226*, 497–509.
30. Moheimani, N.R.; Matsuura, H.; Watanabe, M.M.; Borowitzka, M.A. Non-destructive hydrocarbon extraction from *Botryococcus braunii* BOT-22 (race B). *J. Appl. Phycol.* **2014**, *26*, 1453–1463. [CrossRef]
31. Inoue, S.; Dote, Y.; Sawayama, S.; Minowa, T.; Ogi, T.; Yokoyama, S. Analysis of oil derived from liquefaction of *Botryococcus Braunii*. *Biomass Bioenergy* **1994**, *6*, 269–274. [CrossRef]
32. Aramaki, T.; Watanabe, M.M.; Nakajima, M.; Ichikawa, S. Bench-scale dehydration of a native microalgae culture by centrifugation, flocculation and filtration in Minamisoma city, Fukushima, Japan. *Bioresour. Technol. Rep.* **2020**, *10*, 100414. [CrossRef]
33. Wiedemann, T.; Stahl, W. Experimental investigation of the shrinkage and cracking behaviour of fine participate filter cakes. *Chem. Eng. Process.* **1996**, *35*, 35–42. [CrossRef]
34. Slade, R.; Bauen, A. Micro-algae cultivation for biofuels: Cost, energy balance, environmental impacts and future prospects. *Biomass Bioenergy* **2013**, *53*, 29–38. [CrossRef]

35. Das, S.; Das, S.; Das, I.; Ghangrekar, M.M. Application of bioelectrochemical systems for carbon dioxide sequestration and concomitant valuable recovery: A review. *Mater. Sci. Energy Technol.* **2019**, *2*, 687–696. [CrossRef]
36. Das, S.; Das, S.; Ghangrekar, M.M. Quorum-sensing mediated signals: A promising multi-functional modulators for separately enhancing algal yield and power generation in microbial fuel cell. *Bioresour. Technol.* **2019**, *294*, 122138. [CrossRef]

Article

Electro-Fenton Based Technique to Enhance Cell Harvest and Lipid Extraction from Microalgae

Shuai Zhang [1,2,†], Yuyong Hou [1,†], Zhiyong Liu [1], Xiang Ji [3], Di Wu [3], Weijie Wang [4], Dongyuan Zhang [1], Wenya Wang [5], Shulin Chen [2,*] and Fangjian Chen [1,*]

1 Tianjin Key Laboratory of Industrial Biosystems and Bioprocess Engineering, Tianjin Institute of Industrial Biotechnology, Chinese Academy of Sciences, Tianjin 300308, China; shuai.zhang5@email.wsu.edu (S.Z.); hou_yy@tib.cas.cn (Y.H.); liu_zy@tib.cas.cn (Z.L.); zhang_dy@tib.cas.cn (D.Z.)
2 Department of Biological Systems Engineering, Washington State University, Pullman, WA 99164-6120, USA
3 School of Life Science and Technology, Inner Mongolia University of Science and Technology, Baotou 014010, China; jixiang@imust.cn (X.J.); diwu@hainanu.edu.cn (D.W.)
4 College of Life Science, North China University of Science and Technology, Tangshan 063210, China; weijiewang@ncst.edu.cn
5 College of Life Science and Techenology, Beijing University of Chemical Technology, Beijing 100029, China; wangwy@mail.buct.edu.cn
* Correspondence: chens@wsu.edu (S.C.); chen_fj@tib.cas.cn (F.C.)
† These authors contributed equally to this work.

Received: 15 June 2020; Accepted: 23 July 2020; Published: 24 July 2020

Abstract: Currently, lipid extraction remains a major bottleneck in microalgae technology for biofuel production. In this study, an effective and easily controlled cell wall disruption method based on electro-Fenton reaction was used to enhance lipid extraction from the wet biomass of *Nannochloropsis oceanica* IMET1. The results showed that 1.27 mM of hydroxide radical (HO•) was generated under the optimal conditions with 9.1 mM $FeSO_4$ in a 16.4 mA·cm^{-2} current density for 37.0 min. After the electro-Fenton treatment, the neutral lipid extraction yield of microalgae (~155 mg) increased from 40% to 87.5%, equal to from 12.2% to 26.7% dry cell weight (DCW). In particular, the fatty acid composition remained stable. The cell wall disruption and lipid extraction processes were displayed by the transmission electron microscope (TEM) and fluorescence microscopy (FM) observations, respectively. Meanwhile, the removal efficiency of algal cells reached 85.2% within 2 h after the reaction was terminated. Furthermore, the biomass of the microalgae cultured in the electrolysis wastewater treated with fresh nutrients reached 3 g/L, which is 12-fold higher than that of the initial after 24 days. These finds provided an economic and efficient method for lipid extraction from wet microalgae, which could be easily controlled by current magnitude regulation.

Keywords: lipid extraction; electro-Fenton reaction; cell wall disruption; microalgae

1. Introduction

Microalgae is considered as a sustainable biomass source for producing biofuel, contributing to meet energy demands as well as addressing environmental concerns [1]. Driven by solar energy, microalgae can bio-fix CO_2 (4–14%, *v/v*) [2] into biomass, such as lipids, starch, proteins and pigments, thus abate greenhouse gas (GHG) emissions [3]. However, several economic and technical constraints limit the industrial applications of microalgae-based biofuel, including high costs of production and low lipid extraction yield [4,5]. As the most useful sources for biodiesel production, triacylglycerols are the major components of microalgae lipid, which make up 20–50% of the total lipid [6–11], and disperse in the cytoplasm, bounded by a rigid cell wall. However, the rigid cell wall makes extraction of the lipid a challenging process. Hence, effective extraction techniques are required to extract lipid from microalgae cells.

Currently, several methods have been investigated at the laboratory-scale for lipid extraction, such as organic solvents, supercritical CO_2, subcritical water extraction and milking [12–16]. Among these techniques, organic solvent extraction is a practical method of lipid extraction in industrial applications. Halim et al. and Lee et al. proposed that organic solvents extract lipid from the microalgae cell by diffusing into cytoplasm and dissolved most lipids [16,17]. Lipids undergo counter-diffusion through microalgae cell to the bulk solvent for the downstream process [18]. Organic solvent extraction techniques include both dry and wet routes, depending on the water content of the microalgae. The dry route requires over 90% of the water removal in the microalgae, water evaporation takes up nearly 90.5% of the total energy consumption in the whole process [19,20]. Furthermore, the potential fossil energy ratio (FER) of the wet and dry routes, which shows the ratio of renewable energy input to fossil energy output [20], reaches 1.82 and 2.38, respectively. This indicates that the wet route has higher energy efficiency and more feasibility for piloting [21]. However, cell wall disruption becomes the bottleneck for lipid liberation and extraction of the wet route [22].

The algal cell wall is constructed of complex carbohydrates and glycoproteins and has high mechanical strength and chemical resistance. Pretreatments with mechanical and non-mechanical techniques are usually required for the cell wall disruption. The former typically includes high-pressure homogenization (HPH), high-speed homogenization (HSH), hydrodynamic cavitation, microwaving, and ultrasonication (USN) [23–31]. These processes destroy the rigid cell wall with strong external forces, which causes high energy concerns [32,33]. Grimi et al. reported energy consumption ranging from 12 $kJ \cdot kg^{-1}$ dry weight (DW) for USN treatment of 1.5 $MJ \cdot kg^{-1}$ DW for HPH process in *Nannochloropsis sp.* cell disruption [32]. On the other hand, non-mechanical methods are known for lower energy consumption, with direct physical energy transfer and additional chemical reactions for cell disruption [23]. Non-mechanical methods include acid hydrolysis [34], ionic liquid extraction [35–37], steam explosion [38], and more. However, further challenges include continuous chemical supplementation, lipid degradation and waste liquor treatment, limiting their commercial applications [23,24].

To address these issues, recent research has focused on making the process more cost-effective and controllable, reducing energy consumption below the algae's potential energy storage of 21 $kJ \cdot g^{-1}$, as well as improving the mildness and adaptability of the process and recoverability of products [23,24]. Fenton cell disruption techniques represent a new approach that effectively disrupts the microalgae cell wall. Additionally, the Fenton treatment is a promising technique in effluent disposal. In fact, it can remove over 90% of the chemical oxygen demand (COD) in combined industrial and domestic wastewater for water recycling [39]. These reduce the costs of equipment, maintenance and wastewater treatment, as well as energy consumption. In the Fenton reaction, high oxidative species, HO•, forms through iron catalysis of H_2O_2 decomposition. They have the ability to efficiently degrade microalgae cell walls (3~5 min) in mild temperatures and atmospheric pressures [40,41]. Moreover, iron catalysts needed in Fenton reactions can also be used for microalgae harvest [35]. However, continuous H_2O_2 supplement remains a bottleneck of the Fenton treatment due to safety concerns in the handling and shipping process and higher operation costs [42,43]. Meanwhile, difficult controllability of the Fenton reaction is also a major challenge for industrial applications. The excessive reaction usually leads to lipid degradation and markedly decreases the lipid extraction yield. The common ways to solve this problem are to add organic solvents and inhibitors in order to terminate the reaction [40,41], further raising costs [41,44].

It is highly desirable to develop an effective cell wall breaking method under mild conditions which could be regulated quickly and accurately. Moreover, the intracellular fatty acid profile remains stable. As an electrically driven reaction, the electro-Fenton reaction seems to be able to gradually destroy the cell wall by current controlling till the desired breaking effect is obtained. Thus, this study aimed to explore the use of the electro-Fenton reaction as an alternative technique for microalgae cell wall disruption. In this process, H_2O_2 is continuously generated on the surface of the cathode through cathodic oxygen reduction, which addresses the continuous supplementation of H_2O_2 as well as safety concerns [45]. Furthermore, the electro-Fenton process can be terminated through regulating current and electrolysis time with convenient operation. For this study, we examined the effects of different parameters (e.g., $FeSO_4$ concentration, current density and time) on disruption of microalgae cell wall with response surface methodology (RSM). We used this models in the optimization process to identify the optimal conditions for lipid extraction. Then, the amount of hydroxyl radical, distribution of iron, monitored changes in the cell wall and lipid droplets were measured by High Performance Liquid Chromatography (HPLC), spectrophotometry, a transmission electron microscope (TEM) and a fluorescence microscope (FM), respectively. Subsequently, to further identify the quality of the extracted lipid, gas chromatography analysis was applied. Finally, the effect of the wastewater directly from electrolysis on cell growth was demonstrated.

2. Materials and Methods

2.1. Strain and Culture Conditions

N. oceanica IMET1, the strain with high triacylglycerol (TAG) and eicosapentaenoic acid (EPA) content, was considered as promising feedstock for microalgal biodiesel production and kindly provided by Dr. Yubin Ma from the Qingdao Institute of Bioenergy and Bioprocess Technology, Chinese Academy of Sciences. Seed cultures were maintained in 250 mL Erlenmeyer flasks with 150 mL of seawater BG-11 medium under continuous artificial illumination. In the experiments, 50 mL logarithmic phase seed cells and 350 mL fresh medium were transferred to glass columns (4.1 cm in diameter, 37 cm in height) with 2% (v/v) CO_2. Aeration flow for each column was 10 mL·min^{-1}. Light intensity and temperature were maintained at 50 μM·m^{-2}·s^{-1} and 25 °C with an initial pH of 7.2, respectively. For the dry cell weight (DCW) measurement, the microalgal culture samples (10 mL) were filtered by pre-weighed 0.8 μm microporous filter papers, and then dried overnight at 105 °C. DCW was calculated using the difference between the final and beginning weights of the filter papers [46].

2.2. Neutral Lipid Extraction and Fatty Acid Analysis

The microalgal total neutral lipid and lipid extraction yield from electrolysis were obtained based on Steriti et al.'s and Fajardo et al.'s methods [41,47], respectively. The yield of extracted neutral lipids was calculated using the ratio between the weight of extracted lipids and the weight of total lipids in the microalgae. For the fatty acid methyl ester (FAME) reaction, 1 mL of the KOH-methanol solution (0.5 M) with nonadecanoic acid (0.1 mg·mL^{-1}) was added to the lipid extract at 65 °C for 15 min, and shaken every 5 min. Next, 2 mL of 14% BF3-methanol solution was added at 65 °C for 2 min with continuous shaking. Subsequently, 2 mL of hexane and 1 mL of saturated NaCl solution was employed, and the FAMEs were dissolved in the hexane layer. For reproducibility, triplicates were carried out in each experimental condition.

FAME analysis was carried out using GC (GC2010, Shimadzu, Kyoto, Japan) equipped with a FID detector and a SP-2560 column (100 m × 0.25 mm × 0.2 μm, Supelco, Bellefonte, PA, USA), based on Zhang et al. [48]. An amount of 1 μL methylated sample was injected at 250 °C. The column temperature was initially set at 165 °C and held for 5 min. Subsequently, it was programmed to 180 °C at the rate of 5 min^{-1} and held for 5 min. Next, the temperature was increased from 180 °C to 240 °C at the rate of 5 min^{-1} and held for another 5 min. The temperature of FID detector remained at 260 °C. The flow rate of hydrogen and air were supplied at 40 and 400 mL·min^{-1}, respectively.

2.3. Cell Disruption

Microalgae cells with 26 days cultivation were chosen as the material for electro-Fenton reaction treatment. Amounts of 50 mL cell samples were added into 100 mL glass beakers along with different amounts of $FeSO_4$. Then, two graphite electrodes were dipped into the medium (3 cm^2 effective area) and the distance between them was situated at 2 cm. A magnetic stirrer (350–400 rpm) was employed to enhance mass transfer in the electro-Fenton reaction. Compressed oxygen (2 L·min^{-1}) was fed to the cathode for H_2O_2 production. Finally, electro-Fenton reaction was terminated by turning off the electricity. After electrolysis, microalgae precipitate was collected for lipid extraction according to the Section 2.2.

2.4. Optimization of Electro-Fenton Conditions

Several parameters affecting disruption of microalgae cell wall were evaluated through single-factor experiments by calculating the lipid extraction yield, including electrolysis time, $FeSO_4$ concentration and current density. Triplicate runs were conducted for each combination: $FeSO_4$ (5–15 mM), current density (5–30 mA·cm^{-2}) and electrolysis time (0–60 min). Design expert (version 8.0.6) was used to design the experiments and analyze the data. Experimental designs of the electro-Fenton conditions are shown in Table 1.

Table 1. Experimental design matrix for the optimization of electro-Fenton conditions.

Std. Order	Run Order	Fe^{2+} Conc. (mM)	Current Density (mA·cm^{-2})	Time (min)	Extraction Yield (wt%)
17	1	10.00	17.50	30.00	84.00
11	2	10.00	5.00	60.00	64.97
6	3	15.00	17.50	0.00	43.31
3	4	5.00	30.00	30.00	58.41
10	5	10.00	30.00	0.00	40.36
12	6	10.00	30.00	60.00	54.80
13	7	10.00	17.50	30.00	81.71
16	8	10.00	17.50	30.00	84.99
4	9	15.00	30.00	30.00	46.60
15	10	10.00	17.50	30.00	88.60
14	11	10.00	17.50	30.00	85.64
5	12	5.00	17.50	0.00	46.27
7	13	5.00	17.50	60.00	66.28
1	14	5.00	5.00	30.00	57.75
2	15	15.00	5.00	30.00	56.11
9	16	10.00	5.00	0.00	41.67
8	17	15.00	17.50	60.00	46.92

2.5. Hydroxyl Radical Detection

The amount of HO• formation in the electro-Fenton reaction was monitored by the modified salicylic acid method [49]. The experimental apparatus was the same as that in Section 2.3. Based on the results of optimization, the electro-Fenton reaction was carried out at 9.1 mM $FeSO_4$, 16.4 mA·cm^{-2} current density, 2 L·min^{-1} oxygen feeding at 350–400 rpm in room temperature. In addition, 5 mL salicylic acid-ethanol solution (90 mM) was added into the electro-Fenton reactor, which also contains a 40 mL BG11 culture medium, and the pH of the solution was adjusted to 6.0, which was the same as the final pH value of microalgae medium in electrolysis. For HO• detection, samples were collected at 0, 5, 15, 37 and 60 min, respectively. They were filled through a 0.22 μm luer syringe filter before analysis with HPLC (Shimadzu, LC-20AD). HPLC analysis was performed on an ODS-BD C-18 column (Sinochrom, 4.6 × 260 mm, 5.0 μm) at 35 °C. In the mobile phase, an acetic acid–water–methanol solution (1:79:20) was passed through the column with a flow rate of 1 mL·min^{-1}. The concentrations of salicylic acid and its derivatives were measured with a UV detector at 296 nm. The injection volume was 10 μL each time [49].

2.6. Iron Detection

The iron concentrations in the supernatant and microalgae precipitation were detected with the modified 1,10-phenanthroline method [50]. An amount of 1 mL of microalgae samples collected at day 26 were subjected to electrolysis treatment at the optimized conditions of 0, 5, 15, 37 and 60 min, respectively. The cells without the addition of $FeSO_4$ were used as a control. All experiments were performed in triplicate. Then the experimental cells underwent flocculation for 30 min after electrolysis. Next, the supernatant and precipitate were collected. Before iron detection, each precipitate was added in 50 μL concentrated HCl (37%, v/v), and 1 mL of water mixed by ultrasonic treatment for 5 min to dissolve the precipitated iron. In the detection, 2.5 mL ammonium acetate buffer (mixing 40 g ammonium acetate, 50 mL glacial acetic acid and water to 100 mL) was added into each supernatant and mixture. Then, 1.25 mL hydroxylamine hydrochloride solution (10%) and 2.5 mL 1, 10-phenanthroline (0.5%, m/v) were used to chelate the iron. Sequentially, the volume of solution was supplied to 25 mL with e-pure water. After 10 min, the color development reaction was complete. Finally, samples were measured with a spectrophotometer at 510 nm. The iron detection standard curve was required in each batch.

2.7. Electron Microscopy and Fluorescence Microscope Analysis

To observe the cell wall disruption process, electro-Fenton treatment ranging from 0 to 60 min were performed, respectively. Transmission electron microscopic (TEM) studies were carried out with TEM (JEM-1400; JEOL, Tokyo, Japan) according to Hou et al. [46]. To elucidate the lipid extraction process under electro-Fenton system, the microalgae cells stained by Nile red [51,52] were examined with a fluorescence spectrometer (Leica DM5000B microscope) with BP 516–560 nm excitation filter, 580 nm dichoic mirror and LP 590 nm emission filter (Leica filter cube N2.1) at 25 °C [53].

2.8. Reuse of Electro-Fenton Treated Microalgal Culture Medium for Subsequent Cultivation of Algae

After the reaction was terminated, the cells were settled in tubes for 2 h, the flocculation efficiency was measured by OD680 at different time (*t*) during the settling process [54]. Flocculation efficiency $= (1 - OD_t/OD_0) \times 100\%$, where OD_t is the optical density of the cells at 2 h and OD_0 is the initial optical density of the culture for the settling process. For comprehensive utilization of wastewater from the electro-Fenton reaction, culture medium recycling was carried out. The resulting supernatant was centrifuged for further cultivation with replenished fresh BG-11 nutrients, except for $FeSO_4$. Cultivation conditions were consistent with methods described in Section 2.1. DCW was used to evaluate the effects of wastewater on cell growth.

3. Results and Discussion

3.1. Growth Properties

To get the high biomass and lipid accumulation, we measured the cell growth curve using dry weight analysis, as shown in the Figure 1A. Results showed that this strain reached the stationary stage, with a biomass concentration of 3.10 $g{\cdot}L^{-1}$ after 26 days, and a neutral lipid content of 30.47% (wt/wt, DCW). Before reaching at stationary stage, microalgae began to turn yellow on day 22 (data not shown). This indicated that lipid accumulated gradually in the cells, which was in accordance with with Ma's results [6]. To evaluate the effects of the electro-Fenton treatment on cell disruption, we need to keep microalgae alive and all the organelles intact, considering that the lack of nutrients and accumulation of toxic catabolites induced cell apoptosis. Thus, the cells cultured to the 26th day were collected for further analysis.

3.2. The Effects of Different Factors on Electro-Fenton Treatment for Lipid Extraction

In this study, we employed the electro-Fenton process for the simultaneous cell disruption and harvest of microalgae as wet biomass. Theoretically, oxygen was directly reduced on the surface of cathodes driven by electricity for H_2O_2 production. Next, the resulting H_2O_2 was catalyzed by Fe^{2+} to produce HO• for microalgae cell wall degradation through the Fenton reaction. Furthermore, Fe^{3+} kept obtaining electrons on the cathode, which inhibited the side reaction of oxygen generation. The related mechanism is shown as follows [42]:

$$O_2 + e^- + 2H + \rightarrow H_2O_2 \tag{1}$$

$$Fe^{2+} + H_2O_2 \rightarrow Fe^{3+} + OH^- + HO\bullet \quad (2)$$

$$Fe^{3+} + e^- \rightarrow Fe^{2+} \quad (3)$$

Side reaction in the Fenton process:

$$Fe^{3+} + H_2O_2 \rightarrow Fe^{2+} + 2H^+ + O_2 \quad (4)$$

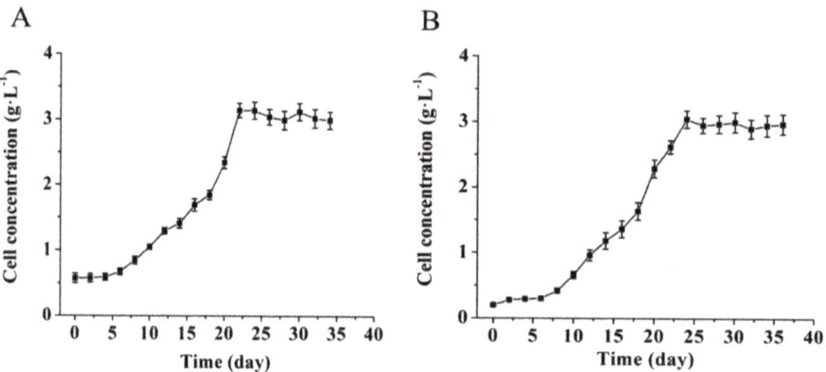

Figure 1. The growth curve of *N. oceanica* IMET1 (**A**. fresh seawater BG-11; **B**. electro-Fenton resulting wastewater).

As the main factors, $FeSO_4$ concentration, current density and electrolysis time thus largely affected lipid extraction yield. The optimum experimental conditions were 10 mM $FeSO_4$, 17.5 mA·cm^{-2} current density and 30 min (Std. order 15 in Table 1). Lipid extraction yield reached the maximum 88.6% (wt%, total neutral lipid). In other electro-Fenton runs, parameters with either lower or higher value decreased the lipid extraction yield. In this process, three factors played important roles in lipid extraction by varying the amount of HO• as well as its contact time with microalgae. Pimentel et al. (2008) found that Fe^{2+} concentration and current density were directly related to the HO• production rate [55]. In addition, Le et al. and Steriti et al. found that the contact time between HO• and microalgae largely affected cell wall degradation and lipid extraction yield [41,56]. In the Fenton reaction treatment, insufficient HO• could not efficiently achieve cell wall disruption. However, excessive HO• caused lipid consumption. Thus, the method of HO• production must be strictly controlled. In this study, we precisely controlled the electro-Fenton reaction through electricity regulation. The lipid extraction yield of untreated *N. oceanica* IMET1 was about 40% (%, total neutral lipid), which was much lower than the electro-treated ones. Therefore, the electro-Fenton treatment played an important role in enhancing lipid extraction.

3.3. Confirmation of Predicted Optimum Condition

The predictive quadratic model for lipid extraction yield were based on actual parameters as follows:

$$Y = -37.83889 + 12.64796 \times X_1 + 3.79849 \times X_2 + 1.91768 \times X_3 - 0.040689 \times X_1 \times X_2 - 0.027345 \times X_1 \times X_3 - 0.00590643 \times X_2 \times X_3 - 0.60049 \times X_1{}^2 - 0.097653 \times X_2{}^2 - 0.021420 \times X_3{}^2;$$

where Y, X_1, X_2 and X_3 were lipid extraction yield (wt%, total neutral lipid), $FeSO_4$ concentration (mM), current density (mA·cm^{-2}) and time (min), respectively. The predicted maximum lipid extraction yield calculated by the predictive model was 86.33% with the optimum conditions, including 9.1 mM $FeSO_4$, 16.4 mA·cm^{-2} current density and 37.0 min electrolysis (Figure 2). This model described the experimental data well. The lipid extraction yield verified under the predicted optimum conditions was 87.53%, which was close to the predicted lipid extraction yield. In the optimal conditions, electro-Fenton enhanced the yield from 40% to 87.53%, equal to from 12.2% (wt%, DCW) to 26.7% (wt%, DCW), which is 16.1% and 53.4% higher than that in Ti_4O_7-based membrane anodic oxidation (23.0 wt%, DCW) and H_2O_2 with the $FeSO_4$-based (17.4% wt%, DCW) method, respectively [41,57]. Meanwhile, compared to the traditional Fenton reaction, the amount of $FeSO_4$ decreased from 24 mM to 9.1 mM and required no additional H_2O_2 [40,41]. This makes the pretreatment free from safety concerns of H_2O_2 in handling and transportation. As a result, electro-Fenton based way seems to be more economically for lipid extraction from wet microalgae.

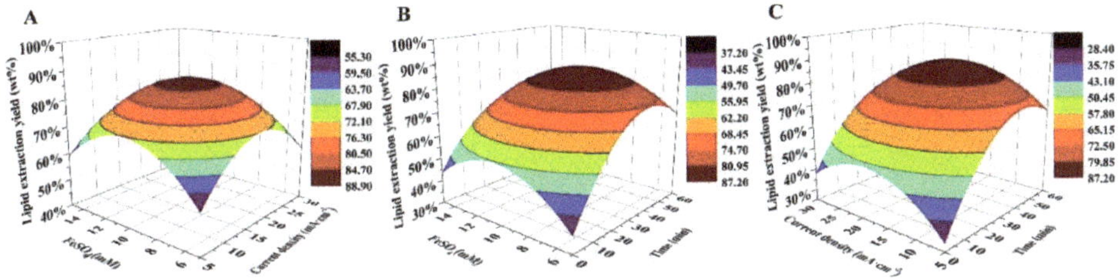

Figure 2. Respond surface curve standing for the interactive effects of $FeSO_4$ concentration, current density and time on the lipid extraction yield: (**A**) effect of $FeSO_4$ concentration and current density; (**B**) effect of $FeSO_4$ and time; (**C**) effect of current density and time.

3.4. Fatty Acid (FA) Composition Analysis during Electro-Fenton Reaction

The quality of the extracted lipid was examined by FAMEs analysis. Previous studies showed that the FAs composition significantly changed under both the $Fe_2(SO_4)_3$-based method and H_2O_2 treatment with or without $FeSO_4$ [35,40,41]. In this study, the identified 15 individual FAMEs were shown in Figure 3. In the control group, the major fatty acids (FAs) of IMET1 were palmitic acid (C16:0) and hexadecenoic acid (C16:1), which accounted for 32.5% and 26.2% of the total FAs, respectively. The third most abundant was oleic acid (C18:1), accounting for 22.9%. Notably, compared with FAs composition in the control, there were no obvious changes in the electro-Fenton treatment group after 1h. This was consistent with a previous report that monounsaturated (MUFA) and saturated fatty acids (SFA) were more resistant to HO• than polyunsaturated fatty acids (PUFA), due to their fewer double bonds [58]. The SFA and MUFA, the major contents in the IMET1, reached 40.5% and 49.4%, respectively. This made it difficult for HO• to change the FA composition through peroxidation and degradation [59,60]. High content of these FAs made IMET1 was suitable to this technology for algae cell wall disruption.

3.5. Microalgae Cell Disruption Mechanism in the Electro-Fenton Reaction

3.5.1. Hydroxyl Radical Generation in the Electro-Fenton Reaction

HO• was continuously produced during the electro-Fenton process. The concentration of HO• was calculated according to this theory that one equivalent of HO• drove one equivalent of hydroxyl group addition on the salicylic acid [49]. Figure 4 presents the line plots for salicylic acid-captured HO• products vs. electrolysis time. The production rates of 2,5-dHBA and 2,3-dHBA were initially kept at 15.76 $\mu M{\cdot}min^{-1}$ and 7.65 $\mu M{\cdot}min^{-1}$, respectively. After 15min, they started to decrease. At the optimal time (37 min), 1.27 mM HO• was formed in the electro-Fenton process. Interestingly, the results showed no obvious accumulation of these products when electricity was turned off (Figure 4). This indicated that the production of HO• could be effectively terminated through electricity regulation. Taking these into account, the electro-Fenton reaction overcomes several problems remaining in the Fenton reactions: first, H_2O_2 was sustainably produced on-site by oxygen feeding, eliminating the need for acquisition, shipment and storage of H_2O_2 [43], and might further decrease treatment costs. Second, the reaction termination and HO• production could be artificially regulated by using electricity without additional inhibitors consumption [40]. This could effectively balance microalgae cell wall degradation and lipid oxidation in the Fenton-based wet cell disruption route. Third, the electro-Fenton reaction decreased co-reactions causing by Ferric ions and increased the efficiency of HO• production. This indicated that the production of HO• could be effectively terminated through electricity regulation.

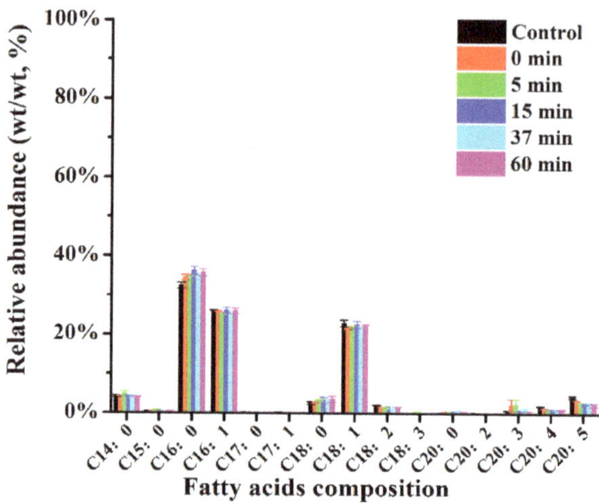

Figure 3. Effect of electrolysis time on the composition of extracted fatty acids from *N. oceanica* IMET1.

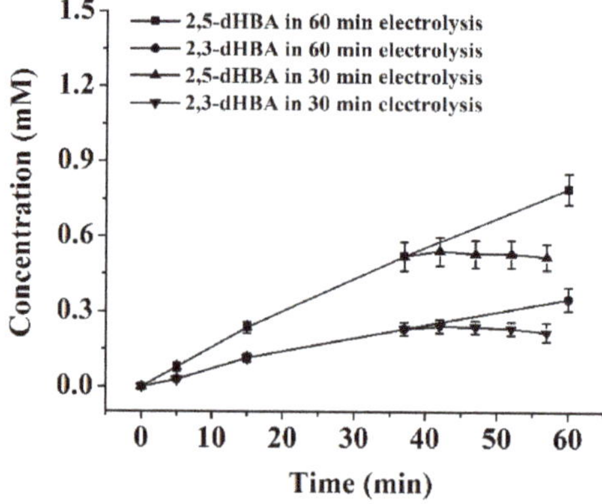

Figure 4. 2, 3-dHBA and 2, 5-dHBA production versus time in electro-Fenton process.

3.5.2. TEM for Microalgae Cell Wall and Cytomembrane Degradation in Electro-Fenton Reaction

The microalgae cell structure was gradually degraded in the electrolysis (Figure 5). Compared with untreated microalgae cells (Figure 5a,b), iron ions were clearly absorbed on the surface of the cell wall, and further diffused to the cytomembrane, even including the lipid body in the $FeSO_4$ group (Figure 5c). However, these microalgae cell structures were kept intact without degradation by the absorbed iron in the 1h treatment (Figure 5d). In addition, we used electrolysis to treat microalgae cells for 15 min. It was discovered that most of microalgae were dead, and some organelles, such as chloroplast, were destroyed (Figure 5e). At the same time, a pore formed by HO• degradation on the cytomembrane (Figure 5f). We presumed that the thickness of cell wall was thicker than that of the cytomembrane, which firstly caused cytomembrane degradation. However, cell wall disruption was the true bottleneck in microalgae lipid wet extraction [12,14]. We found that a 15 min electro-Fenton reaction did not destroy the cell wall (Figure 5d), which was consistent with our previous results showing that 15 min was not the optimized electrolysis time for lipid extraction. In addition, when the electrolysis time reached 37 min, most of the organelles, including the lipid body, were destroyed (Figure 5g). Figure 5h depicted a pore that appeared on the cell wall. This suggested that cell wall disruption occurred in 37 min. Cell wall degradation changed the shape of microalgae cell and resulted in a higher lipid extraction yield. Finally, most of the microalgae cell structure was completely destroyed after 1h of electro-Fenton treatment (Figure 5i). The cell wall and cytomembrane were completely disrupted, and the intracellular contents leaked out (Figure 5j). In the meantime, the lipid was directly exposed to electro-Fenton agents as it diffused into the electrolyte, which caused lipid degradation by HO•. These results were consistent with Steriti et al.'s report [41]. This meant that the disruption degree of the cell wall could be regulated by electrolysis time, which aided the sequential extraction of various products from the microalgal cell.

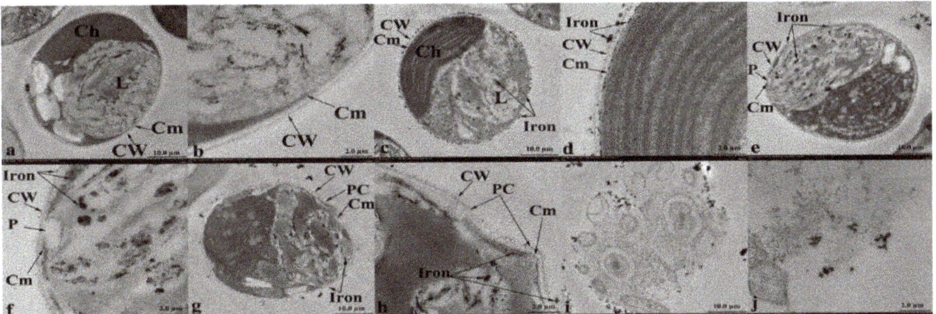

Figure 5. Transmission electron microscope (TEM) image of microalgae. (**a**) fresh microalgae cell without any treatment, (**b**) parts of fresh microalgae cell without any treatment, (**c**) microalgae cell with $FeSO_4$ addition, (**d**) parts of microalgae cell with $FeSO_4$ addition, (**e**) microalgae cell with 15 min electro-Fenton treatment, (**f**) microalgae cell with 15 min electro-Fenton treatment, (**g**) microalgae cell with 37 min electro-Fenton treatment, (**h**) parts of microalgae cell with 30 min electro-Fenton treatment, (**i**) microalgae cell with 1 h electro-Fenton treatment, (**j**) parts of microalgae cell with 1 h electro-Fenton treatment.

3.5.3. Fluorescence Microscope (FM) for Microalgae Observation during Electro-Fenton Reaction

Interestingly, microalgae cells showed obvious coagulation with iron addition, resulting in an efficient sedimentation. In contrast, the untreated microalgae cells dispersed in the culture medium and were separated from each other (Figure 6a). The addition of $FeSO_4$ caused cell aggregation, and the cells were still complete (Figure 6b). The red fluorescence used to track lipid position did not show obvious changes during 1h of $FeSO_4$ treatment. This indicated that cell and lipid degradation did not occur during $FeSO_4$ treatment. Considering that the cost of harvesting is estimated as 20–30% of the total production costs, the treatment might make it more economical for biofuel production. In the next experiment, we used the electro-Fenton reaction to disrupt microalgae cells. The red fluorescence appeared on the whole microalgae cells, which meant that the lipid body was disrupted during the 15 min electro-Fenton reaction (Figure 6c). It was probable that the lipid body membranes absorbed the iron ions (Figure 5e), which catalyzed the resulting H_2O_2 from electrolysis to produce HO• on site and destroyed the membranes. When the reaction time was lengthened to 37 min, red fluorescence began to appear outside of the cells, as shown in the FM image (Figure 6d). This was consistent with our previous results that the cell wall was disrupted (Figure 5h). However, a one-hour electro-Fenton reaction disrupted most cells, and their fragments aggregated together (Figure 6e). A large amount of red fluorescence appeared among the resulting fragments with low intensity. This confirmed that HO• was able to completely destroy the microalgae cell structure. The leaked lipids were constantly exposed to HO• and consumed through oxidation reaction. Luckily, the reaction was accurately controlled, the high flocculation rate, high lipid extraction and low cost was thus obtained easily by the manual controls.

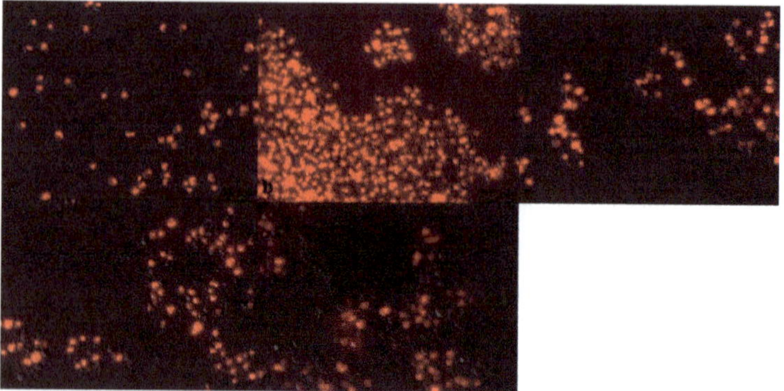

Figure 6. Fluorescence microscopy (FM) image of microalgae. (**a**) Fresh microalgae cell without any treatment, (**b**) microalgae cell with $FeSO_4$ addition, (**c**) microalgae cell with 15 min electro-Fenton treatment, (**d**) microalgae cell with 37 min electro-Fenton treatment, (**e**) microalgae cell with 1 h electro-Fenton treatment.

3.6. Iron Distribution

In the electrolysis process, iron functioned as the catalyst that induced HO• production [55], and also as the coagulant that enhanced the microalgae harvest through microalgae surficial charge neutralization [35,61]. Owning to these functions, iron distribution was further investigated to explain the trend of HO• production and provide the data for iron recovery in the downstream process. As shown in Figure 7a, iron concentration in the supernatant decreased from 322 $mg{\cdot}L^{-1}$ to 22 $mg{\cdot}L^{-1}$, while it increased from 179 $mg{\cdot}L^{-1}$ to 320 $mg{\cdot}L^{-1}$ in the precipitate during 60 min electrolysis. In addition, part of the iron was absorbed on the surface of the cathode through iron reduction. Figure 7b indicates that the iron abundance of supernatant during 60 min electrolysis decreased from 64% to 4%. However, the iron abundance of precipitate and electrodes increased from 33% to 63% and 2% to 33%, respectively. These results indicate that there were three main stages of iron distribution. In the first stage, about 33% of the iron was rapidly adhered on the microalgae through absorption, and 64% of that remained as free ions in the electrolyte. Metal ions could neutralize the negative charges distributed on the surface of the microalgae cell wall [61]. This was also confirmed by the fact that iron dispersed on the cell wall, cytomembrane and lipid droplets of microalgae (Figure 5c). In the second stage, 18% of additional iron attached to the microalgae during 5 min of electrolysis. It appears that the chemicals, such as $Fe(OH)_2$ and $Fe(OH)_3$, which formed during electrolysis, enhanced cell coagulation. In the third stage, iron began to accumulate on the surface of the cathode by iron reduction and continuously absorbed on the microalgae. As a result, the decreased contact area between cathode and oxygen would inhibit the electrolysis efficiency and cell disruption. Furthermore, more than 60% of iron was removed from the supernatant, prompting us to recover iron from precipitate residues after lipid extraction. On the other hand, $FeSO_4$ also functioned as a precipitant for microalgae co-flocculation during electrolysis, the self-flocculation efficiency reached 85.2% after the reaction was terminated (data not shown).

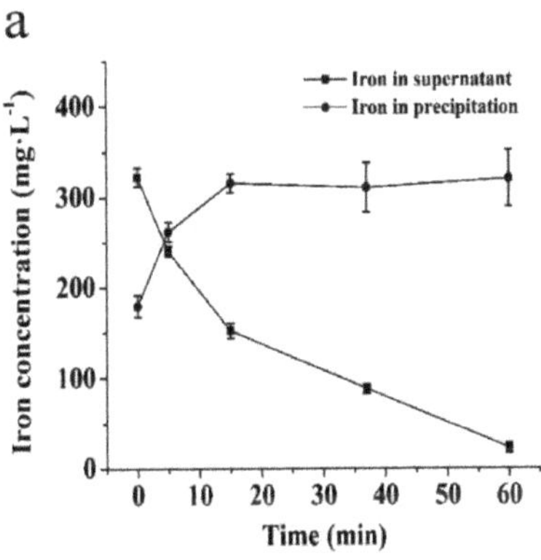

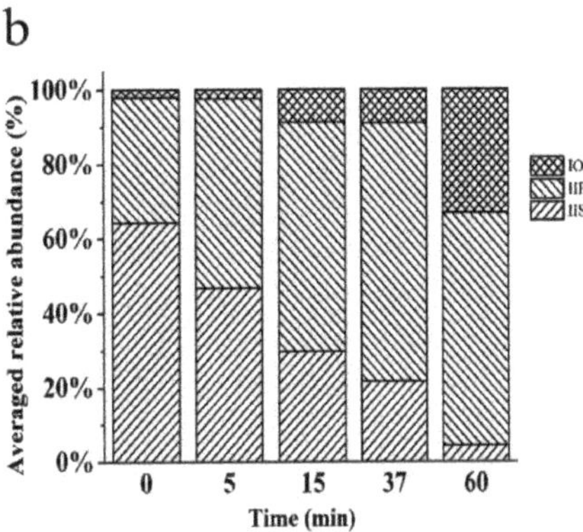

Figure 7. Iron distribution in the supernatant and precipitate (**a**). Averaged relative abundance of iron in each phase (**b**). IOE stands for iron abundance on the surface of electrode. IIP stands for iron abundance in the precipitate. IIS stands for iron abundance in the supernatant.

3.7. Microalgae Cultivation in the Electro-Fenton Wastewater

The electro-Fenton process is well-known as a kind of advanced oxidation process (AOP) in wastewater treatment. We discovered that microalgae could grow well in the resulting wastewater. Figure 1B displays the growth curve of IMET1 in the wastewater. The results show that microalgae could grow in the wastewater at up to 3.0 g·L^{-1} in 24 days cultivation. After that, microalgae reached a stationary stage without obvious biomass accumulation. The growth curve of microalgae cultured with wastewater was similar to fresh medium. This indicated that the quality of wastewater met the requirements for IMET1 growth and did not negatively affect the growth of microalgae.

4. Conclusions

This study investigated a novel microalgae cell disruption technique for *N. oceanica* IMET1 lipid extraction based on the electro-Fenton reaction. The findings showed that the yield of lipid extraction depended on current density, Fe^{2+} concentration and electrolysis time. With the optimal conditions of 16.4 mA·cm^{-2} current density, 9.1 mM Fe^{2+} and 37.0 min electrolysis time, 87.53% of the lipid was extracted (%, total neutral lipids), and there were no obvious changes in intracellular components. After electrolysis, over 60% of the total iron distributed in the microalgae precipitation and on the surface of the electrode, which could be used for iron recovery and contribute towards cell harvest (85.2%). Furthermore, the resulting wastewater containing 88 mg·L^{-1} iron ions met the quality requirements for microalgae cultivation, and no inhibitors clearly affected microalgae growth. This electrochemistry technology holds promise as an efficient approach for microalgae biofuel extraction.

Author Contributions: Concenptualization, S.Z. and Y.H.; Data curation, D.Z.; Formal analysis, S.Z., Z.L. and W.W. (Weijie Wang); Fungding acquisition, F.C.; Investigation, S.Z., Y.H., D.W. and F.C.; Methodology, S.Z. and Y.H.; Project administration, S.C. and F.C.; Software, W.W. (Weijie Wang) and W.W. (Wenya Wang); Supervision, X.J. and W.W. (Wenya Wang); Validation, Z.L. and X.J.; Visualizaiton, D.W. and D.Z.; Writing-original draft, S.Z. and Y.H.; Writing-review & editing, Y.H., S.C. and F.C. All authors have read and agreed to the published version of the manuscript.

Funding: This work was supported partly by the National Key R&D Program of China (2019YFA090460203, 2018YFE0107200, 2016YFD0501405), the Tianjin Synthetic Biotechnology Innovation Capacity Improvement Project (TSBICIP-KJGG-004-05), the National Science Foundation of the United States (Grant No. 1231085), the Key Program for International Cooperation Projects of Sino-Canada (Grant No. 155112KYSB20160030), and the National Natural Science Foundation of China (Grant No. 31570047).

Acknowledgments: The authors thank Dr. Huanhuan Zhai (Tianjin Institute of Industrial Biotechnology, Chinese Academy of Sciences) for analysis of transmission electron microscope (TEM) observations and Dr. Lixian Wang

(Tianjin Institute of Industrial Biotechnology, Chinese Academy of Sciences) for analysis of fluorescence microscopy (FM) observations.

Conflicts of Interest: The authors declare no conflict of interest.

References

1. Demirbas, M.F. Biorefineries for biofuel upgrading: A critical review. *Appl. Energy* **2009**, *86*, S151–S161. [CrossRef]
2. Moreira, D.; Pires, J.C. Atmospheric CO_2 capture by algae: Negative carbon dioxide emission path. *Bioresour. Technol.* **2016**, *215*, 371–379. [CrossRef] [PubMed]
3. Mata, T.M.; Martins, A.A.; Caetano, N.S. Microalgae for biodiesel production and other applications: A review. *Renew. Sustain. Energy Rev.* **2010**, *14*, 217–232. [CrossRef]
4. Concas, A.; Lutzu, G.A.; Pisu, M.; Cao, G. Experimental analysis and novel modeling of semi-batch photobioreactors operated with *Chlorella vulgaris* and fed with 100% (*v*/*v*) CO_2. *Chem. Eng. J.* **2012**, *213*, 203–213. [CrossRef]
5. Concas, A.; Steriti, A.; Pisu, M.; Cao, G. Comprehensive modeling and investigation of the effect of iron on the growth rate and lipid accumulation of *Chlorella vulgaris* cultured in batch photobioreactors. *Bioresour. Technol.* **2014**, *153*, 340–350. [CrossRef]
6. Ma, Y.; Wang, Z.; Yu, C.; Yin, Y.; Zhou, G. Evaluation of the potential of 9 *Nannochloropsis* strains for biodiesel production. *Bioresour. Technol.* **2014**, *167*, 503–509. [CrossRef]
7. Gonçalves, A.L.; Rodrigues, C.M.; Pires, J.C.M.; Simões, M. The effect of increasing CO_2 concentrations on its capture, biomass production and wastewater bioremediation by microalgae and cyanobacteria. *Algal Res.* **2016**, *14*, 127–136. [CrossRef]
8. Wahlen, B.D.; Willis, R.M.; Seefeldt, L.C. Biodiesel production by simultaneous extraction and conversion of total lipids from microalgae, cyanobacteria, and wild mixed-cultures. *Bioresour. Technol.* **2011**, *102*, 2724–2730. [CrossRef]
9. Hu, Q.; Sommerfeld, M.; Jarvis, E.; Ghirardi, M.; Posewitz, M.; Seibert, M.; Darzins, A. Microalgal triacylglycerols as feedstocks for biofuel production: Perspectives and advances. *Plant J.* **2008**, *54*, 621–639. [CrossRef]
10. Ranjan, A.; Patil, C.; Moholkar, V.S. Mechanistic assessment of microalgal lipid extraction. *Ind. Eng. Chem. Res.* **2014**, *49*, 2979–2985. [CrossRef]
11. Sharma, K.K.; Schuhmann, H.; Schenk, P.M. High Lipid Induction in Microalgae for Biodiesel Production. *Energies* **2012**, *5*, 1532–1553. [CrossRef]
12. Wijffels, R.H.; Barbosa, M.J. An Outlook on Microalgal Biofuels. *Science* **2010**, *329*, 796–799. [CrossRef] [PubMed]
13. Schäfer, K. Accelerated solvent extraction of lipids for determining the fatty acid composition of biological material. *Anal. Chim. Acta* **1998**, *358*, 69–77. [CrossRef]
14. Cheng, C.H.; Du, T.B.; Pi, H.C.; Jang, S.M.; Lin, Y.H.; Lee, H.T. Comparative study of lipid extraction from microalgae by organic solvent and supercritical CO_2. *Bioresour. Technol.* **2011**, *102*, 10151–10153. [CrossRef] [PubMed]
15. Herrero, M.; Cifuentes, A.; Ibañez, E. Sub- and supercritical fluid extraction of functional ingredients from different natural sources: Plants, food-by-products, algae and microalgae: A review. *Food Chem.* **2006**, *98*, 136–148. [CrossRef]
16. Lee, A.K.; Lewis, D.M.; Ashman, P.J. Disruption of microalgal cells for the extraction of lipids for biofuels: Processes and specific energy requirements. *Biomass Bioenergy* **2012**, *46*, 89–101. [CrossRef]
17. Halim, R.; Danquah, M.K.; Webley, P.A. Extraction of oil from microalgae for biodiesel production: A review. *Biotechnol. Adv.* **2012**, *30*, 709–732. [CrossRef]
18. Halim, R.; Gladman, B.; Danquah, M.K.; Webley, P.A. Oil extraction from microalgae for biodiesel production. *Bioresour. Technol.* **2011**, *102*, 178–185. [CrossRef]
19. Lardon, L.; Hélias, A.; Sialve, B.; Steyer, J.P.; Bernard, O. Life-cycle assessment of biodiesel production from microalgae. *Environ. Sci. Technol.* **2009**, *43*, 6475. [CrossRef]
20. Pradhan, A.; Shrestha, D.; McAloon, A.; Yee, W.; Haas, M.; Duffield, J.; Shapouri, H. Energy Life-Cycle assessment of soybean biodiesel revisited. *Trans. ASABE* **2009**, *54*, 1031–1039. [CrossRef]

21. Xu, L.; Brilman, D.W.F.; Withag, J.A.M.; Brem, G.; Kersten, S. Assessment of a dry and a wet route for the production of biofuels from microalgae: Energy balance analysis. *Bioresour. Technol.* **2011**, *102*, 5113. [CrossRef]
22. Taher, H.; Al-Zuhair, S.; Al-Marzouqi, A.H.; Haik, Y.; Farid, M. Effective extraction of microalgae lipids from wet biomass for biodiesel production. *Biomass Bioenergy* **2014**, *66*, 159–167. [CrossRef]
23. Günerken, E.; D'Hondt, E.; Eppink, M.H.; Garcia-Gonzalez, L.; Elst, K.; Wijffels, R.H. Cell disruption for microalgae biorefineries. *Biotechnol. Adv.* **2015**, *33*, 243. [CrossRef]
24. Shirgaonkar, I.Z.; Lothe, R.R.; Pandit, A.B. Comments on the mechanism of microbial cell disruption in high-pressure and high-speed devices. *Biotechnol. Prog.* **1998**, *14*, 657–660. [CrossRef]
25. Balasundaram, B.; Pandit, A.B. Selective release of invertase by hydrodynamic cavitation. *Biochem. Eng. J.* **2001**, *8*, 251–256. [CrossRef]
26. Cheng, J.; Sun, J.; Huang, Y.; Feng, J.; Zhou, J.; Cen, K. Dynamic microstructures and fractal characterization of cell wall disruption for microwave irradiation-assisted lipid extraction from wet microalgae. *Bioresour. Technol.* **2013**, *150*, 67. [CrossRef]
27. Cheng, J.; Yu, T.; Li, T.; Zhou, J.; Cen, K. Using wet microalgae for direct biodiesel production via microwave irradiation. *Bioresour. Technol.* **2013**, *131*, 531–535. [CrossRef]
28. Naghdi, F.G.; González, L.M.G.; Chan, W.; Schenk, P.M. Progress on lipid extraction from wet algal biomass for biodiesel production. *Microb. Biotechnol.* **2016**, *9*, 718–726. [CrossRef]
29. Gogate, P.R.; Pandit, A.B. Application of cavitational reactors for cell disruption for recovery of intracellular enzymes. *J. Chem. Technol. Biotechnol.* **2008**, *83*, 1083–1093. [CrossRef]
30. Keris-Sen, U.D.; Sen, U.; Soydemir, G.; Gurol, M.D. An investigation of ultrasound effect on microalgal cell integrity and lipid extraction efficiency. *Bioresour. Technol.* **2014**, *152*, 407–413. [CrossRef]
31. Valizadeh Derakhshan, M.; Nasernejad, B.; Dadvar, M.; Hamidi, M. Pretreatment and kinetics of oil extraction from algae for biodiesel production. *Asia Pac. J. Chem. Eng.* **2014**, *9*, 629–637. [CrossRef]
32. Grimi, N.; Dubois, A.; Marchal, L.; Jubeau, S.; Lebovka, N.I.; Vorobiev, E. Selective extraction from microalgae *Nannochloropsis* sp. using different methods of cell disruption. *Bioresour. Technol.* **2014**, *153*, 254–259. [CrossRef] [PubMed]
33. Kim, D.-Y.; Vijayan, D.; Praveenkumar, R.; Han, J.-I.; Lee, K.; Park, J.-Y.; Chang, W.-S.; Lee, J.-S.; Oh, Y.-K. Cell-wall disruption and lipid/astaxanthin extraction from microalgae: *Chlorella* and *Haematococcus*. *Bioresour. Technol.* **2016**, *199*, 300–310. [CrossRef]
34. Park, J.Y.; Oh, Y.K.; Lee, J.S.; Lee, K.; Jeong, M.J.; Choi, S.A. Acid-catalyzed hot-water extraction of lipids from *Chlorella vulgaris*. *Bioresour. Technol.* **2014**, *153*, 408. [CrossRef]
35. Kim, D.Y.; Oh, Y.K.; Park, J.Y.; Kim, B.; Choi, S.A.; Han, J.I. An integrated process for microalgae harvesting and cell disruption by the use of ferric ions. *Bioresour. Technol.* **2015**, *191*, 469–474. [CrossRef]
36. Teixeira, R.E. Energy-efficient extraction of fuel and chemical feedstocks from algae. *Green Chem.* **2012**, *14*, 419–427. [CrossRef]
37. Vanthoorkoopmans, M.; Wijffels, R.H.; Barbosa, M.J.; Eppink, M.H.M. Biorefinery of microalgae for food and fuel. *Bioresour. Technol.* **2013**, *135*, 142–149. [CrossRef]
38. Nurra, C.; Torras, C.; Clavero, E.; Ríos, S.; Rey, M.; Lorente, E.; Farriol, X.; Salvadó, J. Biorefinery concept in a microalgae pilot plant. Culturing, dynamic filtration and steam explosion fractionation. *Bioresour. Technol.* **2014**, *163*, 136. [CrossRef]
39. Badawy, M.I.; Ali, M.E. Fenton's peroxidation and coagulation processes for the treatment of combined industrial and domestic wastewater. *J. Hazard. Mater.* **2006**, *136*, 961. [CrossRef]
40. Concas, A.; Pisu, M.; Cao, G. Disruption of microalgal cells for lipid extraction through Fenton reaction: Modeling of experiments and remarks on its effect on lipids composition. *Chem. Eng. J.* **2015**, *263*, 392–401. [CrossRef]
41. Steriti, A.; Rossi, R.; Concas, A.; Cao, G. A novel cell disruption technique to enhance lipid extraction from microalgae. *Bioresour. Technol.* **2014**, *164*, 70–77. [CrossRef]
42. Brillas, E.; Sirés, I.; Oturan, M.A. Electro-Fenton process and related electrochemical technologies based on Fenton's reaction chemistry. *Chem. Rev.* **2009**, *109*, 6570–6631. [CrossRef]
43. Qiang, Z.; Chang, J.-H.; Huang, C.-P. Electrochemical generation of hydrogen peroxide from dissolved oxygen in acidic solutions. *Water Res.* **2002**, *36*, 85–94. [CrossRef]

44. Lopes, G.K.; Schulman, H.M.; Hermeslima, M. Polyphenol tannic acid inhibits hydroxyl radical formation from Fenton reaction by complexing ferrous ions. *BBA Biomembr.* **1999**, *1472*, 142–152.
45. Oturan, M.A. An ecologically effective water treatment technique using electrochemically generated hydroxyl radicals for in situ destruction of organic pollutants: Application to herbicide 2, 4-D. *J. Appl. Electrochem.* **2000**, *30*, 475–482. [CrossRef]
46. Hou, Y.; Liu, Z.; Yue, Z.; Chen, S.; Zheng, Y.; Chen, F. CAH1 and CAH2 as key enzymes required for high bicarbonate tolerance of a novel microalga *Dunaliella salina* HTBS. *Enzym. Microb. Technol.* **2016**, *87*, 17–23. [CrossRef]
47. Fajardo, A.R.; Cerdán, L.E.; Medina, A.R.; Moreno, P.A.G.; Grima, E.M. Lipid extraction from the microalga *Phaeodactylum tricornutum*. *Eur. J. Lipid Sci. Technol.* **2007**, *109*, 120–126. [CrossRef]
48. Zhang, K.; Li, H.; Chen, W.; Zhao, M.; Cui, H.; Min, Q.; Wang, H.; Chen, S.; Li, D. Regulation of the Docosapentaenoic Acid/Docosahexaenoic Acid Ratio (DPA/DHA Ratio) in *Schizochytrium limacinum* B4D1. *Appl. Biochem. Biotechnol.* **2016**, *182*, 1–15. [CrossRef]
49. Jen, J.F.; Leu, M.F.; Yang, T.C. Determination of hydroxyl radicals in an advanced oxidation process with salicylic acid trapping and liquid chromatography. *J. Chromatogr. A* **1998**, *796*, 283–288. [CrossRef]
50. Gu, X.; Chen, C.; Zhou, T. Spectrophotometric method for the determination of ascorbic acid with iron (III)-1,10-phenanthroline after preconcentration on an organic solvent-soluble membrane filter. *Anal. Bioanal. Chem.* **1996**, *355*, 94–95. [CrossRef]
51. Alemán-Nava, G.S.; Cuellar-Bermudez, S.P.; Cuaresma, M.; Bosma, R.; Muylaert, K.; Ritmann, B.E.; Parra, R. How to use Nile Red, a selective fluorescent stain for microalgal neutral lipids. *J. Microbiol. Methods* **2016**, *128*, 74–79. [CrossRef]
52. Lieselot, B.; Cedrick, V.; Koen, G.; Charlotte, B.; Koenraad, M.; Imogen, F. Optimization of a Nile Red method for rapid lipid determination in autotrophic, marine microalgae is species dependent. *J. Microbiol. Methods* **2015**, *118*, 152.
53. Greenspan, P.; Mayer, E.P.; Fowler, S.D. Nile red: A selective fluorescent stain for intracellular lipid droplets. *J. Cell. Biol.* **1985**, *100*, 965–973. [CrossRef]
54. Liu, Z.Y.; Liu, C.F.; Hou, Y.Y.; Chen, S.L.; Xiao, D.G.; Zhang, J.K.; Chen, F.J. Isolation and Characterization of a Marine Microalga for Biofuel Production with Astaxanthin as a Co-Product. *Energies* **2013**, *6*, 2759–2772. [CrossRef]
55. Pimentel, M.; Oturan, N.; Dezotti, M.; Oturan, M.A. Phenol degradation by advanced electrochemical oxidation process electro-Fenton using a carbon felt cathode. *Appl. Catal. B Environ.* **2008**, *83*, 140–149. [CrossRef]
56. Le, G.; Li, D.; Feng, G.; Liu, Z.; Hou, Y.; Chen, S.; Zhang, D. Hydroxyl radical-aided thermal pretreatment of algal biomass for enhanced biodegradability. *Biotechnol. Biofuels* **2015**, *8*, 194.
57. Hua, L.; Guo, L.; Thakkar, M.; Wei, D.; Agbakpe, M.; Kuang, L.; Magpile, M.; Chaplin, B.P.; Tao, Y.; Shuai, D. Effects of anodic oxidation of a substoichiometric titanium dioxide reactive electrochemical membrane on algal cell destabilization and lipid extraction. *Bioresour. Technol.* **2016**, *203*, 112–117. [CrossRef]
58. Carocho, M.; Ferreira, I.C. A review on antioxidants, prooxidants and related controversy: Natural and synthetic compounds, screening and analysis methodologies and future perspectives. *Food Chem. Toxicol.* **2013**, *51*, 15–25. [CrossRef]
59. Gutteridge, J.M. Lipid peroxidation and antioxidants as biomarkers of tissue damage. *Clin. Chem.* **1995**, *41*, 1819. [CrossRef]
60. Machlin, L.J.; Bendich, A. Free radical tissue damage: Protective role of antioxidant nutrients. *FASEB J.* **1987**, *1*, 441. [CrossRef]
61. Chen, C.Y.; Yeh, K.L.; Aisyah, R.; Lee, D.J.; Chang, J.S. Cultivation, photobioreactor design and harvesting of microalgae for biodiesel production: A critical review. *Bioresour. Technol.* **2011**, *102*, 71–81. [CrossRef] [PubMed]

Article

Perspective Design of Algae Photobioreactor for Greenhouses—A Comparative Study

Kateřina Sukačová [1,*], Pavel Lošák [2], Vladimír Brummer [2], Vítězslav Máša [2], Daniel Vícha [1] and Tomáš Zavřel [1]

[1] Global Change Research Institute, Academy of Sciences of the Czech Republic, Bělidla 986/4a, 603 00 Brno, Czech Republic; vicha.d@czechglobe.cz (D.V.); zavrel.t@czechglobe.cz (T.Z.)

[2] Institute of Process Engineering, Brno University of Technology, Technická 2896/2, 616 69 Brno, Czech Republic; losak.p@fme.vutbr.cz (P.L.); brummer@fme.vutbr.cz (V.B.); masa@fme.vutbr.cz (V.M.)

* Correspondence: sukacova.k@czechglobe.cz; Tel.: +420-511-440-550

Abstract: The continued growth and evolving lifestyles of the human population require the urgent development of sustainable production in all its aspects. Microalgae have the potential of the sustainable production of various commodities; however, the energetic requirements of algae cultivation still largely contribute to the overall negative balance of many operation plants. Here, we evaluate energetic efficiency of biomass and lipids production by *Chlorella pyrenoidosa* in multi-tubular, helical-tubular, and flat-panel airlift pilot scale photobioreactors, placed in an indoor environment of greenhouse laboratory in Central Europe. Our results show that the main energy consumption was related to the maintenance of constant light intensity in the flat-panel photobioreactor and the culture circulation in the helical-tubular photobioreactor. The specific power input ranged between 0.79 W L^{-1} in the multi-tubular photobioreactor and 6.8 W L^{-1} in the flat-panel photobioreactor. The construction of multi-tubular photobioreactor allowed for the lowest energy requirements but also predetermined the highest temperature sensitivity and led to a significant reduction of *Chlorella* productivity in extraordinary warm summers 2018 and 2019. To meet the requirements of sustainable yearlong microalgal production in the context of global change, further development towards hybrid microalgal cultivation systems, combining the advantages of open and closed systems, can be expected.

Keywords: microalgae; biomass; photobioreactors; power consumption; *Chlorella*; power input; sustainability; lipids; temperature stress; photoinhibition

Citation: Sukačová, K.; Lošák, P.; Brummer, V.; Máša, V.; Vícha, D.; Zavřel, T. Perspective Design of Algae Photobioreactor for Greenhouses—A Comparative Study. *Energies* **2021**, *14*, 1338. https://doi.org/10.3390/en14051338

Academic Editors: José Carlos Magalhães Pires and Ana Luísa Gonçalves

Received: 2 February 2021
Accepted: 25 February 2021
Published: 1 March 2021

Publisher's Note: MDPI stays neutral with regard to jurisdictional claims in published maps and institutional affiliations.

1. Introduction

We are currently experiencing a growing demand for products that do not burden our planet with the co-production of greenhouse gases, as well as for products that respect the principles of sustainability [1]. In line with these requirements, we can see a growing demand for bio-based products [2]. These trends are very well met by the possibilities offered by the use of microalgae, especially in the field of pharmacy [3], food industry [4–6], and industrial lipids production [7,8]. In the pharmaceutical and food industries, a high added value of products entering the market can be often predicted, which opens a wide range of possibilities for microalgae production applications. Lipid production is currently investigated mainly regarding biofuels production, which is, however, still not competitive enough compared to the existing technologies [9,10], even though the algae biodiesel is of comparable quality with other biodiesels [11]. Intensive research on microalgae cultivation, system optimization, and other aspects of the supply chain is currently being carried out using the most advanced algorithms [12]. Microalgae have also high potential for wastewater and industrial waste gases treatment [9,13].

Microalgae come in many forms and one of the indisputable advantages is the possibility of cultivating adapted strains according to the chosen applications. In addition, the current technological level allows for the subsequent cultivation of selected strains in order to perform specific tasks. For this purpose, a molecular sieve [14] or genome editing tools [15] can be used.

The presented article focuses on the usage of closed photobioreactors (PBRs) technology for microalgal cultivation in Central European region. The climate conditions of a temperate zone predetermine the need to use the PBR technology as the environment does not provide stable sunlight and a suitable temperature for a sufficient part of the year to use open-air cultivation systems. Even though the climate conditions are far from optimal, the use of closed PBRs in temperate zone makes sense. Unlike in open-air cultivation systems, the use of PBRs prevents or reduces contamination (by bacteria, protozoa, or unwanted algal species [2]). Cultivation in PBRs also allows for better control of the cultivation process, including the maintenance of balanced growth conditions (nutrients, light, temperature) for effective biomass production [16]. Because of these advantages, it can be assumed that PBRs will remain a necessary technology in the algae cultivation industry as open-air cultivation systems and their modifications cannot offer a sufficient product quality. The future challenges of algae cultivation technology will be to improve the construction of PBRs, to reduce their energy consumption, and to optimize cultivation processes, together with the optimization of the algae strains according to specific applications.

This work provides a comparison of three different PBRs designs (geometry) in the context of biomass and lipid production and energy consumption. The choice of cultivation strategy strongly depends on the operation site [17]. The advantages of open-air cultivation systems over closed systems [18] are lower investment costs [19] and lower energy consumption [20]; however, only in the areas with high solar exposure. Acién Fernández et al. [21] reported the average yearly energetic consumption of raceway pond around 1 W/m^3 compared to the energetic consumption of tubular PBR about 500 W/m^3. On the contrary, the disadvantages of the open systems are the higher chance of contamination and lower biomass production compared to the closed PBRs [16,21,22]. A closed PBRs technology generally allows for better regulation and control of the main biotic and abiotic growth factors [1,18,23].

Closed PBRs can be operated indoors, using natural or artificial lighting, or outdoors. Outdoor applications will require less energy for lighting; however, depending on the location, this benefit may be eliminated by the need to cool the PBRs to avoid overheating [17]. Wang et al. [18] highlighted that an efficient cooling method, such as water spraying, may increase a cost especially for large scale operation.

Many authors have studied specific aspects of efficient production in individual PBRs types, including light and temperature management, energetic management or optimization of cultivation parameters [2,24–26]. However, most of the studies focus on one specific PBR type [24,26] or compare results from different cultivation conditions [2,17,25]. Only few studies evaluate PBRs design during parallel PBRs operation under identical environmental conditions. For instance, Wolf et al., [27] compared five specific microalgae production systems (high-rate ponds, low and high flat panel, low and high tubular PBRs) under sub-tropical Australian conditions that are well suited for microalgal production. The work of Wolf et al. [27] described the relationship between temperature, heat load and PBR geometry in the context of biomass production and highlighted the importance of the surface-to-volume ratio for maximizing biomass production in PBRs.

The aim of this study is the comparison of biomass and lipid production during parallel cultivation in three specific PBRs and evaluation of impact of different design (geometry) on energy consumption as a crucial factor for evaluating cultivation cost. The Multi-Tubular airlift PBR and Helical-Tubular airlift PBRs were placed in an indoor environment of greenhouse laboratory that ensured a sufficient supply of natural light through the glass ceiling and walls and provided protection against cool temperatures in the spring or autumn. Moreover, Flat Panel airlift PBR with full control of cultivation conditions was included in the study and placed in the same greenhouse laboratory. To the best of our knowledge, the presented results evaluate, for the first-time, energy consumption and biomass and lipid productivity in parallel microalgae cultivation in the

Central European region, which allows us to directly assess the impact of PBR design on cultivation process and costs.

2. Materials and Methods

2.1. Cultivation Conditions and Experimental Design

Chlorella pyrenoidosa Chick (IPPAS C2, denoted as *C. pyrenoidosa* hereafter), was kindly provided by Dr. Maria Sinetova (Institute of Plant Physiology, Moscow, Russia). The stock cultures were maintained in BG-11 medium [28] at 25 °C, under light intensity of 70 µmol m^{-2} s^{-1} in a cultivation chamber (Algaetron, Photon System Instruments, Drásov, Czech Republic) with CO_2 enriched air (final concentration of 2%).

The experiments were performed throughout two consecutive years, 2018 and 2019. *C. pyrenoidosa* was cultivated in three photobioreactors of specific construction: Multi-Tubular Airlift photobioreactor (MTA PBR, working volume 60 L), Helical-Tubular Photobioreactor (H-T PBR, working volume 200 L) and Flat-Panel Airlift Photobioreactor (FPA PBR, working volume of 25 L, see Figure 1 and Section 2.2 for details).

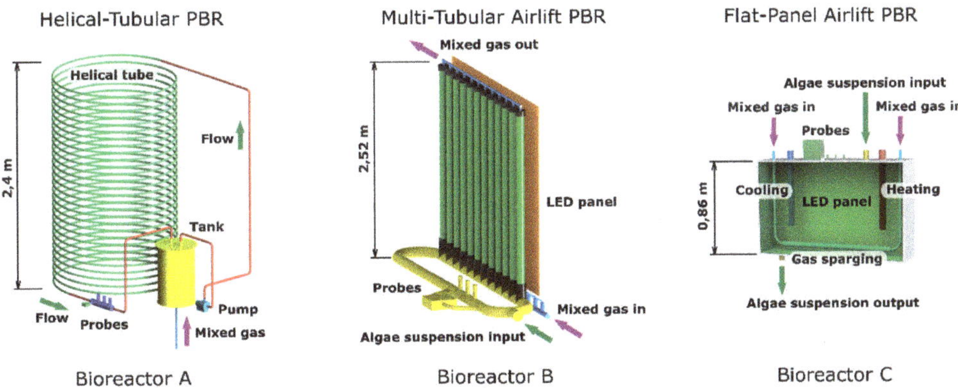

Figure 1. Schematic representation of design of the three photobioreactors used in this study: Helical-tubular PBR (**Bioreactor A**), Multi-tubular airlift PBR (**Bioreactor B**) and Flat panel PBR (**Bioreactor C**). Green parts of each PBR represent sunlight illuminated zones, yellow parts in Helical-Tubular Photobioreactor (H-T PBR), and Multi-Tubular Airlift photobioreactor (MTA PBR) represent dark zones. Total volumes of H-T PBR, MTA PBR, and Flat-Panel Airlift Photobioreactor FPA PBR were 200 L, 60 L and 25 L, respectively. LED panel in FPA PBR was placed at the back side of the cultivation tank.

Two types of experiments were performed: *C. pyrenoidosa* was cultivated in batch cultures in full BG-11 medium and in the N-limited BG-11 medium where NO_3^- concentration was decreased to 150 mg/L (10 times lower compared to the original BG-11 medium). This concentration was identified as optimal for lipid accumulation [29]. The cultures for the inoculation of MTA and H-T PBRs were prepared in FPA PBR and the biomass content of inoculum cultures was 1.5 g L^{-1} (dry weight). Inoculum volume was 8 L for MTA PBR (16% working volume) and 15 L for H-T PBR (15% working volume). Cultures in both MTA PBR and H-T PBR were initiated simultaneously to ensure identical environmental conditions (light and temperature) for both cultivations. Both MTA PBR and H-T PBR were illuminated solely by solar irradiance, and the temperature was monitored but not controlled. Both photobioreactors were aerated by CO_2-enriched air (final CO_2 concentration of 3%) with flow rate of 0.4 L min^{-1} $L_{culture}^{-1}$ (20 L min^{-1} for MTA PBR and 40 L min^{-1} for H-T PBR).

In the FPA PBR, the cultivation conditions were controlled as follows: white light was set to 400 µmol m^{-2} s^{-1} (provided by LED panel at back side of the PBR, in addition to day light penetrating front side), temperature was set to 25 °C and 3% CO_2 concentration in the air mixture was bubbled through the culture with flow rate of

2 L min^{-1} (0.08 L min^{-1} $L_{culture}{}^{-1}$, according Sukačová et al. [30]). All cultivations lasted for eight days.

2.2. Photobioreactor Units

MTA PBR consisted of 12 transparent PVC tubes with a diameter of 0.05 m and a height of 2.52 m, arranged in a vertical position. The distance between the tubes was 0.06 m. Each vertical tube contained an inlet port for the air/CO_2 mixture at the bottom part. At the bottom of the PBR, the transparent tubes were inserted in polyurethane interconnected T-shape tubes. The upper part of the PBR consisted of interconnected T-shape tubes in which the transparent tubes were fixed and that allowed for headspace presence over the cultures (10–15% of total volume). The total volume (V) of MTA PBR was 60 L. The illuminated surface area (SA) was 4.68 m^2, which resulted in the SA/V ratio 78 m^2 m^{-3}.

H-T PBR consisted of a spiral coiled tube (illuminated part of PBR) and a tank with a pump that secured the culture circulation (dark part of PBR). The spiral coiled tube (diameter of 0.034 m) with working volume of 170 L was made of transparent fluorinated ethylene propylene. The outer diameter of the H-T PBR was 1.7 m and the height was 2.4 m. The position of the transparent spiral tubes was secured by fixing the tubes to a metal frame placed inside the PBR. The tubes were connected to the dark part of PBR with a working volume of 30 L from which the algal suspension was pumped to the upper part of the PBR to circulate through the illuminated part back to the tank. The illuminated surface area (SA) was 25 m^2. The total volume of H-T PBR was 200 L. The SA/V ratio was 125 m^2 m^{-3}. Flow rate of algal suspension was 6 L min^{-1}. Both the MTA PBR and H-T PBR were equipped by sensors that allowed for the real time monitoring of culture temperature, pH, optical density (at both 680 nm and 720 nm), and steady-state fluorescence parameters (F_T, F_M, $Y_{(II)}$).

The flat panel PBR consisted of closed glass cultivation tank (25 L) fitted in a metal frame (outer dimensions of the PBR were 0.86 × 1.03 × 0.17 m). The aeration element was placed horizontally at the bottom part inside the tank. The PBR was equipped with light regulation and temperature control. The temperature control consisted of a heating element (not turned on throughout the entire experimental period) and a custom-made cooling system connected to the chilled water distribution pipes and a solenoid valve controlling the chilled water inlet into the cooling element of the FPA PBR. The PBR monitoring system allowed for measuring the same parameters as in MTA PBR and H-T PBR. Additional FPA PBR accessories are described in Sukačová et al. [30].

The aeration of all PBRs was secured by a gas mixing system GMS 150 (Photon Systems Instruments, Drásov, Czech Republic). PBRs were placed in an indoor greenhouse laboratory that ensured a supply of natural light through the glass ceiling and glass walls in South Moravia Region of the Czech Republic (49.3373761 N, 16.4751258 E). Inside the greenhouse, the temperature was not regulated.

2.3. Analytical Procedures

2.3.1. Determination of Dry Weight, Total Lipid Content and Productivity

The dry weight (DW) of algal biomass was detected gravimetrically. First, 1 mL of *C. pyrenoidosa* culture was centrifuged (10,000× *g*, MegaStar 3.0R, VWR, Wayne, PA, USA) and pellet freeze-dried (ScanVac CoolSafe, LaboGene, Lillerød, Denmark) in pre-weighed Eppendorf tubes. The tubes were then weighted using XA105DR analytical balances (Mettler Tolledo, Greifensee, Switzerland). For each sampling, DW determinations were performed in triplicates.

To derive the culture productivity parameters, the dry weight data were interpolated by a logistic function:

$$N_t = \frac{N_0\, K}{N_0 + (K - N_0)\, e^{-rt}}, \tag{1}$$

where N_0 represents the population size at the beginning of the cultivation, N_t represents the population size in time t, K represents the maximum population size, and r represents the population growth rate constant. To derive the culture doubling time in the exponential growth phase, the initial two days of the logistic interpolations were subjected to further analysis:

$$T_D = \frac{ln_2}{\left(\frac{ln\,(\frac{N_{t2}}{N_{t1}})}{t_2 - t_1}\right)}, \tag{2}$$

where T_D represents the doubling time and N_{t1} and N_{t2} represent the biomass content as derived from the logistic function in times t_1 and t_2, respectively.

The total lipid content was determined by a gravimetric method based on Folch et al. [31] on ice. The freeze-dried biomass (15 mg for each sample, measured on XA105DR analytical balances) was extracted with a 2:1 (v/v) chloroform-methanol mixture under ultrasound bath (SonoPlus Ultrasonic Homogenizer, Bandelin, Berlin, Germany: amplitude 100%, pulse 5 s, break 0.2 s) containing the ice-cooled water for 30 min. The biomass-chemical mixture was additionally vortexed with sea sand for 30 s (3 cycles) at room temperature (25 °C) using an amalgamator (Silamat S6, Ivoclar Vivadent AG). The mixture was further centrifuged at 2000× g for 10 min at 4 °C (MegaStar 3.0R, VWR, Wayne, PA, USA) which allowed the chemical mixture to be separated into two layers. The upper layer (methanol-water phase) was removed by a pipette and discarded. The lower chloroform layer containing lipids was stored in pre-weighed tubes. The extraction steps were repeated three times, and the chloroform layers with lipids were collected in the same tubes. Finally, the chloroform phase was evaporated in a vacuum desiccator at 45 °C under vacuum (Concentrator Plus, Eppendorf, Hamburg, Germany). The lipid residues in the tubes were weighed using XA105DR analytical balances.

The biomass and lipid productivities were determined from logistic interpolation of the dry weight measurements (according to Equation (1)), and from the lipid content data as measured after 8 days of cultivation. The effect of temperature on the biomass productivity was quantified by calculating the temperature coefficient Q_{10}:

$$Q_{10} = \left(\frac{P_2}{P_1}\right)^{\frac{10\ ^\circ C}{T_2 - T_1}}, \tag{3}$$

where P_1 and P_2 are biomass productivities at temperatures T_1 and T_2, respectively.

2.3.2. Measurement of Photosynthetic Activity

To further probe the effect of temperature on *C. pyrenoidosa* physiology, photosynthesis-irradiance (P-I) curves were measured by Multi-Color-PAM fluorometer (Walz, Effeltrich, Germany), using *Light Curves* protocol. The light curves consisted of 21 individual steps with stepwise increased white actinic light intensities between 0–4500 µmol (photons) $m^{-2}\ s^{-1}$, each step took 30 s. Prior to the measurement, no culture treatment or dilution was applied, and besides the first (dark) step, no extra dark acclimation was applied. For each light curve, the temperature was controlled by Temperature Control Unit US-T (Walz) to correspond with the actual temperature as measured in each PBR in time of sample withdrawal.

To derive the parameters of photosynthesis efficiency, the light curves were interpolated by a function according to [32]:

$$ETR = ETR_{mPot}\left(1 - e^{-\frac{\alpha\, PAR}{ETR_{mPot}}}\right) e^{-\frac{\beta\, PAR}{ETR_{mPot}}}, \tag{4}$$

where *ETR* is electron transport rate, ETR_{mPot} is maximal potential *ETR* capacity without photoinhibition included, *PAR* is photosynthetically active irradiance, α is the initial slope of *ETR* increase under low *PAR*, and β is photoinhibition factor, with the same units as α. The parameters α, β and ETR_{mPot} resulted directly from light curves fitting. To derive ETR_{max}, the maximal *ETR* capacity with photoinhibition included, the following equation was further applied:

$$ETR_{max} = ETR_{mPot} \left(\frac{\alpha}{\alpha + \beta}\right) \left(\frac{\beta}{\alpha + \beta}\right)^{\frac{\beta}{\alpha}} \tag{5}$$

2.3.3. Elemental Composition and Calorific Value

Biomass samples obtained after 8 days of cultivation were subjected to an elemental analysis. A proximate analysis was carried out—the dry matter together with ash content at 550 °C of the original samples was measured, and subsequently the samples were fully dried. The elemental analysis (CHNS, Vario Macro Cube analyser, Elementar Analysensysteme GmbH, Langenselbold, Germany) was performed on dried samples, and the Higher heating Values (HHV) of dried samples were measured (calorimeter Parr 6100, Parr Instrument Company, Moline, IL, USA). Lower Heating Values (LHV) at 20 °C were calculated according to HHV values, hydrogen and water content. Oxygen values were calculated to 100% balance (concentrations of Cl and F were neglected). To verify the results, the HHV values were also calculated from the results of the elemental analysis using the average values of 10 different empirical equations for calculation of HHV from the measured CHNS composition [33]. All results were provided for the original sample, the dry matter of the sample, and the volatile matter of the sample, and were further used for the calculations of biomass calorific values.

2.3.4. Measurement of Electric Input Power

The energy consumption of each PBR can be expressed as the sum of the electric input power P (W) of individual components. All three PBRs include a gas mixing and monitoring system. This default configuration is complemented by the pump for the MTA and H-T PBR and LED lighting and cooling for the FPA PBR.

The measurement methodology consisted of several consecutive steps. First, the electric input power of individual components was measured for the three selected PBRs. The total input power (P_{TOT}) was then calculated as the sum of these partial values as follows:

$$P_{TOT\_H\text{-}T} = P_{GMS\_H\text{-}T} + P_{MU\_H\text{-}T} + P_{CP\_H\text{-}T}, \tag{6}$$

$$P_{TOT\_MTA} = P_{GMS\_MTA} + P_{MU\_MTA} + P_{CP\_MTA}, \tag{7}$$

$$P_{TOT\_FPA} = P_{GMS\_FPA} + P_{MU\_FPA} + P_{LED\_FPA} + P_{COOL\_FPA}, \tag{8}$$

where P is the electric input power and the indexes represent the gas mixing system (GMS), the monitoring unit (MU), LED lighting (LED), LED cooling (COOL), and the circulating pump (CP) in MTA, H-T and FPA PBRs. Subsequently, the total electric input power was measured at full operation for each PBR, and the results were compared. A confidence interval was determined for the measured values. The confidence interval includes the standard deviation of the measurement and the accuracy of the measuring device. Each PBR has a different working capacity, so the specific electric input power P_S related to one liter of working volume ($W\ L^{-1}$) was finally calculated. During the measurement, the stability of the electrical network was checked by voltage and frequency. The electric input power was measured by the wattmeter analyzer LUTRON DW-6092 with the accuracy of 1%.

2.4. Statistical Analysis

To determine the effect of PBR design on the biomass and lipid content, the data were analyzed by one-way ANOVA, followed by Tukey's HSD post-hoc test ($p < 0.05$). Data normality was verified using the Shapiro–Wilk test, which showed that all data followed normal distribution ($p > 0.05$). Since number of replicates was small, the data normality was also verified by inspecting distribution histograms. Homogeneity of variances was verified by the Cochran's C–Hartley–Bartlett test ($p > 0.05$). The analysis was performed using Statistica software (TIBCO Software Inc., Palo Alto, CA, USA).

3. Results

3.1. Biomass Productivity

The photobioreactors were operated in two successive years of 2018 and 2019. The microalga *C. pyrenoidosa* was cultivated in two distinct growth media: in full BG-11 medium (under nutrient sufficiency) in 2018 and in BG-11 medium with 10-times reduced nitrogen content (under N limitation) in 2019. Growth of biomass under nutrient sufficient conditions in MTA, H-T and FPA PBRs is summarized in Figure 2. In MTA PBR, the slowest growth of *C. pyrenoidosa* was detected in July when the biomass content increased from 0.31 to 0.52 g L^{-1} after 8 days of cultivation. The fastest growth was detected in October when the biomass content increased from 0.36 to 1.90 g L^{-1} after 8 days (doubling time in exponential phase was T_D = 1.7 days). In H-T PBR, the slowest growth was detected in September when biomass increased from 0.13 to 1.29 g L^{-1} after 8 days (T_D = 2.2 days), while the fastest growth (T_D = 1.2 days) was determined in June; however, with a very similar biomass increase as in September (biomass increased from 0.08 to 1.28 g L^{-1} after 8 days). The difference in growth rates between June and September was related to potentially higher maximal populations size K, as derived from data interpolation by the logistic function (K = 1.27 and 4.50 g L^{-1} for June and September, respectively, for details of K determination see Equation (1)). The lowest variation in growth throughout 2018 was detected in FPA PBR: T_D was detected as 1.52–1.94 days in January and May, respectively. The biomass increased from 0.24 ± 0.07 to 1.84 ± 0.11 g L^{-1}) after 8 days (Figure 2).

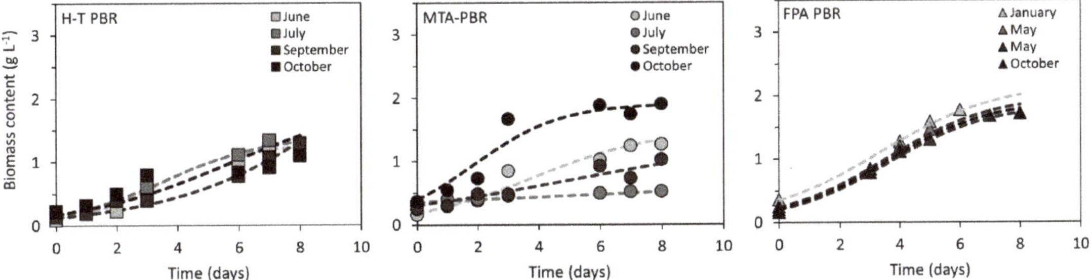

Figure 2. Growth of *C. pyrenoidosa* in H-T PBR, MTA PBR, and FPA PBR under nutrient sufficiency (as determined throughout the year 2018). Measured data points from individual cultivations were interpolated by logistic function (Equation (1)) using least squares fitting method. Data of *C. pyrenoidosa* growth under N limitation are summarized in Supplementary Figure S1.

The biomass accumulation under N-limited conditions is summarized in Figure S1. The average biomass content after 8 days of cultivation in N-limited medium achieved 1.81 ± 0.87 g L^{-1} for MTA PBR, 1.27 ± 0.35 g L^{-1} for H-T PBR and 1.35 ± 0.08 g L^{-1} for FPA PBR.

The highest variance in biomass productivity under both nutrient sufficiency and N limitation was found in MTA PBR, throughout both years of 2018 and 2019. In H-T PBR, the variability was reduced compared to MTA PBR, and the lowest variability was determined in FPA PBR with the possibility of temperature control. As discussed further in Sections 3.3 and 3.4, temperature was identified as the main factor responsible for the variability in biomass production. The average biomass productivity under nutrient sufficient conditions was 107 ± 72 mg L^{-1} d^{-1} in MTA PBR, 147 ± 6 mg L^{-1} d^{-1}

in H-T PBR and 199 ± 6 mg L^{-1} d^{-1} in FPA PBR. Under N limitation, the productivities were 202 ± 109 mg L^{-1} d^{-1} in MTA PBR, 143 ± 30 mg L^{-1} d^{-1} in H-T PBR and 148 ± 6 mg L^{-1} d^{-1} in FPA PBR (Table 1). The highest determined biomass productivities achieved for respective PBR types were 178 mg L^{-1} d^{-1} in H-T PBR (July 2019), 279 mg L^{-1} d^{-1} in MTA PBR (September 2019) and 207 mg L^{-1} d^{-1} in FPA PBR (January 2018). However, the effect of N availability, the same as the effect of PBR design on the biomass productivity, was not found (Table 1, ANOVA: $p > 0.05$).

Table 1. Biomass and lipid accumulation in *C. pyrenoidosa* cultures cultivated under nutrient sufficiency (full BG-11 medium) and under N limitation (BG-11 medium with 10 times lower N content) in the Multi-tubular airlift photobioreactor (MTA PBR), Helical-tubular photobioreactor (H-T PBR) and Flat panel photobioreactor (FPA PBR, for details on PBRs construction see Figure 1). The data represent mean values ± standard deviations, n = 2–4. In both 2018 and 2019, no significant differences in biomass or lipid accumulation and production were found between individual photobioreactors (ANOVA followed by Tukey's HSD test: $p > 0.05$). N.D.: not determined.

Parameters (Determined after 8 Days of Cultivation)	Nutrient Sufficiency			N-Limitation		
	MTA PBR	H-T PBR	FPA PBR	MTA PBR	H-T PBR	FPA PBR
Biomass content (g L^{-1})	1.17 ± 0.58	1.32 ± 0.08	1.84 ± 0.12	1.81 ± 0.87	1.27 ± 0.35	1.35 ± 0.08
Biomass productivity (mg L^{-1} d^{-1})	107 ± 72	147 ± 6	199 ± 7	202 ± 109	143 ± 30	148 ± 6
Lipid content (%)	34 ± 9	26 ± 3	N.D.	44 ± 2	41 ± 5	47 ± 3
Lipid productivity	34 ± 19	38 ± 5	N.D.	87 ± 44	60 ± 21	70 ± 2

3.2. *Lipid Productivity*

Lipid accumulation under both nutrient sufficient and N-limited conditions during cultivation in three photobioreactors is summarized in Table 1. In full BG-11 medium, the average lipid content was established as 35 ± 9% (of dry weight) in MTA PBR and 26 ± 3% in H-T PBR. Under N limitation, the lipid content was 44 ± 2% in MTA PBR, 41 ± 5% in H-T PBR and 47 ± 3% in FPA PBR (Table 1). The lipid content was not significantly different between the three PBRs (ANOVA: $p > 0.05$). However, the lipid content in N-limited medium was significantly higher compared to full BG-11 (ANOVA: $p < 0.05$).

Lipid productivity under nutrient sufficiency was 34 ± 19 mg L^{-1} d^{-1} in MTA PBR and 35 ± 8 mg L^{-1} d^{-1} in H-T PBR. Under N-limitation, lipid productivity was 87 ± 44 mg L^{-1} d^{-1} in MTA PBR, 60 ± 21 mg L^{-1} d^{-1} in H-T PBR and 70 ± 2 mg L^{-1} d^{-1} in FPA PBR (Table 1). Lipid productivity was not different between photobioreactors (ANOVA: $p > 0.05$), but it increased significantly under N-limitation (ANOVA: $p < 0.05$).

The elemental composition and calorific values of *C. pyrenoidosa* biomass under N-limitation are described in Table 2. The N content ranged from 3.15 ± 0.05% in FPA PBR to 4.24 ± 0.85% in HT PBR. The calorific values ranged between 24.0 ± 1.85 MJ kg^{-1} in H-T PBR and 25.2 ± 1.4 MJ kg^{-1} in FPA PBR. Neither the measured elemental composition parameters nor the calorific values were significantly different between the three photobioreactors (ANOVA: $p > 0.05$).

Table 2. Calorific values and the elemental analysis of *C. pyrenoidosa* biomass produced under nitrogen limited conditions. The data represent averages ± standard deviations, n = 2–3. No differences between photobioreactors were identified (ANOVA followed by Tukey's HSD test: $p > 0.05$).

	H-T PBR	MTA PBR	FPA PBR
Carbon (%)	54.2 ± 1.6	56.0 ± 1.2	57.8 ± 0.6
Hydrogen (%)	7.7 ± 0.4	7.8 ± 0.2	8.0 ± 0.06
Nitrogen (%)	4.2 ± 0.9	4.1 ± 0.1	3.2 ± 0.05
Sulphur (%)	0.36 ± 0.03	0.38 ± 0.04	0.3 ± 0.01
Oxygen (%)	29.4 ± 1.3	27.9 ± 1.3	27.2 ± 1.2
Calorific values (MJ kg^{-1})	24.0 ± 1.9	25.2 ± 1.1	25.2 ± 1.4

3.3. Temperature

The daily temperature of *C. pyrenoidosa* cultures in MTA and H-T PBRs varied between 18 °C and 40 °C (during the monitored period). In FPA PBR, the temperature was regulated and kept constant at 25 °C. Daily temperature profiles during June, July, September, and October 2018 in MTA and H-T PBRs are shown in Figure 3. The daily temperature courses showed similar trends between months: the lowest temperatures were recorded at night until 8 a.m., and maximal temperatures were detected in the afternoon between 2–4 p.m. During the whole monitoring period of June–October 2018, higher temperatures were measured in MTA PBR compared to H-T PBR: while the maximum culture temperature of 40 °C was recorded in several individual days during June, July and September 2018 in MTA PBR, maximal temperature in the H-T PBR did not exceed 38 °C. Minimal culture temperatures of 18 °C were measured in June in both MTA and H-T PBRs. Overall, as indicated in Figure 3, a temperature difference between day and night was higher in MTA PBR compared to H-T PBR.

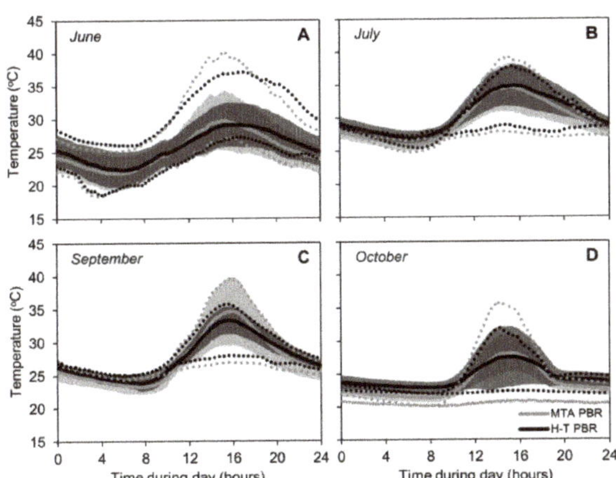

Figure 3. Temperature profiles in H-T PBR (black lines) and MTA PBR (grey lines) in June (**A**), July (**B**), September (**C**), and October 2018 (**D**). The full lines represent averages from 8 days of cultivation in each photobioreactor, the dark and light grey error intervals represent standard deviations for H-T PBR and MTA PBR, respectively. The doted lines represent temperature profiles in days when minimal and maximal temperatures during the light phase of the day were detected (within each month). Temperature in FPA PBR was regulated to 25 °C.

The productivity of C. *pyrenoidosa* cultures showed a higher temperature sensitivity in MTA PBR compared to H-T PBR. Temperature coefficient Q_{10} (see Equation (3)) was identified as 0.08 in MTA-PBR between 25–30 °C, whereas in H-T PBR it was identified as 0.56 between 27–30 °C (Figure 4). This suggests, in addition to the increased sensitivity of MTA PBR to high temperatures, a shift in temperature optima between the photobioreactors. In MTA PBR, the biomass productivity was decreasing between 25–30 °C, which suggests a temperature optimum equal or lower to 25 °C. On the other hand, the productivity in the H-T PBR increased between 25–27 °C and decreased between 27–31 °C (Figure 3), which suggests optimal temperature for culture productivity of 27 °C.

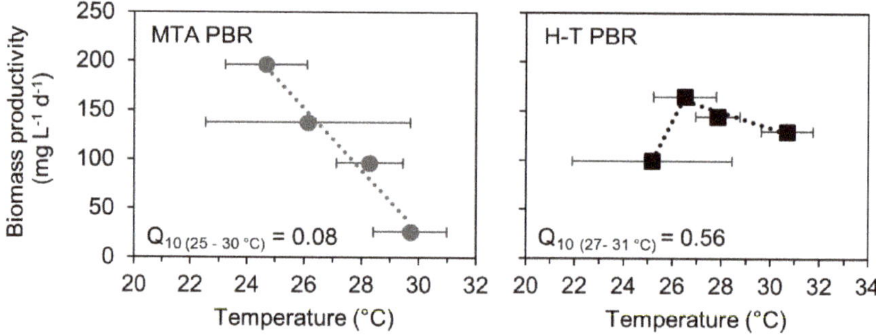

Figure 4. Temperature dependence of biomass productivity in MTA PBR (left panel) and H-T PBR (right panel) under nutrient sufficiency. Each point represents average daily temperature as measured during 8 days of C. *pyrenoidosa* cultivation in June, July, September and October 2018, error bars represent standard deviations. Biomass productivity was determined from a single point measurement at 8th cultivation day, data of C. *pyrenoidosa* growth are summarized in Figure 2. Temperature coefficient Q_{10} was calculated according to Equation (2).

3.4. *Photosynthetic Activity*

The photosynthetic activity was monitored in June, July, September and October 2018 during 6th and 8th day of each cultivation, each day between 6 a.m. and 6 p.m., with a temporal resolution of two hours. The highest photosynthesis efficiency and capacity in C. *pyrenoidosa* cultures were determined in June and October when the daily temperature maxima did not exceed 36 °C (Figure 5). In July and September, when the temperature rose to 38 °C, both photosynthetic efficiency and photosynthetic capacity decreased. The maximal photosynthesis drops in the daytime were measured in both July and September at 4 p.m., and the drop was higher in MTA PBR compared to H-T PBR. The reduction of photosynthesis efficiency was accompanied by a slight increase of photoinhibition factor β (see Equation (4) for details). However, we noted that during the photosynthesis efficiency measurement, the cultures were light-limited rather than photo-inhibited, as reflected by the courses of light curves showing that both ETR_{max} (see Equation (5) for details) and β were detected at actinic light intensities of 2000 μmol photons m^{-2} s^{-1} or higher (Supplementary Figure S1), whereas light available for both photobioreactors did not exceed 1000 μmol photons m^{-2} s^{-1} in any of the measured days (Figure 5).

Rather minor differences in photosynthetic performance between both H-T and MTA PBRs correlate with the results of biomass and lipid productivity, as summarized in Table 1. Higher sensitivity of the MTA PBR to increased temperature is consistent with reduced productivity under the increasing temperature in this photobioreactor, as reflected by coefficient Q_{10} (Figure 4, see Section 3.3 for details).

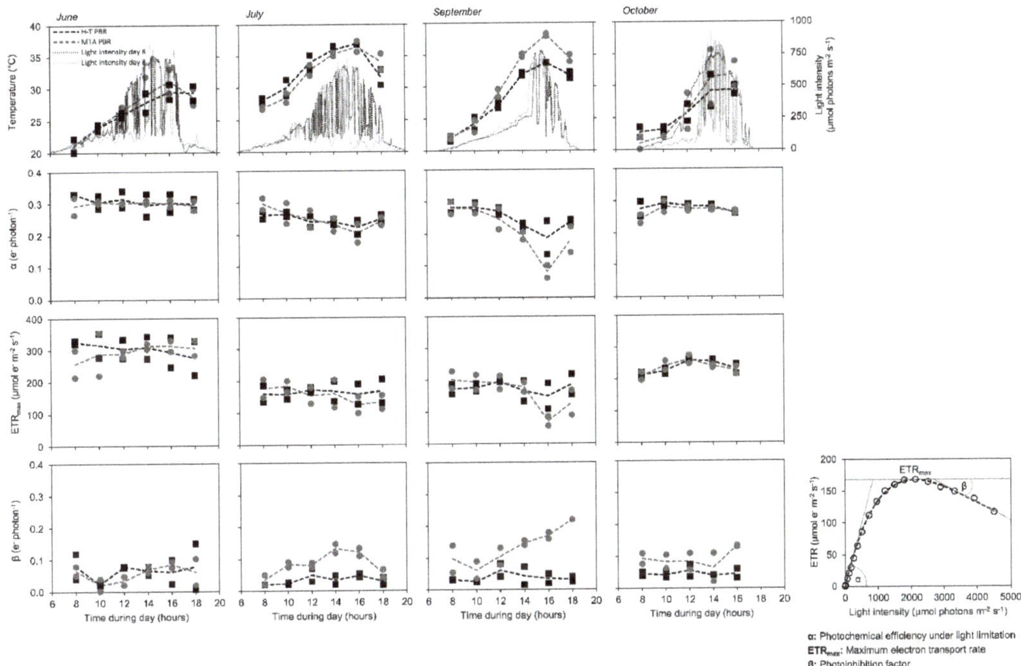

Figure 5. Light intensity and temperature profiles (upper row panels) and photosynthesis efficiency parameters (bottom rows panels) during 6th and 8th day of *C. pyrenoidosa* cultivation in June, July, September, and October 2018 (left to right columns) in H-T PBR (black lines, black squares) and MTA PBR (grey lines, grey circles). The points represent measured data throughout the light phase of each 6th and 8th cultivation day, the dashed lines represent particular averages. The photosynthesis efficiency parameters, schematically represented in the bottom right panel, were determined according to Equations (4) and (5). All photosynthesis-irradiance curves from which the parameters were derived are summarized in Supplementary Figure S2.

3.5. Electric Input Power of the Tested PBRs

All three PBRs included two basic components: gas mixing system and monitoring unit which allowed for real-time measurement of temperature, pH, optical density and fluorescence signals. The electric input power of these components was P_{GMS_FPA} = 17.2 W for gas mixing system and P_{MU} = 25.3 W for monitoring system.

The FPA PBR was supplemented with LED lighting with an average electric input power of P_{LED_FPA} = 105.6 W and three fans for cooling LED lights with an electric input power of P_{COOL_FPA} = 21.5 W. H-T PBR was equipped with a circulating pump with $P_{CP\_H\text{-}T}$ = 516.5 W. A smaller pump with P_{CP_MTA} = 7.3 W was installed in MTA PBR as an external part of the monitoring system.

Knowing the partial power input of all components, the total power input (P_{TOT}) of the three PBRs was estimated (according to Equations (6)–(8)) as: $P_{TOT\_H\text{-}T} = 561 \pm 9$ W for H-T PBR, $P_{TOT\_MTA} = 48 \pm 2$ W for MTA PBR and $P_{TOT\_FPA} = 169 \pm 3$ W for FPA PBR. The corresponding daily electricity consumption was 13.5 kWh for H-T PBR, 1.1 kWh for MTA PBR and 4.1 kWh for FPA PBR.

However, since the working volumes of H-T, MTA, and FPA PBRs were not identical, for direct electric input power comparison it was necessary, to further normalize the specific values per L culture. As summarized in Figure 6, the highest specific electric input power P_S was identified in FPA PBR (working volume 25 L) as $P_{S\_FPA} = 6.8 \pm 0.1$ W L^{-1}. Such a high value is related specially to LED lighting and cooling which accounts for about 75% of the total FPA input power (Figure 6). In H-T PBR (working volume 200 L), $P_{S\_H\text{-}T}$

was identified as 2.8 ± 0.1 W L^{-1}, and to the power consumption was mainly related to the operation of a circulation pump which accounted for more than 92% of the total H-T PBR's specific electric input power. The lowest consumption was determined in the MTA PBR (of working volume 60 L) with P_{S_MTA} = 0.79 ± 0.03 W L^{-1}. In addition to the basic components (gas mixing and monitoring systems), the MTA PBR was only equipped with a small external pump which secured a culture flow through the monitoring system.

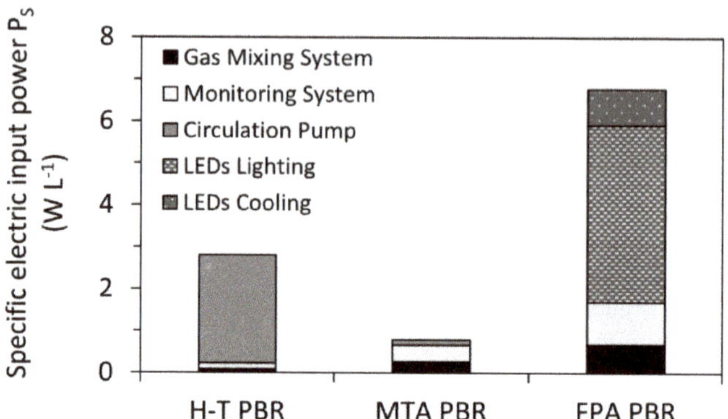

Figure 6. Specific electric input power P_S (W L^{-1}) as measured for H-T PBR, MTA PBR, and FPA PBR. Distribution of individual components within each PBR is summarized in Equations (6)–(8). Note that the circulation pump was not part of FPA PBR and, on the contrary, LEDs lighting and cooling were specific for FPA only.

4. Discussion

4.1. Biomass and Lipid Productivity

In this study, three photobioreactors with different geometries were compared in the context of biomass and lipid production and power consumption required for operation. Moreover, the photosynthetic activity of microalgal cells during the cultivation process was studied in order to compare sensitivity to environmental stress in various PBRs. The average biomass production of 0.20 g L^{-1} d^{-1} in MTA PBR (Table 1) was similar to the *Chlorella* production as reported by both de Morais and Costa [34] in vertical tubular PBR and Huang et al. [35] in multi-column airlift PBR, and it was higher compared to biomass yield of 0.032 g L^{-1} d^{-1} in a pilot-scale sequential-baffled column PBR as reported by Lam et al. [36]. On the other hand, the biomass production in MTA PBR was lower than the values reported for other tubular PBRs, such as outdoor inclined tubular PBR operated under solar irradiation (biomass production 0.3–1.47 g L^{-1} d^{-1}, [37]), small volume (0.6 L) bubbled tubular PBR operated under artificial illumination (*Chlorella* sp. production of 0.5–1.2 g L^{-1} d^{-1}, [38]), or the horizontal tubular system in a sub-tropical Australia climate with *Chlorella* sp. production of 0.43 g L^{-1} d^{-1} [27]. Indeed, a comparison with other works reveals that distinct biomass production can be achieved in various PBRs systems and that specific improvements of the PBR design (such as inclination of PBR tubes or insertion of static mixers inside the tubes, [37]) can lead to a significant production increase. However, the goal of this work was not maximizing the production, but rather providing solid data on direct evaluation of energetic efficiency of various PBR types. In addition, it should be noted that, as suggested by Wolf et al. [27], a comparison of different PBRs systems is often limited due to the use of different strains and exposure to distinct climate and operational conditions.

Data on *Chlorella* production in H-T PBRs are rather scarce. The average production in H-T PBR in this study (0.15 g L^{-1} d^{-1}) was three times higher than the production described by Scragg et al. [39] for *C. vulgaris* and *C. emersonii*, where a lower flow rate of microalgal suspension and lower illumination were used. Similarly, our results are higher than productions reported for microalga *Tetraselmis* sp. in Biocoil (helical tubular) PBR [23], where, in addition to differences in PBRs design, also higher temperature fluctuations were reported.

The average biomass production of 0.20 g L^{-1} d^{-1} achieved in FPA PBR was nearly identical to the production reported for *C. sorokiniana* [27] and higher than the production reported for *C. zofingiensis* [40]. On the other hand, it was lower than the *Chlorella* production in flat panel PBRs improved by baffles insertion, where the productions varied between 0.6–1.1 g L^{-1} d^{-1} [41,42]. An important parameter related to the culture productivity is the culture depth that determines the length of the light path. The culture depth of our FPA PBR was 7 cm, whereas the light path of PBRs in Zhang et al. [42] and Degen et al. [41] was 2.5 cm and 3 cm, respectively. The exceptional *Chlorella* production of 2.64 g L^{-1} d^{-1} was described for flat panel PBRs with baffles and the light path of 1.5 cm [41]. These results suggest that the construction of FPA as used here was far from optimal. However, same as in case of MTA PBR and H-T PBR, the focus of this study was a comparison of energetic requirements rather than yield maximization. From this perspective, rather low energetic efficiency of biomass production is not surprising (95–810 kWh kg^{-1}, only energy for cultivation considered, energy required for biomass harvesting and drying not included to calculations). A direct comparison with previous studies shows that the productivity increase or input power reduction can lead to significant energetic efficiency increase, up to 12 kWh kg^{-1} in tubular PBR or 2.4 kWh kg^{-1} in raceway pond [21]. It should be noted here that total energy requirements higher than 8 kWh kg^{-1} of dry algal biomass were reported as not sustainable [43].

Also, the initial biomass concentration is related tightly to the biomass productivity. During our experiments we used nearly ten times lower initial biomass concentration than recommended by Holdmann et al. [24] which can be considered as another factor related to rather low biomass and lipid yields.

Lipid yields in *C. pyrenoidosa* under nitrogen limitation (60–87 mg L^{-1} d^{-1}) were in the range of the previously reported yields in other *Chlorella* strains in laboratory conditions (41–106 mg L^{-1} d^{-1}; [30,44–47]). In line with this, the maximal lipid production in MTA PBR (118 mg L^{-1} d^{-1}) was only 11% lower than the maximum values reported up to date [48]. Also, lipid yields of different microalgae strains cultivated in (pilot scale) flat panel PBRs (36–91 mg L^{-1} d^{-1}) correspond well with our results [49–51], same as the lipid content (41–47%) and calorific values (24–25 MJ kg^{-1}, [30,39,52]). Considering the suboptimal cultivation conditions in our setup, the comparative lipid yields suggest possible lipid production limitation in the previous studies.

4.2. PBR Design and Susceptibility to Temperature Fluctuations

As shown in Figure 3, biomass productivity was strongly related to temperature. The results of photosynthetic activity measurement (Figure 4) further suggest that: (1) temperature was the main stress factor for *C. pyrenoidosa*, and (2) the PBR geometry allowed for a shift in *C. pyrenoidosa* temperature optimum from 27 °C in H-T PBR to $\leq$25 °C in MTA PBR. The increased temperature sensitivity of *C. pyrenoidosa* cultures in MTA PBR is likely related to the surface-to-volume ratio (see next paragraph for details) and the amount of irradiance received by both PBRs. The MTA PBR was exposed to direct solar irradiance over its whole illuminated area, which included also light (of weaker intensity) reflected by the back side plate. In contrast, in H-T PBR, from the nature of helical tubes stacked on the top of each other (see Figure 1 and Section 2.2 for details), the illuminated part consisted of the directly illuminated zone that received full sunlight irradiance and the indirectly illuminated zone with reduced light availability. We note that size of the dark zones in both MTA and H-T PBRs represented 10–15% of the working volumes and the effect of dark zone size on productivities can be there for considered similar for both PBRs.

The relation between temperature, heat load and PBR geometry was studied previously [27], and special importance in terms of microalgae productivity was assigned to the surface-to-volume ratio (SA/V ratio). Optimal SA/V ratio for maximization of microalgal production was determined as 43–73 m^2 m^{-3} [27]. Higher SA/V ratio led to increased heat load and resulted in susceptibility to temperature fluctuations, algal stress and reduced growth rates. MTA PBR in this study had the SA/V ratio of 78 m^2 m^{-3} and H-T PBR had the SA/V ratio of 125 m^2 m^{-3}. However, as shown in Figure 4, culture in H-T PBR had lower temperature sensitivity compared to MTA PBR. This result can be assigned to the specific geometry of H-T PBR which allowed for reduced heat transfer in partially shaded coils, and it suggests that the optimal value of SA/V ratio can be specific for different PBRs designs.

Several ways that can prevent PBR from overheating have been described. Water spraying is considered as an efficient method of PBR cooling; however, water pumping is connected with additional costs [18]. More recent approach preventing cultures from overheating is represented by utilization of spectrally selective glass or plastic material that filter the infrared part of (sun) light spectra [2,27]. Operation of PBRs in 2018 and 2019 was affected by extremely warm and sunny weather (see temperature profiles in Figure 3). As a consequence, the culture temperature in many days exceeded 35 °C. According to Wolf et al. [27], such a temperature acts as a stress factor for *Chlorella* and leads to a significant growth reduction. This is consistent with temperature optimum of 25 °C and growth reduction observed already at 29 °C, as identified by Sukačová et al. [29]. The lowest biomass productivity in the MTA PBR in summers 2018 and 2019 (Figure 2 and S1) as well as both the reduced photosynthesis efficiency and the increased sensitivity to photoinhibition under temperatures close to 40 °C (Figure 5) are in full agreement with the previous results, and document stress in *C. pyrenoidosa* cultures during our cultivations.

4.3. Energetic Requirements for Chlorella Production

The energetic requirements for microalgal biomass production have been described in a number of works [17,25,53]. Acién Fernández et al. [21] reported an average power consumption of 500 W m^{-3} for tubular large scale PBRs, which is lower than the consumption identified for H-T PBR (2800 W m^{-3}) and MTA PBR (790 W m^{-3}) here. According to our evaluation, the MTA PBR had at least 8.5 times lower energy consumption per L culture than FPA PBR that was operated under fully controlled cultivation conditions, including illumination by LED and active cooling (however, the cooling power input was not considered in the energy consumption, see the next paragraph for details). The average biomass productivity in FPA PBR was not different from MTA PBR, and the maximum production was even 25% higher in MTA PBR compared to FPA PBR. Similarly, the operation of FPA PBR consumed at least 2.5 times more energy than H-T PBR but the average biomass production was again not different between these two PBRs. Here, the maximum measured production was 15% higher in FPA PBR compared to H-T PBR.

From the perspective of energy efficiency of biomass production, maintaining constant conditions (including artificial light and cooling/heating) seems disadvantageous. Lam et al. [36] reported 69% reduced energy cost for outdoor cultivation in pilot-scale sequential-baffled column PBR using solar irradiance when compared to the identical system with artificial light. Itoiz et al. [22] also highlighted 85% reduction in energy requirements for the outdoor system compared to the indoor production. Also PBR cooling/heating requires substantial amount of energy. In our work, the FPA PBR was actively cooled during the summers 2018 and 2019 by a chilled water (see Section 2.2 for details) to maintain constant temperature of 25 °C. Since the chilled water distribution pipes were common for the whole greenhouse facility in which the PBRs were placed, it was not possible to separate cooling requirements of the FPA PBR from other systems. However, as shown on the example of FPA plant operated in southern Europe (Toscana, Italy), active cooling can cost up to 12% of the annual energy consumption in warm climate [54]. From this perspective, the effort towards reduction of the cooling energy costs [55] is only logical.

The concept of this work was based on placement of PBRs in the greenhouse that allowed for usage of solar irradiance for microalgal cultivation and simultaneously protected the cultivation system during periods of colder and rainy weather that may occur from spring to autumn in the temperate climate zone of Central Europe. On the other hand, the cultivation process in PBRs was affected by extremely warm and sunny weather in the summers of 2018 and 2019. This extraordinary hot weather is expected to become a standard in next decades and rising average temperatures that are well documented worldwide [56,57] may lead to considering relocation of microalgae cultivations outside of greenhouses in the forthcoming period.

5. Conclusions

We did not determine significant differences in biomass and lipid production among the studied PBRs. MTA PBR showed the lowest energy consumption but was more temperature sensitive than H-T PBR, even though the maximum achieved biomass production was found in MTA PBR during colder period of October 2018. Our results suggest MTA PBR as the most appropriate design for further development, especially in the context of energy efficiency. The possibilities of biomass and lipid production increase are related to the prevention of overheating (for instance by usage of infrared irradiance filtering) and the optimization of initial biomass concentration and culture depth/thickness. Our results show the need of reconsidering the concept of algae cultivation inside greenhouses in the central European region. For the development of feasible yearlong microalgal cultivation in Central Europe, it will be necessary to develop hybrid systems for microalgal cultivation, combining the advantages of the open system in summer and the closed one in winter.

Supplementary Materials: The following are available online at https://www.mdpi.com/1996-1073/14/5/1338/s1, Figure S1: Growth of *C. pyrenoidosa* in H-T PBR, MTA PBR and FPA PBR under N-limitation (as determined throughout summer–autumn 2019), Figure S2: Photosynthesis-irradiance (P-I) curves measured during 6th day and 8th day of *C. pyrenoidosa* cultivation in June, July, September and October 2018.

Author Contributions: Conceptualization, K.S., V.M., T.Z.; methodology, K.S., P.L., V.B., V.M., D.V., T.Z.; formal analysis, K.S., V.B., T.Z.; investigation, K.S., P.L., V.B., V.M., D.V., T.Z.; data curation, V.M., T.Z.; writing—original draft preparation, K.S., P.L., V.B., V.M., T.Z.; writing—review and editing, K.S., P.L., V.B., V.M., T.Z.; visualization, V.M., T.Z.; supervision, T.Z.; project administration K.S., D.V. All authors have read and agreed to the published version of the manuscript.

Funding: This research was funded by the Ministry of Education, Youth and Sports of the Czech Republic under the OP RDE grant number CZ.02.1.01/0.0/0.0/16_026/0008413 'Strategic Partnership for Environmental Technologies and Energy Production' and by Czech Science Foundation (GA ČR, grant number 18–24397S).

Institutional Review Board Statement: Not applicable.

Informed Consent Statement: Not applicable.

Data Availability Statement: Not applicable.

Conflicts of Interest: The authors declare no conflict of interest.

References

1. Brennan, L.; Owende, P. Biofuels from microalgae—A review of technologies for production, processing, and extractions of biofuels and co-products. *Renew. Sustain. Energy Rev.* **2010**, *14*, 557–577. [CrossRef]
2. Nwoba, E.G.; Parlevliet, D.A.; Laird, D.W.; Alameh, K.; Moheimani, N.R. Light management technologies for increasing algal photobioreactor efficiency. *Algal Res.* **2019**, *39*, 101433. [CrossRef]
3. Manirafasha, E.; Ndikubwimana, T.; Zeng, X.; Lu, Y.; Jing, K. Phycobiliprotein: Potential microalgae derived pharmaceutical and biological reagent. *Biochem. Eng. J.* **2016**, *109*, 282–296. [CrossRef]
4. Batista, A.P.; Niccolai, A.; Fradinho, P.; Fragoso, S.; Bursic, I.; Rodolfi, L.; Biondi, N.; Tredici, M.R.; Sousa, I.; Raymundo, A. Microalgae biomass as an alternative ingredient in cookies: Sensory, physical and chemical properties, antioxidant activity and in vitro digestibility. *Algal Res.* **2017**, *26*, 161–171. [CrossRef]
5. Adarme-Vega, T.C.; Thomas-Hall, S.R.; Schenk, P.M. Towards sustainable sources for omega-3 fatty acids production. *Curr. Opin. Biotechnol.* **2014**, *26*, 14–18. [CrossRef]
6. Sathasivam, R.; Radhakrishnan, R.; Hashem, A.; Abd_Allah, E.F. Microalgae metabolites: A rich source for food and medicine. *Saudi J. Biol. Sci.* **2019**, *26*, 709–722. [CrossRef] [PubMed]

7. Gouveia, L.; Oliveira, A.C. Microalgae as a raw material for biofuels production. *J. Ind. Microbiol. Biotechnol.* **2009**, *36*, 269–274. [CrossRef]
8. Richmond, A.; Hu, E.Q. *Handbook of Microalgal Culture: Applied Phycology and Biotechnology*; John Wiley & Sons, Ltd.: Oxford, UK, 2013.
9. Yadav, G.; Dash, S.K.; Sen, R. A biorefinery for valorization of industrial waste-water and flue gas by microalgae for waste mitigation, carbon-dioxide sequestration and algal biomass production. *Sci. Total Environ.* **2019**, *688*, 129–135. [CrossRef]
10. Garcia, R.; Figueiredo, F.; Brandão, M.; Hegg, M.; Castanheira, É.; Malça, J.; Nilsson, A.; Freire, F. A meta-analysis of the life cycle greenhouse gas balances of microalgae biodiesel. *Int. J. Life Cycle Assess.* **2020**, *25*, 1737–1748. [CrossRef]
11. Kokkinos, N.; Lazaridou, A.; Stamatis, N.; Orfanidis, S.; Mitropoulos, A.C.; Christoforidis, A.; Nikolaou, N. Biodiesel production from selected microalgae strains and determination of its properties and combustion specific characteristics. *J. Eng. Sci. Technol. Rev.* **2015**, *8*, 1–6. [CrossRef]
12. Teng, S.Y.; Yew, G.Y.; Sukačová, K.; Show, P.L.; Máša, V.; Chang, J. Microalgae with artificial intelligence: A digitalized perspective on genetics, systems and products. *Biotechnol. Adv.* **2020**, *44*, 1–17. [CrossRef]
13. Sukačová, K.; Kočí, R.; Žídková, M.; Vítěz, T.; Trtílek, M. Novel insight into the process of nutrients removal using an algal biofilm: The evaluation of mechanism and efficiency. *Int. J. Phytoremed.* **2017**, *19*, 909–914. [CrossRef] [PubMed]
14. Chen, J.Z.; Wang, S.; Zhou, B.; Dai, L.; Liu, D.; Du, W. A robust process for lipase-mediated biodiesel production from microalgae lipid. *RSC Adv.* **2016**, *6*, 48515–48522. [CrossRef]
15. Zhang, Y.-T.; Jiang, J.-Y.; Shi, T.-Q.; Sun, X.-M.; Zhao, Q.-Y.; Huang, H.; Ren, L.-J. Application of the CRISPR/Cas system for genome editing in microalgae. *Appl. Microbiol. Biotechnol.* **2019**, *103*, 3239–3248. [CrossRef]
16. Chisti, Y. Biodiesel from microalgae. *Biotechnol. Adv.* **2007**, *25*, 294–306. [CrossRef]
17. Schade, S.; Meier, T. A comparative analysis of the environmental impacts of cultivating microalgae in different production systems and climatic zones: A systematic review and meta-analysis. *Algal Res.* **2019**, *40*, 101485. [CrossRef]
18. Wang, B.; Lan, C.Q.; Horsman, M. Closed photobioreactors for production of microalgal biomasses. *Biotechnol. Adv.* **2012**, *30*, 904–912. [CrossRef] [PubMed]
19. Jorquera, O.; Kiperstok, A.; Sales, E.A.; Embiruçu, M.; Ghirardi, M.L. Comparative energy life-cycle analyses of microalgal biomass production in open ponds and photobioreactors. *Bioresour. Technol.* **2010**, *101*, 1406–1413. [CrossRef] [PubMed]
20. Pawar, S.B. Process Engineering Aspects of Vertical Column Photobioreactors for Mass Production of Microalgae. *ChemBioEng Rev.* **2016**, *3*, 101–115. [CrossRef]
21. Acién Fernández, F.G.; Fernández Sevilla, J.M.; Molina Grima, E. Photobioreactors for the production of microalgae. *Rev. Environ. Sci. Biotechnol.* **2013**, *12*, 131–151. [CrossRef]
22. Itoiz, E.S.; Fuentes-Grünewald, C.; Gasol, C.M.; Garcés, E.; Alacid, E.; Rossi, S.; Rieradevall, J. Energy balance and Environmental impact analysis of microalgal biomass production for biodieselgeneration in a photobioreactor pilot plant. *Biomass Bioenergy* **2012**, *39*, 324–335. [CrossRef]
23. Raes, E.J.; Isdepsky, A.; Muylert, K.; Borowitzka, M.A.; Moheimani, N.R. Comparison of growth of *Tetraselmis* in a tubular photobioreactor (Biocoil) and a raceway pond. *J. Appl. Phycol.* **2014**, *26*, 247–255. [CrossRef]
24. Holdmann, C.; Schmid-Staiger, U.; Hirth, T. Outdoor microalgae cultivation at different biomass concentrations—Assesment of different daily and seasonal light scenarios by modeling. *Algal Res.* **2019**, *38*, 101405. [CrossRef]
25. Pegallapati, A.K.; Arudchelvam, Y.; Nirmalakhandan, N. Energetic performance of photobioreactors for algal cultivation. *Environ. Sci. Technol. Lett.* **2014**, *1*, 2–7. [CrossRef]
26. Endres, C.H.; Roth, A.; Brück, T.B. Modeling Microalgae Productivity in Industrial-Scale Vertical Flat Panel Phototbioreactors. *Environ. Sci Technol.* **2018**, *52*, 5490–5498. [CrossRef]
27. Wolf, J.; Stephens, E.; Steinbusch, S.; Yarnold, J.; Ross, I.L.; Steinweg, C.; Doebbe, A.; Krolovitsch, C.; Müller, S.; Jakob, G.; et al. Multifactorial comparison of photobioreactor geometries in parallel microalgae cultivations. *Algal Res.* **2016**, *15*, 187–201. [CrossRef]
28. Stanier, R.Y.; Kunisawa, R.; Mandel, M.; Cohen-Bazire, G. Purification and properties of unicellular blue-green algae (Order Chroococcales). *Bacteriol. Rev.* **1971**, *35*, 171–205. [CrossRef]
29. Sukačová, K.; Búzová, D.; Červený, J. Biphasic optimization approach for maximization of lipid production by the microalga Chlorella pyrenoidosa. *Folia Microbiol.* **2020**, *65*, 901–908. [CrossRef] [PubMed]
30. Sukačová, K.; Búzová, D.; Trávníček, P.; Červený, J.; Vítězová, M.; Vítěz, T. Optimization of microalgal growth and cultivation parameters for increasing bioenergy potential: Case study using the oleaginous microalga *Chlorella pyrenoidosa Chick (IPPAS C2)*. *Algal Res.* **2019**, *40*, 101519. [CrossRef]
31. Folch, J.; Lees, M.; Sloane Stanley, G.H. A simple method for the isolation and purification of total lipides from animal tissues. *J. Biol. Chem.* **1957**, *226*, 497–509. [CrossRef]
32. Platt, T.; Gallegos, C.L.; Harrison, W.G. Photoinhibition of photosynthesis in natural assemblages of marine phytoplankton. *J. Mar. Res.* **1980**, *38*, 687–701.
33. Friedl, A.; Padouvas, E.; Rotter, H.; Varmuza, K. Prediction of heating values of biomass fuel from elemental composition. *Anal. Chim. Acta* **2005**, *544*, 191–198. [CrossRef]
34. de Morais, M.G.; Costa, J.A.V. Carbon dioxide fixation by Chlorella kessleri, C. vulgaris, Scenedesmus obliquus and Spirulina sp. cultivated in flasks and vertical tubular photobioreactors. *Biotechnol. Lett.* **2007**, *29*, 1349–1352. [CrossRef]

35. Huang, J.; Ying, J.; Fan, F.; Yang, Q.; Wang, J.; Li, Y. Development of a novel multi-column airlift photobioreactor with easy scalability by means of computational fluid dynamics simulations and experiments. *Bioresour. Technol.* **2016**, *222*, 399–407. [CrossRef]
36. Lam, M.K.; Lee, K.T. Cultivation of Chlorella vulgaris in a pilot-scale sequential-baffled column photobioreactor for biomass and biodiesel production. *Energy Convers. Manag.* **2014**, *88*, 399–410. [CrossRef]
37. Ugwu, C.U.; Ogbonna, J.C.; Tanaka, H. Improvement of mass transfer characteristics and productivities of iclined tubular photobioreactors by installation of internal static mixers. *Appl. Microbiol. Biotechnol.* **2002**, *58*, 600–607. [CrossRef] [PubMed]
38. Rodolfi, L.; Chini Zittelli, G.; Bassi, N.; Padovani, G.; Biondi, N.; Bonini, G.; Tredici, M.R. Microalgae for oil: Strain selection, induction of lipid synthesis and outdoor mass cultivation in a low-cost photobioreactor. *Biotechnol. Bioeng.* **2009**, *102*, 100–112. [CrossRef] [PubMed]
39. Scragg, A.H.; Illman, A.M.; Carden, A.; Shales, S.W. Growth of microalgae with increased calorific values in a tubular bioreactor. *Biomass Bioenergy* **2002**, *23*, 67–73. [CrossRef]
40. Feng, P.; Deng, Z.; Hu, Z.; Fan, L. Lipid accumulation and growth of Chlorella zofingienis in flat plate photobioreactors outdoors. *Bioresour. Technol.* **2011**, *102*, 10577–10584. [CrossRef]
41. Degen, J.; Uebele, A.; Retze, A.; Schmid-Staiger, U.; Walter, T. A novel airlift photobioreactor with baffles for improved light utilization through the flashing light effect. *J. Biotechnol.* **2001**, *92*, 89–94. [CrossRef]
42. Zhang, Q.H.; Wu, X.; Xue, S.Z.; Wang, Z.H.; Yan, C.H.; Cong, W. Hydrodynamic Characteristics and Microalgae Cultivation in a Novel Flat-Plate Photobioreactor. *Biotechnol. Prog.* **2013**, *29*, 127–134. [CrossRef]
43. Martinez-Guerra, E.; Gude, V.G. Energy aspects of microalgal biodiesel production. *AIMS Energy* **2016**, *4*, 347–362. [CrossRef]
44. Ördög, V.; Stirk, W.A.; Bálint, P.; Aremu, A.O.; Okem, A.; Lóvázs, C.; Molnár, Z.; van Staden, J. Effect of temperature and nitrogen concentration on lipid productivity and fatty acid composition in three Chlorella strains. *Algal Res.* **2016**, *16*, 141–149. [CrossRef]
45. Kakarla, R.; Choi, J.W.; Yun, J.H.; Kim, B.H.; Heo, J.; Lee, S.; Cho, D.H.; Ramanam, R.; Kim, H.S. Application of high-salinity stress for enhancing the lipid productivity of Chlorella sorokiniana HS1 in a two-phase process. *J. Microbiol.* **2018**, *56*, 56–64. [CrossRef] [PubMed]
46. He, Q.; Yang, H.; Wu, L.; Hu, C. Effect of light intensity on physiological changes, carbon allocation and neutral lipid accumulation in oleaginous microalgae. *Bioresour. Technol.* **2015**, *191*, 219–228. [CrossRef] [PubMed]
47. Sajjadi, B.; Chen, W.-Y.; Raman, A.A.A.; Ibrahim, S. Microalgae lipid and biomass for biofuel production: A comprehensive review on lipid enhancement strategies and their effects on fatty acid composition. *Renew. Sustain. Energy Rev.* **2018**, *97*, 200–232. [CrossRef]
48. Yeh, K.L.; Chang, J.S. Nitrogen starvation strategies and photobioreactor design for enhancing lipid content and lipid production of a newly isolated microalgae Chlorella vulgaris ESP-31: Implication of biofuel. *Biotechnol. J.* **2011**, *6*, 1358–1366. [CrossRef]
49. Piligaev, A.V.; Sorokina, K.N.; Samoylova, Y.V.; Parmon, V.N. Lipid production by microalga Micractinium sp. IC-76 in a flat panel photobioreactor and its transesterification with cross-linked enzyme aggregates of *Burkholderia cepacia* lipase. *Energy Convers. Manag.* **2018**, *156*, 1–9. [CrossRef]
50. Sarayloo, E.; Simsek, S.; Unlu, Y.S.; Cevahir, G.; Erkey, C.; Kavakli, I.H. Enhancement of the lipid productivity and fatty acid methyl ester profile of Chlorella vulgaris by two rounds mutagenesis. *Bioresour. Technol.* **2018**, *250*, 764–769. [CrossRef]
51. Khichi, S.S.; Anis, A.; Ghosh, S. Mathematical modeling of light energy flux balance in flat panel photobioreactor for Botryococcus braunii growth, CO_2 biofixation and lipid production under varying light regimes. *Biochem. Eng. J.* **2018**, *134*, 44–56. [CrossRef]
52. Illman, A.; Scragg, A.; Shales, S. Increase in Chlorella strains calorific values when grown in low nitrogen medium. *Enzyme Microb. Technol.* **2000**, *27*, 631–635. [CrossRef]
53. Slade, R.; Bauen, A. Micro-algae cultivation for biofuels: Cost, energy balance, environmental impacts and future prospects. *Biomass Bioenergy* **2013**, *53*, 29–38. [CrossRef]
54. Tredici, M.R.; Rodolfi, L.; Biondi, N.; Bassi, N.; Sampietro, G. Techno-economic analysis of microalgal biomass production in a 1-ha Green Wall Panel (GWP®) plant. *Algal Res.* **2016**, *19*, 253–263. [CrossRef]
55. Nwoba, E.G.; Parlevliet, D.A.; Laird, D.W.; Alameh, K.; Moheimani, N.R. Pilot-scale self-cooling microalgal closed photobioreactor for biomass production and electricity generation. *Algal Res.* **2020**, *45*, 1–11. [CrossRef]
56. Tripathy, A.; Tripathi, D.K.; Chauhan, D.K.; Kumar, N. Paradigms of climate change impacts on some major food sources of the worls: A review on current knowledge and future prospects. *Agric. Ecosyst. Environ.* **2016**, *216*, 356–373. [CrossRef]
57. Barati, B.; Lim, P.E.; Gan, S.Y.; Poong, S.W.; Phang, S.M.; Beardall, J. Effect of elevated temperature on the physiological responses of marine Chlorella strains from different latitudes. *J. Appl. Phycol.* **2018**, *30*, 1–13. [CrossRef]

Review

Immobilising Microalgae and Cyanobacteria as Biocomposites: New Opportunities to Intensify Algae Biotechnology and Bioprocessing

Gary S. Caldwell [1,*], Pichaya In-na [2], Rachel Hart [1], Elliot Sharp [3], Assia Stefanova [4], Matthew Pickersgill [2,5], Matthew Walker [1], Matthew Unthank [3], Justin Perry [3] and Jonathan G. M. Lee [2]

1 School of Natural and Environmental Sciences, Newcastle University, Newcastle upon Tyne NE1 7RU, UK; R.Hart1@newcastle.ac.uk (R.H.); M.J.Walker@newcastle.ac.uk (M.W.)
2 School of Engineering, Newcastle University, Newcastle upon Tyne NE1 7RU, UK; p.in-na@newcastle.ac.uk (P.I.-n.); m.pickersgill@newcastle.ac.uk (M.P.); Jonathan.Lee@newcastle.ac.uk (J.G.M.L.)
3 Department of Applied Sciences, Northumbria University, Newcastle upon Tyne NE1 8ST, UK; Elliot.b.sharp@northumbria.ac.uk (E.S.); matthew.unthank@northumbria.ac.uk (M.U.); justin.perry@northumbria.ac.uk (J.P.)
4 School of Architecture, Planning & Landscape, Newcastle University, Newcastle upon Tyne NE1 7RU, UK; a.stefanova@newcastle.ac.uk
5 Northumbrian Water Ltd., Bran Sands, Tees Dock Road, Middlesbrough TS6 6UE, UK
* Correspondence: gary.caldwell@newcastle.ac.uk; Tel.: +44-(0)1912086660

Abstract: There is a groundswell of interest in applying phototrophic microorganisms, specifically microalgae and cyanobacteria, for biotechnology and ecosystem service applications. However, there are inherent challenges associated with conventional routes to their deployment (using ponds, raceways and photobioreactors) which are synonymous with suspension cultivation techniques. Cultivation as biofilms partly ameliorates these issues; however, based on the principles of process intensification, by taking a step beyond biofilms and exploiting nature inspired artificial cell immobilisation, new opportunities become available, particularly for applications requiring extensive deployment periods (e.g., carbon capture and wastewater bioremediation). We explore the rationale for, and approaches to immobilised cultivation, in particular the application of latex-based polymer immobilisation as living biocomposites. We discuss how biocomposites can be optimised at the design stage based on mass transfer limitations. Finally, we predict that biocomposites will have a defining role in realising the deployment of metabolically engineered organisms for real world applications that may tip the balance of risk towards their environmental deployment.

Keywords: bioreactor; carbon capture; carbon dioxide; eutrophication; immobilization; latex polymers; process intensification; wastewater

Citation: Caldwell, G.S.; In-na, P.; Hart, R.; Sharp, E.; Stefanova, A.; Pickersgill, M.; Walker, M.; Unthank, M.; Perry, J.; Lee, J.G.M. Immobilising Microalgae and Cyanobacteria as Biocomposites: New Opportunities to Intensify Algae Biotechnology and Bioprocessing. *Energies* **2021**, *14*, 2566. https://doi.org/10.3390/en14092566

Academic Editor: Jaakko Puhakka

Received: 1 April 2021
Accepted: 28 April 2021
Published: 29 April 2021

Publisher's Note: MDPI stays neutral with regard to jurisdictional claims in published maps and institutional affiliations.

1. Introduction

Two of the three dominant mass microalgae and cyanobacteria (hereon microalgae) cultivation systems (ponds and photobioreactors) focus on maintaining the cells as a colloidal suspension, equivalent to the microalgae living within the planktonic state, i.e., free floating within the water column with minimal physical cell–cell or cell–substratum interactions. The third main cultivation system (biofilm bioreactors) exploits surface attachment, equivalent to the cells living within the benthic or substratum-associated state, defined by more or less continuous cell–cell and cell–substratum interactions (Figure 1a,b).

There are pros and cons for each approach, particularly when attempting to culture at industrial scale. Ponds are advantageous in terms of their simplicity (both to build and operate) and their low capital costs, but they consume large tracts of land, are inefficient with water use [1] and, if used for remediation services (e.g., wastewater treatment), the cells, i.e., the active biomass, are subject to washout from the process (hydraulic retention time, i.e., how long cultures are retained within the pond, is a critical operational parameter [2]). Further, their dependence on ambient light and temperature, combined with their vulnerability to contamination from non-target organisms (predators, pathogens, competitors) makes achieving consistent performance challenging [3–7].

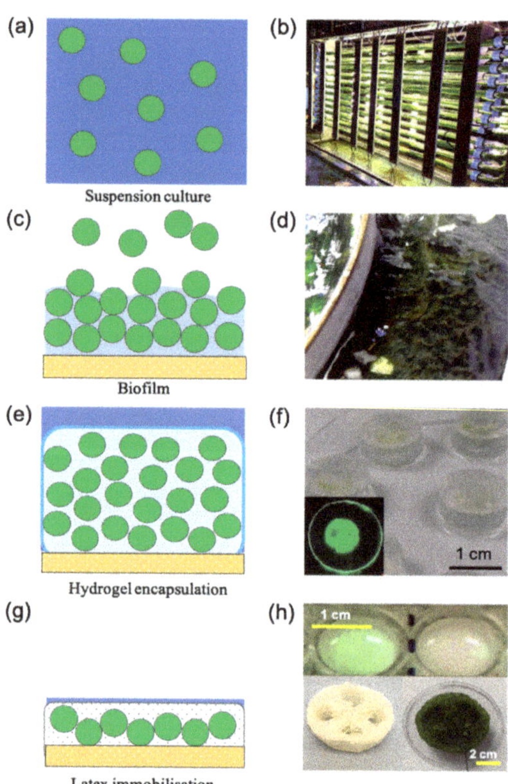

Figure 1. Comparing cell distribution and approaches for cell retention in mass microalgae cultivation: (**a**) Schematic of suspension culture—typical of ponds, raceways and photobioreactors. The cells are free floating in the growth medium, maintaining spatial separation through electrochemical repulsion which limits cell density; (**b**) artificially illuminated tubular photobioreactor using suspension cultivation for wastewater treatment; (**c**) open biofilm culture—cells attach to a substratum and biofilm cohesion is maintained through the natural production of extracellular polymeric substances (EPS). Biofilms are prone to failure leading to biomass loss; (**d**) cyanobacteria biofilm within a small raceway used for wastewater remediation; (**e**) encapsulated biofilms—cells are embedded within an artificial EPS, typically a hydrogel. Encapsulated biofilms are vulnerable to failure from desiccation of the hydrogel and subsequent loss of integrity; (**f**) microalgae encapsulated within kappa-carrageenan. Inset shows the chlorophyll fluorescence of the cells using imaging pulse amplitude modulated fluorometry; (**g**) latex biocomposites—cells are immobilised within materials other than hydrogels, such as latex; and (**h**) cyanobacteria in suspension as a biocoating in wet latex (upper left), the latex without cells is shown for comparison (upper right). The biocoating will be applied to a loofah sponge scaffold (lower left is uncoated, lower right is coated). Biocomposites have longer service lives that other immobilisation systems and, depending on the nature of the binder, can deliver orders of magnitude improvements in performance compared with the other cultivation systems.

Photobioreactors ameliorate many of the drawbacks of open ponds (reduced land and water consumption, improved control of culture conditions, and substantially reduced threat from non-target organisms) culminating in greater biomass yield [8,9]; however, these gains come at the cost of higher capital and operating costs [10].

Both approaches share some drawbacks, such as challenges around maximising carbon dioxide (CO_2) mass transfer [11]. However, other shortcomings are more explicitly linked with the colloidal suspension, notably limitations in cell density and difficulty in harvesting. The planktonic state has evolved partly to minimise intraspecific resource competition, particularly for light and nutrients (including carbon). Microalgae typically have negative cell surface charges (measured as the zeta-potential) which drives cell–cell repulsive forces and establishes stable colloidal suspensions. This allows each cell to occupy its own space (c.f. sphere of influence) within the water column, together with the light and nutrients associated with that space. Further, by limiting physical cell–cell interactions, the microalgae (many of which have no means of independent motility) reduce the risk of floc formation (flocculation) which would otherwise increase the cell's sinking rate (Stoke's Law) from sunlit surface waters (photic zone), thereby compromising photosynthesis and threatening cell survival.

In stark contrast, algae that constitute biofilms have made a virtue out of a necessity. Rather than investing in means to maintain spatial separation, benthic microalgae proactively engage in intimate associations with the substratum, as well as actively encouraging close association with both con- and heterospecifics. This benthic niche negates the risk of sinking from the photic zone (the cell is attached to a stable surface), thus enabling the cells to focus on beneficial cell–cell interactions rather than treating these antagonistically. Nevertheless, the electrochemical repulsive forces must still be overcome. This is achieved through the production of copious extracellular polymeric substances (EPS) which function both as glue and medium through which the cells may move (many benthic microalgae are motile, particularly diatoms, often using EPS production and release to facilitate gliding movement).

Whereas the evolutionary benefit of overcoming the existential threat of sinking is recognised, it does come at a cost. Biofilm dwelling cells compromise on the guaranteed access to 'wrap around' space and resource beholden of their planktonic siblings. Whereas the living space along the xy-plane is only limited by the dimensions of the substratum, occupation of the z-plane is strictly limited by light penetration and EPS cohesiveness, particularly under shear stress. This commonly results in biofilm failure (Figure 1b), which may be accidental or deliberate (many biofilm organisms maintain the capacity to escape the biofilm as a means of dispersal).

Biofilm or attached microalgae cultivation, both as monospecific and mixed cultures, have been developed as alternatives to suspension culture [12,13], particularly for wastewater treatment applications. Biofilm formation can be initiated by adhering a concentrated algae cell paste to a solid supporting material such as glass fiber reinforced plastic [14], filter paper [15], cellulose ester membrane [16], polyester [17,18], polypropylene, [19], polyvinyl chloride, polyacrylonitrile [20], polytetrafluoroethylene [21], polycarbonate, cellulose acetate/nitrate, or polyethylene membranes [22], polyurethane [23,24], nylon or stainless steel mesh [25] and even concrete [26]. Natural fibres such as cotton, chamois cloth and loofah sponge (aka Luffa plant) have also been trialled [27–29] and have successfully outperformed suspension culture controls [30,31]. Subsequently, the cells produce EPS to assist surface attachment, forming a stable matrix [32].

Biofilm cultivation may be divided into three types: (1) permanently immersed in a liquid medium, (2) biofilms that alternate between gaseous and liquid phases, and (3) permeated biofilms wherein liquid medium is delivered through the substratum [33] (Figure 2). The productivity of two phase systems can greatly exceed that of raceways [34], although they generally require rotation or a rocking motion to expose the biofilm to both gaseous and liquid phases [35,36], although this may be difficult to scale up. Most studies with permeated biofilms have arranged their systems vertically to minimise the footprint and thereby the land requirement.

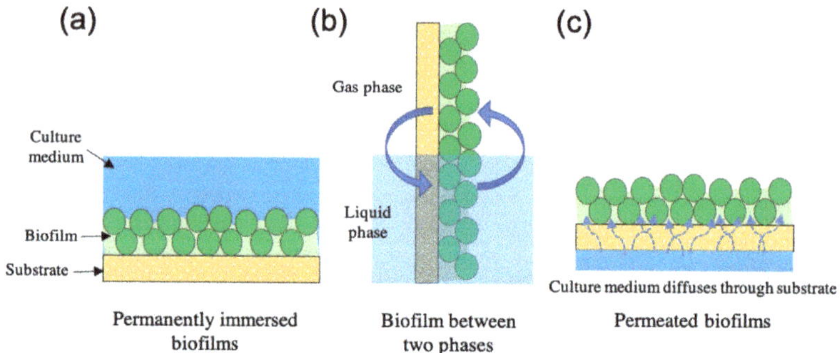

Figure 2. Illustration of: (**a**) a permanently immersed biofilm in which the cells have continuous access to water and nutrients. A degree of mixing of the growth medium is required to prevent nutrient and CO_2 exhaustion at the boundary between biofilm and water (solid–liquid interface); (**b**) a biofilm between two phases whereby rotational or oscillatory motion of the biofilm is used to ensure the biofilm is sequentially exposed to liquid and gas phases—this ameliorates the need to mix the growth medium but presents challenges around the impacts of shear stress on biofilm cohesion; and (**c**) a permeated biofilm in which capillary forces are utilised to wick growth medium through the solid support on which the biofilm is attached. This approach requires neither mixing or motion but is dependent on the selection of an appropriate porous and hydrophilic substratum [13].

Biofilms are intuitively attractive from a mass cultivation perspective as they consume less water (EPS is strongly hydrophilic, conferring good gelling properties and promoting the biofilm to remain hydrated), the cells have evolved to grow in high density situations (enabling the footprint of any culture operation to be reduced), light penetration is more easily controlled, CO_2 mass transfer is improved, and biomass harvesting and dewatering are much easier. However, technical challenges remain; notably the prevalent contamination threat from non-target organisms, the potential for the biofilm to excessively desiccate, and of course the risk that the biofilm may fail as part of the natural biofilm lifecycle. There is the added complication that not all microalgae are amenable to (or even capable of) biofilm formation, i.e., the genetic ties to the planktonic existence are so deep rooted that other than through radical intervention (e.g., gene editing) such species must be deemed non-starters. Or must they?

There is a "Fourth Way" [37] to mass microalgae cultivation that involves the deliberate immobilisation of microalgae within 'engineered biofilms' or 'living biocomposites' [38–40] (Figure 1c,d). A number of studies have developed biocoatings (a binder containing live microbes) and biocomposites (a biocoating applied to a supporting structural material) with different microorganisms for a range of environmental applications including biofuel production, gas and chemical synthesis, environmental remediation, and as biosensors [41–44]. Many have successfully reported that biocoatings and biocomposites can intensify biological process and performance relative to suspension cultures [45–47]. In this article, we explore the rationale for, and approaches to immobilised cultivation, focusing on latex-based polymer immobilisation, and discuss how biocomposites can be optimised at the design stage based on mass transfer limitations.

2. Immobilised Cultivation

There are six immobilisation types: (1) affinity immobilisation, (2) adsorption, (3) covalent coupling, (4) confinement in a liquid-liquid emulsion, (5) capture behind a semi-permeable membrane, and (6) entrapment within polymers [40] (Figure 3). These techniques can be separated into passive and active methods. Passive immobilisation utilises the natural attachment ability of microorganisms to natural or synthetic surfaces, while active immobilisation uses artificial techniques including flocculent agents, chemical attachment, and gel/polymer entrapment [48,49]. Affinity immobilisation is a very mild method and is based on complementary biomolecular interactions which do not involve drastic reactions and no chemical exposure. The method is often used for purification or separation of biomolecule mixtures and a desorption step is required to extract compounds from the immobilised substrate [50]. Adsorption immobilisation is a reversible process involving cells that strongly adhere to the sorbent. Covalent coupling is a well-known immobilisation technique for enzymes, but not for living cells because cell division can lead to cell leakage from loose bonding. Confinement in liquid-liquid emulsions is an aqueous method in which phase separation occurs from two different water-soluble polymers based on their surface properties [40]. For semi-permeable membranes, the cells are immobilised into the membrane and this technique is often used for biosensor fabrication. However, this method causes excessive accumulation of biomass growth on the substrate, which leads to pressure build up and damages the membrane [40]. Entrapment and encapsulation in polymers are the most common immobilisation methods, in which the cells are captured in a matrix made from synthetic polymers (e.g., acrylamide, polyurethanes, polyvinyl, polystyrene), proteins (e.g., gelatine, collagen, egg white) or natural polysaccharides (e.g., agars, carrageenan, alginates) [49]. These two techniques have been widely used to immobilise many microalgae species on various polymers for wastewater applications.

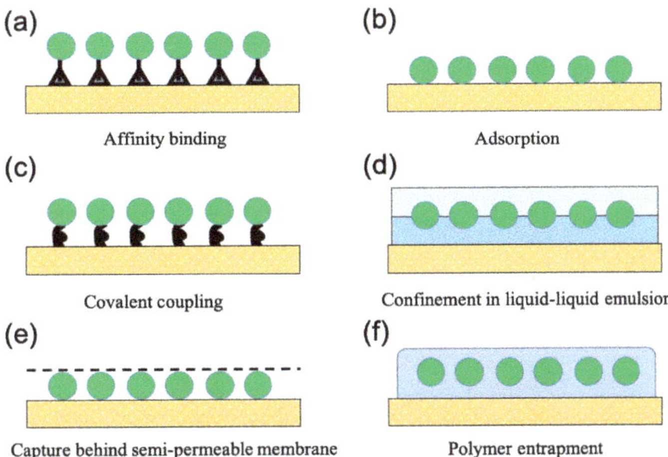

Figure 3. Illustration of the six main approaches to immobilisation: (**a**) affinity binding, which is based on complementary biomolecular interactions; (**b**) adsorption, which is a process involving cells that strongly adhere to a sorbent; (**c**) covalent coupling, which is effective for enzymes but not for cells; (**d**) confinement in liquid-liquid emulsion, which exploits phase separation between two different water-soluble polymers based on their surface properties; (**e**) capture behind a semi-permeable membrane, in which cells are immobilised into the membrane, however this method is prone to failure of the membrane in response to high pressure from excessive accumulation of biomass; and, (**f**) polymer entrapment, in which cells are embedded in a matrix comprising synthetic or natural polymers.

3. Effects of Immobilisation on Microalgae

Biocomposites include one or more discontinuous phases that provide solid support, and a continuous phase creating a matrix around the discontinuous [51]. Biocomposites comprising natural fibres (plant, animal, and mineral materials) are widely used due to their high tensile strength, low weight, and resistance to degradation [52,53]. Biocomposites retain biomass whilst allowing the exchange of molecules across a semi-permeable membrane [39]. It is important to note that biocomposites are different from biofilm reactors. The matrix in which cells are held should also act as protection from non-target organisms whilst still providing room for cellular maintenance and growth [54]. Biocomposites should support greater biomass within a smaller surface area, supported by improved cell retention [47]. This in turn confers greater flexibility in bioreactor design compared with suspension or biofilm counterparts.

Immobilising microalgae can improve photosynthetic rate, growth, and pigment and lipid content compared with suspension cultures [49,55,56]. Some studies have reported toxicity of some polymer immobilising techniques, with chemical forces and interactions between the matrix and the cell wall causing significant stresses on both the material and the entrapped microorganisms [39]. Mallick [40] summarised the fundamental elements for effective immobilisation and argues for ideal properties of the matrix. Immobilisation should retain cell viability and metabolic activity at high cell densities (through photosynthesis if using phototrophs), with low levels of cell loss from the matrix. To achieve this, the immobilising matrix should be non-toxic, appropriately transparent, chemically stable when exposed to growth media, and be mechanically robust.

Many immobilisation studies have focussed on entrapment within gel-like matrices such as alginate, carrageenans or chitosan [57–59]. There are several problems with gel immobilisation, notably cell leaching due to matrix breakdown over time. Furthermore, as gels are often up to 90% water, they are prone to rapid desiccation resulting in poor mechanical properties when rehydrated [60].

Alginates are permeable and transparent mannuronic and guluronic acid polymers from brown macroalgae which are cheap, nontoxic, and easy to process [61]. The microorganisms do not experience extreme physicochemical conditions during the immobilisation process [39]. Despite the benefits, the polymeric structure cannot be maintained when high phosphate concentrations or high cation levels (e.g., K^+ and Mg^{2+}) are present [62]. Similarly, sodium alginate beads degrade when exposed for more than two weeks to wastewater with high phosphorous and nitrogen levels [63]. However, their mechanical resistance can be doubled if mixed with 5–10 kDa chitosan [64]. Carrageenans are polysaccharides extracted from red algae, and are known to support microalgae growth and metabolism [65,66]. Iota carrageenan forms clear elastic gels in the presence of calcium salts, and is thermo-reversible at 50–55 °C [67]. Lambda-carrageenan only forms high viscosity solutions [68] and cannot be used for microalgae biocoatings. Kappa-carrageenan can be mixed with potassium salts to produce strong and rigid set gels or with calcium salts to form brittle gels [68]; and although opaque, they can be made transparent with the addition of sugar [67]. Chitosan, a linear amino polysaccharide of β-D-glucosamine [39], is a promising microalgae immobilising agent [69–71]. It is insoluble in neutral and alkaline solutions which allows the gel structure to be formed at room temperature [72].

The use of gel-hardeners such as $CaCl_2$ can compromise bioprocess efficiency and may result in cell loss [73]. Furthermore, increasing the thickness of a chitosan or alginate matrix to prevent cell leaching will only reduce the mass transfer of target nutrients such as N, P, and CO_2 [74]. The use of cross-linkers in hydrogel systems can cause intracellular damage to microalgae; most likely due to the presence of glutaraldehyde [49].

4. Latex Polymer Immobilisation

Latexes are widely used in the printing and coatings industries to ensure strong adhesiveness, and greater flexibility and porosity of the products [74,75]. Latex coatings can be made with high precision and with controllable thickness using Mayer rod drawdown. The technique is limited to coating thicknesses of >10 μm although it can be used to produce multi-layered coatings [76]. Convective assembly is an alternative method enabling the creation of layer-by-layer coatings with rapid, well-ordered, and scalable fabrication [77].

Latex films form in three stages; (1) consolidation—the evaporation of water to form a packed latex particle network, (2) compaction—particle deformation begins to eliminate void space between latex particles, and (3) coalescence—polymer diffusion is initiated between particles in contact with one another to form the complete film (Figure 4) [78]. The first immobilisation of microorganisms (yeast and *E. coli*) using latex polymers was conducted in 1991 [79]. The mixture was coated onto an activated carbon particle mesh and calcium carbonate was mixed into the biocoating to generate porosity. Since then, several studies have immobilised many microorganisms (mostly bacteria) with latex polymers onto different solid supports [47,74,80]. *Synechococcus* PCC6301 immobilised with different latex emulsions on a carbon electrode maintained nearly 100% of its photosynthetic activity upon rehydration [81]. Photosynthetic microorganisms have also been immobilised onto filter paper using acrylate copolymer latex polymers to produce artificial leaf biocomposites for hydrogen production with a service life of over 1000 h. The specific photosynthetic rate was enhanced by up to 10 times compared to the suspension controls [82,83].

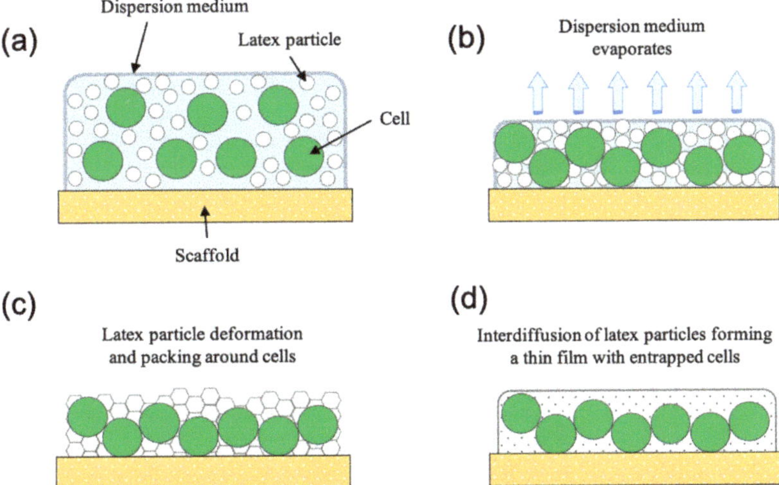

Figure 4. Illustration of the formation of latex films on a surface: (**a**) the latex-cell suspension is deposited onto the scaffold surface. At this stage the cells and latex particles have some scope for movement; (**b**) as the dispersion medium evaporates the cells and latex particles become confined to a thin layer and the immobilisation process has commenced; (**c**) following the evaporation of the dispersion medium the latex particles deform and pack around the cells, both attaching them to the scaffold and creating a protective film at the atmosphere boundary; and, (**d**) immobilisation is complete following the interdiffusion of the latex particles.

In addition to being low cost and with easy access to non-toxic waterborne latex emulsions [46,75], there are numerous biophysical advantages of using latex for cell immobilisation. High cell loadings can be used (500–1000 fold greater than for suspension culture), coinciding with good longevity of the microbes (>1000 h after re-hydration [76,84,85]), particularly if osmoprotectants (e.g., glycerol) are added [47,74,77], and mass transfer rates can be improved due to the capacity to produce very thin coatings.

This is not to say that latex based immobilisation is without issues—film formation and therefore subsequent efficacy of the biocomposites can be affected due to the coating formulation, glass transition temperature (Tg), particle size and distribution, particle morphology, drying and humidity, substrate choice, and biomass loading [80,86–89]. The drying temperature can have significant effects on latex coalescence on porous substrates, where rapid absorption of the aqueous phase into the substrate may result in cracking of the latex [90]. There is also concern regarding gas and liquid mass transfer to immobilised microbes which may limit cellular productivity. There is an inverse relationship between film thickness and mass transfer, therefore film thickness should be minimised whilst ensuring microbes remain immobilised yet metabolically active [85,91]. Therefore, a balance must be struck between latex thickness (which will impact cell retention) versus cells remaining metabolically active with mass transfer of CO_2 and nutrients via the growth media not being limited.

There are methods to mitigate these, for example, latexes that do not require drying to adhere to paper, thus reducing osmotic stress [92], or application methods that deliver cell monolayers [93] that reduce self-shading without compromising cell viability. Another intriguing alternative is arrested coalescence using non-film forming particles (bimodal blends) mixed into the coating formulation to increase the porosity of coatings, which have been used with fungi and bacteria [77,94]. However, unless the particles are sufficiently transparent, bimodal blends may decrease light transmission. In addition, the non-film forming particles have to be non-toxic, inert (non-reactive to latex), ideally smaller sizes than the immobilised cells, and larger than latex particles [47]. However, this approach may increase the total capital cost of the biocomposites and may risk secondary pollution if non-biodegradable additives are used, e.g., microplastics.

There is an additional and as yet poorly understood factor, the biological response of the immobilised organism. The breadth of microalgae diversity that has been trialled for immobilisation is very small, generally reflecting model organisms and those considered as laboratory and industry 'workhorses'. Additionally, species that have been immobilised (e.g., *Synechococcus elongatus, Chlorella vulgaris, Nannochloropsis oculata, Dunialella salina*) tend to be among those considered as structurally robust and capable of withstanding the physical and chemical stressors associated with film formation and the consequent low-water existence. An acid test for ubiquitous tolerance for immobilisation would be to use more fragile species, notably among the dinoflagellates. Intuitively, we would expect low successes during immobilisation; however, once the cells are immobilised (assuming no damage) the structural support afforded by the latex may paradoxically extend cell longevity. Naturally, this is speculative, but it is an intriguing target for future work. Further, the capacity of microalgae to perform once immobilised is not universal. Indeed, we have documented variation in tolerance to immobilisation both across and within species [85,95]. The precise reasons for these variable responses are not known and will required detailed transcriptomic and metabolomic investigation.

5. Bioinspiration from Lichen

Many materials have been assessed for their suitability to support biofilm and biocoatings, although these have mostly been synthetic materials which may pose sustainability concerns. Natural, or even repurposed materials should improve the overall sustainability of biocomposites. Examples include recycled textiles which could function as 2D scaffolds [96]. However, it is important to consider rugosity when selecting textiles as the increased number of surface microstructures increases the hydrophobicity which may affect the formation of a uniform film [97]. Porosity ultimately impacts the homogenous formation of latex films, with more porous structures resulting in faster wicking leading to non-uniform film formation [98]. When utilising woven fabrics, the diameter and spacing of the weave affects the size of the inter-yarn pores which can increase the swelling capacity of the textile [99]. Additionally, highly porous fabrics such as cotton, have high levels of pore-collapse due to structural fragility leading to alterations in the number and distribution of pores [100]. Conversely, the pore size of non-woven fabrics is affected by fibre density, with less dense fibres having larger pores [101] which may lead to poorer microbial retention.

However, from biomedical applications it is well documented that the behaviour of cells grown on 2D surfaces differ to those in 3D matrices [102,103]. 3D scaffolds sustain improved cell proliferation and metabolic activity [104]. The ability to develop 3D structures supporting metabolically active cells permits the development of customisable culture systems [105]. However, before the nascent microalgae biocomposite field can afford serious consideration of 3D printed fabrication, we must advance the development of more affordable and easily accessible options. To progress from 2D to 3D biocomposite systems [95,106], we have drawn inspiration from lichens—ancient composite organisms comprising fungi and photobionts (microalgae and/or cyanobacteria). More than 12% of Earth's land mass is lichen covered [107]. The photobionts live beneath a thin fungal layer (cortex) (Figure 5) which protects them from extreme environmental fluctuations (particularly desiccation). The fungus, which does not harm their photobionts but does influence their growth and cell turnover rates, benefits from the excess carbohydrates produced by the embedded photobionts. Most lichens can tolerate drought, extreme temperatures, can survive under nutrient scarcity and, where necessary can hibernate as part of a dormancy state [108]. These self-sustained microecosystems demonstrate that algae/cyanobacteria can live long and stable lives and operate under minimal quantities of water.

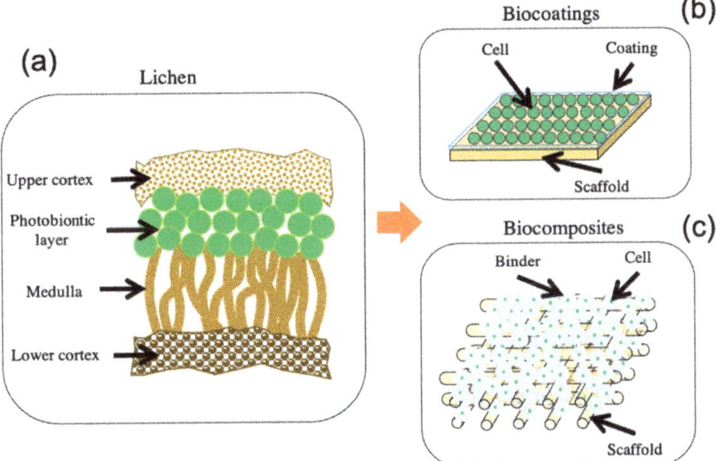

Figure 5. (**a**) Lichens are composite organisms comprising a fungus and a photobiont (microalgae and/or cyanobacteria). The fungal cortex secures and protects the photobiont which donates excess carbon (photosynthate) to its fungal host. The lichen structure provided the inspiration to evolve 2D biocomposites, wherein cells are deposited as a monolayer onto a flat scaffold such as paper (**b**) into 3D biocomposites, wherein cells are deposited on highly porous 3D scaffolds such as loofah sponge with the aid of non-toxic polymer binders, e.g., latex (**c**).

In this context, we have adopted the fibrous skeleton of *Luffa*—a member of the Cucurbitaceae family—commonly referred to as luffa or loofah sponge, as an exemplar sustainable, biodegradable 3D scaffold for algae biocomposites (Figure 6). The highly porous high surface area loofah structure (circa 950 m^2 m^{-3} with >80% void space) supports excellent gas exchange, facilitates reasonable light transmission and, with its hydrophilic nature is effective at retaining moisture within the structure [85,109]. We have demonstrated that loofah-based microalgae and cyanobacteria biocomposites can operate for many weeks as a means to biological CO_2 capture without marked reduction in performance and with negligible maintenance requirements. In particular, the pairing of cyanobacteria with latex binders yielded a carbon capture potential to rival any existing algae-based system. Further, preliminary techno-economic analysis revealed that a scaled system would have a lower annualised CO_2 avoidance cost than the closest algae comparator (a biofilm photobioreactor), with the added benefit of substantially reduced water and energy consumption [In-na et al. unpublished]. Subsequently, we have achieved substantial performance improvement by further optimising our biocomposite formulation (specifically that of the latex binder) [In-na, Sharp et al., unpublished]. Considering that the pre-optimised system was already rivalling the best performing (and long established) algae suspension and biofilm photobioreactors for CO_2 capture, we see great scope for the development and roll-out of biocomposites across the gamut of microalgae and cyanobacteria biotechnology applications.

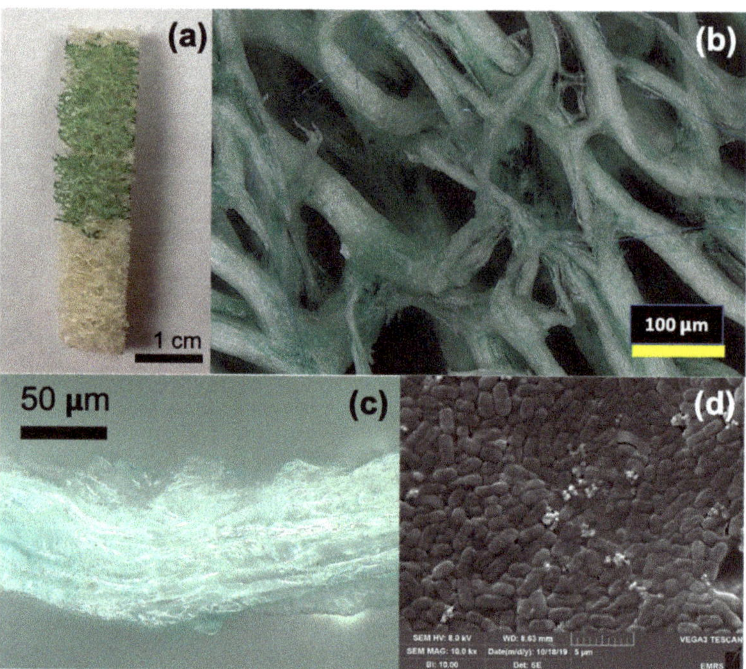

Figure 6. An example of a laboratory scale loofah-cyanobacteria biocomposite. (**a**) Prepared loofah strip coated with cyanobacteria (*Synechococcus elongatus*) enriched latex. (**b**) Higher magnification image of the coated loofah strands demonstrating the high surface area and high porosity of the scaffold. (**c**) Closer focus on an individual coated loofah strand. The latex coating is evidenced by the reflection of light. (**d**) SEM image of an *S. elongatus* biocoating demonstrating the close packed nature of the cells with the thin film latex biocoating.

6. Biocomposites as Process Intensification

Despite photobioreactors increasing the productivity and yield of suspension-based microalgae cultivation, scale-up inefficiencies continue to limit its economic feasibility. Process intensification (PI), conceptualised in the mid-1990′s, is often used to reduce the physical size of operations while achieving set production objectives [110], and is applicable to mass microalgae culture [111]. Moving to immobilised minimal water cultivation would be a clear PI step.

Posten [112] defined photobioreactors in terms of four-phases: (1) solid phase—cells; (2) liquid phase—culture medium; (3) gas phase—air or CO_2 enriched air; and (4) radiation phase—light. This classification remains appropriate for biocomposites, albeit on differing spatial (and perhaps even temporal) scales, and certainly with a shift in emphasis between the phases. Consider the solid phase. Posten's definition includes only the cells. This comparison holds true for biocomposites; albeit biocomposites present substantially greater cell densities than encountered in suspension culture. However, in a biocomposite the solid phase is dominant whereas the liquid phase dominates suspension cultivation and we must not only consider the cells but must also accommodate the influence of the scaffold and the binder. Figure 7 presents a scenario within a cyanobacteria biocomposite. If we consider water and nutrient transport, these molecules must traverse the liquid–solid interface separating the wetted film and the scaffold, to then encounter the solid–solid interface that defines the join between the scaffold and the binder. Clearly, were either scaffold or binder to be hydrophobic this would present a considerable barrier to water and solute transport. A key design consideration is therefore the surface properties of the biocomposite materials. A second solid–solid interface (the boundary between the binder and the cell wall) must then be crossed. The situation is more complex from a CO_2 mass transfer perspective. CO_2 has two main routes into the biocomposite; either transported as HCO_3^- along with water, requiring mass transfer from the gas phase into the liquid phase across a gas–liquid interface, and then following the path as described above, or by direct transfer from the gas phase across the binder where it may react with water to form HCO_3^- either within the binder or at the binder–cell interface. Both paths necessitate a degree of binder porosity. In both cases, these are predominantly physical interactions. There is added complexity once the inputs (water, nutrients, CO_2, and HCO_3^-) encounter the cell, whereby further interfaces must be crossed, e.g., cell wall/plasma membrane, in addition to other intracellular barriers, e.g., carboxysome wall. However, in this respect biology delivers effective solutions such as transmembrane proteins (assuming that the organism remain viable) and the CO_2 concentrating mechanism [113].

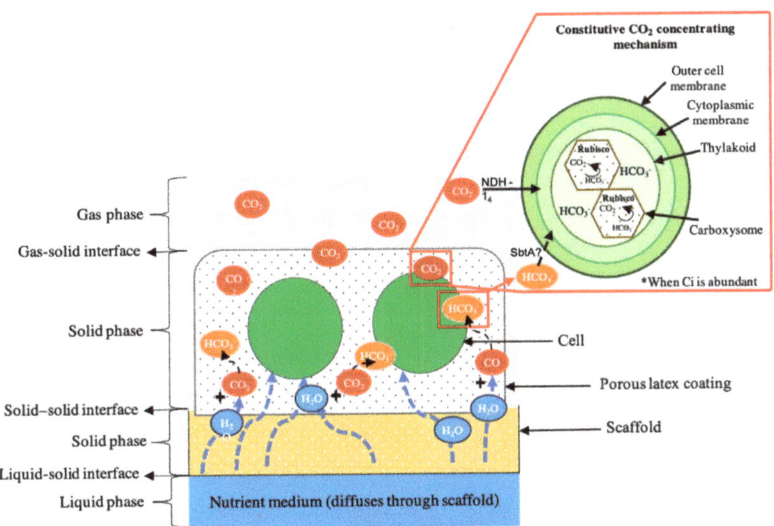

Figure 7. Illustrating the complexities of polymer entrapment to develop living and metabolically functional cyanobacteria biocomposites, with interactions with the CO_2 concentrating mechanism to overcome photorespiration. The biocomposite design brief should specify the effective transport of water and nutrient solutes through capillary forces (wicking). The inclusion of hydrophilic scaffolds enables low moisture operation. The binder requires porosity without compromising integrity. Capacity for mass transfer of CO_2, O_2, water and nutrient solutes is essential to support the cell.

Light transmission is a further critical design consideration. Based on cell growth, photobioreactors may be considered as comprising three distinct light transfer zones. Firstly, as light passes through the photobioreactor wall the intensity can exceed the cell's ability to photosynthesise, having an inhibitory effect (photoinhibition). Light then reaches a maintenance zone in which cells can balance light and nutrient resources for sustained growth. Through self-shading, a reduced light transmission zone propagates towards the middle of the photobioreactor, reducing growth [114]. These issues can be ameliorated at the photobioreactor design stage; for example, access to light may be altered by changing the tube or plate diameter, by altering the construction materials or by regulating the quantity and quality of light, i.e., through artificial illumination. However, post-construction, the available options are more limited. Mixing is a mainstay of operational adjustment, facilitating homogenous light exposure and access to nutrients and CO_2. When we consider these issues for biocomposites, at first glance it would appear that our capacity to fine tune the system is curtailed. For instance, biocomposites are default thin-film systems, thereby the capacity to regulate light transmission by changing culture vessel dimensions is limited. Equally, mixing is not an option. Intuitively, the cells within biocomposites (particularly those at the light facing surface) should be extremely vulnerable to photoinhibition and, given the high cell loadings that define biocomposites, there is a real risk that cells immobilised deeper within the biocomposite structure will be light limited, leaving little margin for error in designing the biocomposite equivalent of the photobioreactor maintenance zone. However, these potential issues can be designed out with comparative ease; indeed, it is possible to craft interesting opportunities out of apparent adversity.

Firstly, it is essential that any binder allows adequate transmission of photosynthetically active radiation (wavelength: 400–700 nm), otherwise photosynthetic performance will be compromised. Latex, particularly when applied as a thin film, supports very good light transmission (Figure 8a). In situations where potentially damaging irradiances may be encountered it is possible to combat this either by altering the latex formulation (Figure 8b), or by the inclusion of reflective particles within the binder. Whereas there may be some loss of efficiency during low light periods, this may be an acceptable trade-off in particular situations to manage photoinhibition. In relation to self-shading driven light limitation, this is easily addressed by using very thin binder applications, potentially to the extent of creating cell monolayers. This avoids self-shading entirely and, paradoxically, favours the adoption of low light conditions. Such a configuration is particularly suited to indoor cultivation using artificial light and may translate into substantial energy savings.

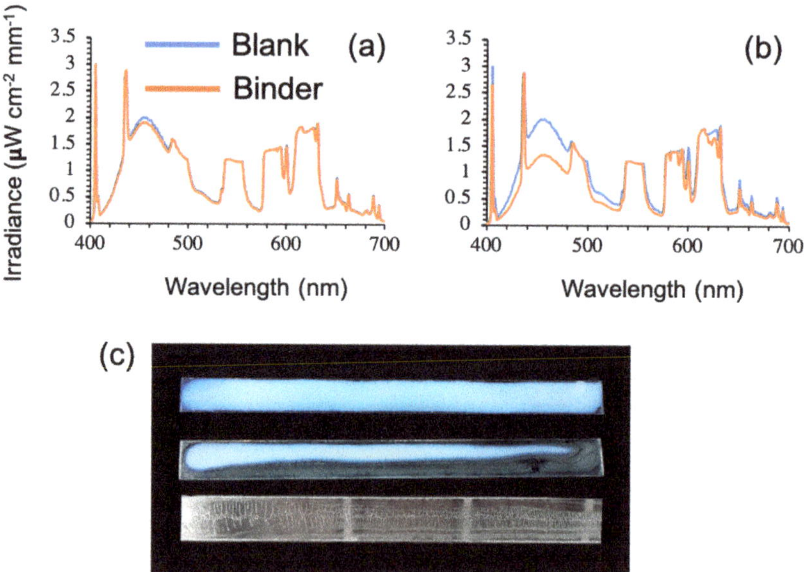

Figure 8. Latex based binders efficiently transmit photosynthetically active radiation to the cell (**a**), yet they may also be designed to filter certain wavelengths (**b**) such that photoinhibition may be managed in high light environments, (**c**) latex film formation on glass, with the upper image showing freshly applied wet latex (opaque), progressively becoming transparent as it dries (middle image), forming a fully transparent film (bottom image) [In-na, Sharp et al. unpublished].

7. Immobilisation for Synthetic Biology

There remains a significant gap in understanding the organisms' response to immobilisation. To progress the technology to commercial application, an increase in system productivity and reproducibility is required. Systems biology is providing greater insight into cellular and metabolic processes, and can be exploited to metabolically engineer microorganisms for enhanced performance [115]. Synthetic biology combines genetic engineering, systems biology, and computational modelling to design biological parts and systems, or redesign existing systems for greater productivity; tailoring the organism to work as effectively as possible, e.g., towards greater production of biochemicals, and by improving photosynthesis by enhancing light harvesting efficiency and CO_2 fixation [116,117]. Wastewater treatment already relies, to some degree, on photosynthetic microbes for bioremediation; this could be intensified by enhancing the robustness of target synthetic microbes, simultaneously increasing CO_2 capture [118]. Enhanced biosorption and biotransformation of nutrients or pollutants from wastewater will lead to higher treatment efficiency, but this is still in its infancy and limited to laboratory scale experiments. However, there are concerns about the use of synthetic microbes [119]. Accidental release could have significant, permanent effects on ecosystems. With large-scale cultivation, the escape of synthetic strains is inevitable, thus care should be taken to ensure the strain cannot establish outside of a controlled environment [120]. Containment within photobioreactors provides less risk as reactor design could feature mitigation measures such as sterilisation prior to release of wastewater, but open-ponds must have enhanced containment such as catchment areas, filters, or UV irradiation, and monitoring strategies when containing synthetic strains [121]. Physical controls can be extremely costly, so biocontainment is recommended by engineering strains to have specific requirements for survival that would not be naturally found.

The development of risk assessments is complex as the fitness of the synthetic strain, all native species of photosynthetic microbes, and all environmental perturbations must be accounted for if models for risk assessment are to have any real meaning [120]. The immobilisation of synthetic strains potentially alleviates some of the concerns about the accidental release by acting as another physical control. When immobilised onto loofah there was just 0.61% release of *S. elongatus* PCC 7942 after 72 h [85]. If both physical and genetic controls were applied, the risk of release to environment could be significantly reduced.

8. Conclusions

We have set out the rationale for immobilisation culture of microalgae as biocomposites within a process intensification framework, with a focus on latex-based systems, particularly for applications that demand prolonged environmental exposure (e.g., carbon capture), are vulnerable to biomass washout (e.g., wastewater bioremediation), or that require secure retention of the cells (e.g., synthetic biology). The choice of organism to immobilise should be driven by the end use of the process. In this article we have focused on applications for bioremediation (air and water), but the technology can be used for bioproducts synthesis, and we see great scope for roll-out as biosensors, particularly given the biocomposites' longevity and low levels of maintenance. Several improvements can be made to increase the performance of biocomposites, aside from targeted species selection. Further study on latex formulations can be used to optimise cell viability, gas mass transfer and photosynthetic performance. Parameters such as cell loading, total solids content of the binder, light intensity, light cycle, and nutrient concentration can greatly influence CO_2 uptake but have not been adequately explored. For example, increasing cell loading can lead to higher competition for nutrients meaning nutrient concentration has to be increased. Design criteria for large-scale and longer-term use of biocomposites should consider the following key points: (1) the need for a biologically safe coating method that delivers a uniform coating throughout the scaffold; (2) the spatial arrangement of the biocomposite should not hinder light penetration to the cells that lie in the inner parts of the biocomposite; (3) the need for a nutrient delivery system that assists/augments the capillary action of loofah without compromising the structural integrity of the binder should be considered; and (4) the containment infrastructure for the biocomposites (if needed) should be cheap but robust enough to support long term operation as well as provide transparency for light penetration.

Author Contributions: All authors contributed to the conceptualisation of the article, writing—original draft preparation, G.S.C., P.I.-n., R.H., J.G.M.L.; writing—review and editing, E.S., M.P., M.W., M.U., J.P. All authors have read and agreed to the published version of the manuscript.

Funding: G.S.C. and R.H. received funding from EnviResearch, P.I. was supported by an International Postgraduate Scholarship (Reference number: 400345497) from the School of Engineering, Newcastle University; J.P. and E.S. were funded by the Engineering and Physical Sciences Research Council EPSRC (EP/R020957/1), M.P. was funded by Northumbrian Water Ltd., and A.S. was supported through Research England as part of the Hub for Biotechnology in the Built Environment.

Institutional Review Board Statement: Not applicable.

Informed Consent Statement: Not applicable.

Data Availability Statement: The data presented in this study are available on request from the corresponding author.

Conflicts of Interest: The authors declare no conflict of interest. The funders had no role in the writing of the manuscript, or in the decision to publish.

References

1. Chisti, Y. Biodiesel from microalgae. *Biotechnol. Adv.* **2007**, *25*, 294–306. [CrossRef]
2. Supriyanto Noguchi, R.; Ahamed, T.; Rani, D.S.; Sakurai, K.; Nasution, M.A.; Wibawa, D.S.; Demura, M.; Watanabe, M.M. Artificial neural networks model for estimating growth of polyculture microalgae in an open raceway pond. *Biosyst. Eng.* **2019**, *177*, 122–129. [CrossRef]
3. Chandra, R.; Amit Ghosh, U.K. Effects of various abiotic factors on biomass growth and lipid yield of *Chlorella minutissima* for sustainable biodiesel production. *Environ. Sci. Pollut. Res.* **2019**, *26*, 3848–3861. [CrossRef] [PubMed]
4. Flynn, K.J.; Greenwell, H.C.; Lovitt, R.W.; Shields, R.J. Selection for fitness at the individual or population levels: Modelling effects of genetic modifications in microalgae on productivity and environmental safety. *J. Theor. Biol.* **2010**, *263*, 269–280. [CrossRef]
5. Hidasi, N.; Belay, A. Diurnal variation of various culture and biochemical parameters of *Arthrospira platensis* in large-scale outdoor raceway ponds. *Algal Res.* **2018**, *29*, 121–129. [CrossRef]
6. Sutherland, D.L.; Howard-Williams, C.; Turnbull, M.H.; Broady, P.A.; Craggs, R.J. Seasonal variation in light utilisation, biomass production and nutrient removal by wastewater microalgae in a full-scale high-rate algal pond. *J. Appl. Phycol.* **2014**, *26*, 1317–1329. [CrossRef]
7. Unnithan, V.V.; Unc, A.; Smith, G.B. Mini-review: A priori considerations for bacteria–algae interactions in algal biofuel systems receiving municipal wastewaters. *Algal Res.* **2014**, *4*, 35–40. [CrossRef]
8. Ugwu, C.; Aoyagi, H.; Uchiyama, H. Photobioreactors for mass cultivation of algae. *Bioresour. Technol.* **2008**, *99*, 4021–4028. [CrossRef] [PubMed]
9. Delrue, F.; Setier, P.-A.; Sahut, C.; Cournac, L.; Roubaud, A.; Peltier, G.; Froment, A.-K. An economic, sustainability, and energetic model of biodiesel production from microalgae. *Bioresour. Technol.* **2012**, *111*, 191–200. [CrossRef] [PubMed]
10. Richardson, J.W.; Johnson, M.D.; Outlaw, J.L. Economic comparison of open pond raceways to photo bio-reactors for profitable production of algae for transportation fuels in the Southwest. *Algal Res.* **2012**, *1*, 93–100. [CrossRef]
11. de Godos, I.; Mendoza, J.L.; Acien, F.G.; Molina, E.; Banks, C.J.; Heaven, S.; Rogalla, F. Evaluation of carbon dioxide mass transfer in raceway reactors for microalgae culture using flue gases. *Bioresour. Technol.* **2014**, *153*, 307–314. [CrossRef]
12. Wang, J.; Liu, W.; Liu, T. Biofilm based attached cultivation technology for microalgal biorefineries-A review. *Bioresour. Technol.* **2017**, *244 Pt 2*, 1245–1253. [CrossRef]
13. Mantzorou, A.; Ververidis, F. Microalgal biofilms: A further step over current microalgal cultivation techniques. *Sci. Total Environ.* **2019**, *651 Pt 2*, 3187–3201. [CrossRef]
14. Shen, Y.; Chen, C.; Chen, W.; Xu, X. Attached culture of *Nannochloropsis oculata* for lipid production. *Bioprocess. Biosyst. Eng.* **2014**, *37*, 1743–1748. [CrossRef] [PubMed]
15. Liu, T.; Wang, J.; Hu, Q.; Cheng, P.; Ji, B.; Liu, J.; Chen, Y.; Zhang, W.; Chen, X.; Chen, L.; et al. Attached cultivation technology of microalgae for efficient biomass feedstock production. *Bioresour. Technol.* **2013**, *127*, 216–222. [CrossRef]
16. Rincon, S.M.; Romero, H.M.; Aframehr, W.M.; Beyenal, H. Biomass production in *Chlorella vulgaris* biofilm cultivated under mixotrophic growth conditions. *Algal Res.* **2017**, *26*, 153–160. [CrossRef]

17. Zhang, Q.; Li, X.; Guo, D.; Ye, T.; Xiong, M.; Zhu, L.; Liu, C.; Jin, S.; Hu, Z. Operation of a vertical algal biofilm enhanced raceway pond for nutrient removal and microalgae-based byproducts production under different wastewater loadings. *Bioresour. Technol.* **2018**, *253*, 323–332. [CrossRef] [PubMed]
18. Xu, X.-Q.; Wang, J.-H.; Zhang, T.-Y.; Dao, G.-H.; Wu, G.-X.; Hu, H.-Y. Attached microalgae cultivation and nutrients removal in a novel capillary-driven photo-biofilm reactor. *Algal Res.* **2017**, *27*, 198–205. [CrossRef]
19. Martín-Girela, I.; Curt, M.D.; Fernández, J. Flashing light effects on CO_2 absorption by microalgae grown on a biofilm photobioreactor. *Algal Res.* **2017**, *25*, 421–430. [CrossRef]
20. Martins, S.C.S.; Martins, C.M.; Fiúza, L.M.C.G.; Santaella, S.T. Immobilization of microbial cells: A promising tool for treatment of toxic pollutants in industrial wastewater. *Afr. J. Biotechnol.* **2013**, *12*, 4412–4418.
21. Hamano, H.; Nakamura, S.; Hayakawa, J.; Miyashita, H.; Harayama, S. Biofilm-based photobioreactor absorbing water and nutrients by capillary action. *Bioresour. Technol.* **2017**, *223*, 307–311. [CrossRef] [PubMed]
22. Schultze, L.K.P.; Simon, M.-V.; Li, T.; Langenbach, D.; Podola, B.; Melkonian, M. High light and carbon dioxide optimize surface productivity in a twin-layer biofilm photobioreactor. *Algal Res.* **2015**, *8*, 37–44. [CrossRef]
23. Travieso, L.; Benitez, F.; Weiland, P.; Sanchez, E.; Dupeyron, R.; Dominguez, A.R. Experiments on immobilization of microalgae for nutrient removal in wastewater treatments. *Bioresour. Technol.* **1996**, *55*, 181–186. [CrossRef]
24. Thepenier, C.; Gudin, C. Immobilization of *Porphyridium cruentum* in polyurethane foams for the production of polysaccharide. *Biomass* **1985**, *7*, 225–240. [CrossRef]
25. Lee, S.H.; Oh, H.M.; Jo, B.H.; Lee, S.A.; Shin, S.Y.; Kim, H.S.; Lee, S.H.; Ahn, C.Y. Higher biomass productivity of microalgae in an attached growth system, using wastewater. *J. Microbiol. Biotechnol.* **2014**, *24*, 1566–1573. [CrossRef] [PubMed]
26. Ozkan, A.; Kinney, K.; Katz, L.; Berberoglu, H. Reduction of water and energy requirement of algae cultivation using an algae biofilm photobioreactor. *Bioresour. Technol.* **2012**, *114*, 542–548. [CrossRef]
27. Hoh, D.; Watson, S.; Kan, E. Algal biofilm reactors for integrated wastewater treatment and biofuel production: A review. *Chem. Eng. J.* **2016**, *287*, 466–473. [CrossRef]
28. Gross, M.; Henry, W.; Michael, C.; Wen, Z. Development of a rotating algal biofilm growth system for attached microalgae growth with in situ biomass harvest. *Bioresour. Technol.* **2013**, *150*, 195–201. [CrossRef]
29. Christenson, L.B.; Sims, R.C. Rotating algal biofilm reactor and spool harvester for wastewater treatment with biofuels by-products. *Biotechnol. Bioeng.* **2012**, *109*, 1674–1684. [CrossRef]
30. Akhtar, N.; Iqbal, J.; Iqbal, M. Removal and recovery of nickel(II) from aqueous solution by loofa sponge-immobilized biomass of *Chlorella sorokiniana*: Characterization studies. *J. Hazard.* **2004**, *108*, 85–94. [CrossRef] [PubMed]
31. Akhtar, N.; Saeed, A.; Iqbal, M. *Chlorella sorokiniana* immobilized on the biomatrix of vegetable sponge of *Luffa cylindrica*: A new system to remove cadmium from contaminated aqueous medium. *Bioresour. Technol.* **2003**, *88*, 163–165. [CrossRef]
32. Xiao, R.; Zheng, Y. Overview of microalgal extracellular polymeric substances (EPS) and their applications. *Biotechnol. Adv.* **2016**, *34*, 1225–1244. [CrossRef] [PubMed]
33. Berner, F.; Heimann, K.; Sheehan, M. Microalgal biofilms for biomass production. *J. Appl. Phycol.* **2014**, *27*, 1793–1804. [CrossRef]
34. Gross, M.; Wen, Z. Yearlong evaluation of performance and durability of a pilot-scale Revolving Algal Biofilm (RAB) cultivation system. *Bioresour. Technol.* **2014**, *171*, 50–58. [CrossRef] [PubMed]
35. Blanken, W.; Janssen, M.; Cuaresma, M.; Libor, Z.; Bhaiji, T.; Wijffels, R.H. Biofilm growth of *Chlorella sorokiniana* in a rotating biological contactor based photobioreactor. *Biotechnol. Bioeng.* **2014**, *111*, 2436–2445. [CrossRef] [PubMed]
36. Johnson, M.B.; Wen, Z. Development of an attached microalgal growth system for biofuel production. *Appl. Microbiol. Biotechnol.* **2010**, *85*, 525–534. [CrossRef]
37. Ouspensky, P.D. *The Fourth Way: A Record of Talks and Answers to Questions Based on the Teaching of G.I. Gurdjieff*; Routledge & Kegan Paul: London, UK, 1957.
38. Pinck, S.; Etienne, M.; Dossot, M.; Jorand, F.P.A. A rapid and simple protocol to prepare a living biocomposite that mimics electroactive biofilms. *Bioelectrochemistry* **2017**, *118*, 131–138. [CrossRef]
39. de-Bashan, L.E.; Bashan, Y. Immobilized microalgae for removing pollutants: Review of practical aspects. *Bioresour. Technol.* **2010**, *101*, 1611–1627. [CrossRef] [PubMed]
40. Mallick, N. Biotechnological potential of immobilized algae for wastewater N, P and metal removal: A review. *BioMetals* **2002**. [CrossRef] [PubMed]
41. Berger, R.G. Biotechnology of flavours—The next generation. *Biotechnol. Lett.* **2009**, *31*, 1651. [CrossRef] [PubMed]
42. Carballeira, J.D.; Quezada, M.A.; Hoyos, P.; Simeó, Y.; Hernaiz, M.J.; Alcantara, A.R.; Sinisterra, J.V. Microbial cells as catalysts for stereoselective red–ox reactions. *Biotechnol. Adv.* **2009**, *27*, 686–714. [CrossRef] [PubMed]
43. Demain, A.L. Biosolutions to the energy problem. *J. Ind. Microbiol. Biotechnol.* **2009**, *36*, 319–332. [CrossRef] [PubMed]
44. Rao, N.N.; Lütz, S.; Würges, K.; Minör, D. Continuous biocatalytic processes. *Org. Process. Res. Dev.* **2009**, *13*, 607–616. [CrossRef]
45. End, N.; Schöning, K.-U. Immobilized biocatalysts in industrial research and production. In *Immobilized Catalysts: Solid Phases, Immobilization and Applications*; Kirschning, A., Ed.; Springer: Berlin/Heidelberg, Germany, 2004; pp. 273–317.
46. Flickinger, M.C.; Fidaleo, M.; Gosse, J.; Polzin, K.; Charaniya, S.; Solheid, C.; Lyngberg, O.K.; Laudon, M.; Ge, H.; Schottel, J.L.; et al. *Engineering Nanoporous Bioactive Smart Coatings Containing Microorganisms: Fundamentals and Emerging Applications*; ACS Symposium Series; American Chemical Society: Washington, DC, USA, 2009.

47. Flickinger, M.C.; Schottel, J.L.; Bond, D.R.; Aksan, A.; Scriven, L.E. Painting and printing living bacteria: Engineering nanoporous biocatalytic coatings to preserve microbial viability and intensify reactivity. *Biotechnol. Prog.* **2007**, *23*, 2–17. [CrossRef]
48. Cohan, Y. Biofiltration—The treatment of fluids by microorganisms immobilized into the filter bedding material: A review. *Bioresour. Technol.* **2001**, *77*, 257–274. [CrossRef]
49. Moreno-Garrido, I. Microalgae immobilization: Current techniques and uses. *Bioresour. Technol.* **2008**, *99*, 3949–3964. [CrossRef]
50. Magdeldin, S.; Moser, A. *Affinity Chromatography: Principles and Applications*; InTech: Rijeka, Croatia, 2012.
51. Chandramohan, D.; Marimuthu, K. A review on natural fibers. *Int. J. Res. Rev. Appl. Sci.* **2011**, *8*, 194–206.
52. Gurunathan, T.; Mohanty, S.; Nayak, S.K. A review of the recent developments in biocomposites based on natural fibres and their application perspectives. *Compos. Part A Appl. Sci. Manuf.* **2015**, *77*, 1–25. [CrossRef]
53. Holzmeister, I.; Schamel, M.; Groll, J.; Gbureck, U.; Vorndran, E. Artificial inorganic biohybrids: The functional combination of microorganisms and cells with inorganic materials. *Acta Biomater.* **2018**. [CrossRef]
54. Perullini, M.; Rivero, M.M.; Jobbágy, M.; Mentaberry, A.; Bilmes, S.A. Plant cell proliferation inside an inorganic host. *J. Biotechnol.* **2007**, *127*, 542–548. [CrossRef] [PubMed]
55. Mallick, N.; Rai, L.C. Removal and assessment of toxicity of Cu and Fe to *Anabaena doliolum* and *Chloreila vulgaris* using free and immobilized cells. *World J. Microbiol. Biotechnol.* **1992**, *8*, 110–114.
56. Yashveer, S. Photosynthetic activity, and lipid and hydracarbon production by alginate-immobilized cells of *Botryococcus* in relation to growth phase. *J. Microbiol. Biotechnol.* **2003**, *13*, 687–691.
57. Pannier, A.; Soltmann, U.; Soltmann, B.; Altenburger, R.; Schmitt-Jansen, M. Alginate/silica hybrid materials for immobilization of green microalgae *Chlorella vulgaris* for cell-based sensor arrays. *J. Mater. Chem. B* **2014**, *2*, 7896–7909. [CrossRef] [PubMed]
58. Desmet, J.; Meunier, C.F.; Danloy, E.P.; Duprez, M.-E.; Hantson, A.-L.; Thomas, D.; Cambier, P.; Rooke, J.C.; Su, B.-L. Green and sustainable production of high value compounds via a microalgae encapsulation technology that relies on CO_2 as a principle reactant. *J. Mater. Chem. A* **2014**, *2*, 20560–20569. [CrossRef]
59. Lode, A.; Krujatz, F.; Brüggemeier, S.; Quade, M.; Schütz, K.; Knaack, S.; Weber, J.; Bley, T.; Gelinsky, M. Green bioprinting: Fabrication of photosynthetic algae-laden hydrogel scaffolds for biotechnological and medical applications. *Eng. Life Sci.* **2015**, *15*, 177–183. [CrossRef]
60. Malik, S.; Hagopian, J.; Mohite, S.; Lintong, C.; Stoffels, L.; Giannakopoulos, S.; Beckett, R.; Leung, C.; Ruiz, J.; Cruz, M.; et al. Robotic extrusion of algae-laden hydrogels for large-scale applications. *Glob. Chall.* **2020**, *4*, 1900064. [CrossRef] [PubMed]
61. Eroglu, E.; Smith, S.M.; Raston, C.L. Application of various immobilization techniques for algal bioprocesses. In *Biomass and Biofuels from Microalgae. Biofuel and Biorefinery Technologies*; Moheimani, N., McHenry, M., de Boer, K., Bahri, P., Eds.; Springer: Berlin/Heidelberg, Germany, 2015; Volume 2, pp. 19–44.
62. Kuu, W.Y.; Polack, J.A. Improving immobilized biocatalysts by gel phase polymerization. *Biotechnol. Bioeng.* **1983**, *25*, 1995–2006. [CrossRef] [PubMed]
63. Faafeng, B.A.; van Donk, E.; Källqvist, S.T. In situ measurement of algal growth potential in aquatic ecosystems by immobilized algae. *Appl. Phycol.* **1994**, *6*, 301–308. [CrossRef]
64. Serp, D.; Cantana, E.; Heinzen, C.; Stockar, U.V.; Marison, I.W. Characterization of an encapsulation device for the production of monodisperse alginate beads for cell immobilization. *Biotechnol. Bioeng.* **2000**, *70*, 41–53. [CrossRef]
65. Lau, P.S.; Tam, N.F.Y.; Wong, Y.S. Effect of carrageenan immobilization on the physiological activities of *Chlorella vulgaris*. *Bioresour. Technol.* **1998**, *63*, 115–121. [CrossRef]
66. Chevalier, P.; de la Noüe, J. Wastewater nutrient removal with microalgae immobilized in carrageenan. *Enzym. Microb. Technol.* **1985**, *7*, 621–624. [CrossRef]
67. McHugh, D.J. A guide to the seaweed industry. In *FAO Fisheries Technical Paper*; FAO: Rome, Italy, 2003; p. 441.
68. Tavassoli-Kafrani, E.; Shekarchizadeh, H.; Masoudpour-Behabadi, M. Development of edible films and coatings from alginates and carrageenans. *Carbohydr. Polym.* **2016**, *137*, 360–374. [CrossRef] [PubMed]
69. Aguilar-May, B.; del Pilar Sánchez-Saavedra, M.; Lizardi, J.; Voltolina, D. Growth of *Synechococcus* sp. immobilized in chitosan with different times of contact with NaOH. *J. Appl. Phycol.* **2007**, *19*, 181–183. [CrossRef]
70. Mallick, N.; Rai, L.C. Removal of inorganic ions from wastewaters by immobilized microalgae. *World J. Microbiol. Biotechnol.* **1994**, *10*, 439–443. [CrossRef]
71. Ekins-Coward, T.; Boodhoo, K.V.K.; Velasquez-Orta, S.; Caldwell, G.; Wallace, A.; Barton, R.; Flickinger, M.C. A microalgae biocomposite-integrated spinning disk bioreactor (SDBR): Toward a scalable engineering approach for bioprocess intensification in light-driven CO_2 absorption applications. *Ind. Eng. Chem. Res.* **2019**, *58*, 5936–5949. [CrossRef]
72. Vorlop, K.D.; Klein, J. Entrapment of microbial cells in chitosan. In *Methods in Enzymology*; Academic Press: Cambridge, MA, USA, 1987; Volume 135, pp. 259–268.
73. Castro-Ceseña, A.B.; del Pilar Sánchez-Saavedra, M.; Ruíz-Güereca, D.A. Optimization of entrapment efficiency and evaluation of nutrient removal (N and P) of *Synechococcus elongatus* in novel core-shell capsules. *J. Appl. Phycol.* **2016**, *28*, 2343–2351. [CrossRef]
74. Cortez, S.; Nicolau, A.; Flickinger, M.C.; Mota, M. Biocoatings: A new challenge for environmental biotechnology. *Biochem. Eng. J.* **2017**, *121*, 25–37. [CrossRef]
75. Guy, A. The science and art of paint formulation. *Chem. Phys. Coat.* **2004**, 317–346. [CrossRef]
76. Lyngberg, O.K.; Stemke, D.J.; Schottel, J.L.; Flickinger, M.C. A single-use luciferase-based mercury biosensor using *Escherichia coli* HB101 immobilized in a latex copolymer film. *J. Ind. Microbiol. Biotechnol.* **1999**, *23*, 668–676. [CrossRef] [PubMed]

77. Jenkins, J.S.; Flickinger, M.C.; Velev, O.D. Deposition of composite coatings from particle-particle and particle-yeast blends by convective-sedimentation assembly. *J. Colloid Interface Sci.* **2012**, *380*, 192–200. [CrossRef] [PubMed]
78. Price, K.; Wu, W.; McCormick, A.V.; Francis, L.F. Measurements of stress development in latex coatings. In *Protective Coatings: Film Formation and Properties*; Wen, M., Dušek, K., Eds.; Springer International Publishing: Cham, Switzerland, 2017; pp. 225–240.
79. Bunning, T.J.; Lawton, C.W.; Klei, H.E.; Sundstrom, D.W. Physical property improvements of a pellicular biocatalyst. *Bioprocess. Eng.* **1991**, *7*, 71–75. [CrossRef]
80. Flickinger, M.C.; Bernal, O.I.; Schulte, M.J.; Broglie, J.J.; Duran, C.J.; Wallace, A.; Mooney, C.B.; Velev, O.D. Biocoatings: Challenges to expanding the functionality of waterborne latex coatings by incorporating concentrated living microorganisms. *J. Coat. Technol. Res.* **2017**, *14*, 791–808. [CrossRef]
81. Martens, N.; Hall, E.A.H. Immobilisation of photosynthetic cells based on film-forming emulsion polymers. *Anal. Chim. Acta* **1994**, *292*, 49–63. [CrossRef]
82. Bernal, O.I.; Pawlak, J.J.; Flickinger Michael, C. Microbial paper: Cellulose fiber-based photo-absorber producing hydrogen gas from acetate using dry-stabilized *Rhodopseudomonas palustris*. *BioResources* **2017**, *12*, 4013–4030. [CrossRef]
83. Bernal, O.I.; Mooney, C.B.; Flickinger, M.C. Specific photosynthetic rate enhancement by cyanobacteria coated onto paper enables engineering of highly reactive cellular biocomposite "leaves". *Biotechnol. Bioeng.* **2014**, *111*, 1993–2008. [CrossRef] [PubMed]
84. Lyngberg, O.K.; Ng, C.P.; Thiagarajan, V.; Scriven, L.E.; Flickinger, M.C. Engineering the microstructure and permeability of thin multilayer latex biocatalytic coatings containing *E. coli*. *Biotechnol. Prog.* **2001**, *17*, 1169–1179. [CrossRef]
85. In-na, P.; Umar, A.A.; Wallace, A.D.; Flickinger, M.C.; Caldwell, G.S.; Lee, J.G.M. Loofah-based microalgae and cyanobacteria biocomposites for intensifying carbon dioxide capture. *J. Co2 Util.* **2020**, *42*, 101348. [CrossRef]
86. Wicks, J.Z.W.; Jones, F.; Peppas, S. Organic coatings: Science and technology volume 1: Film formation, components and appearance. *Dry. Technol.* **1993**, *11*, 1477. [CrossRef]
87. Reyes, Y.; Campos-Terán, J.; Vázquez, F.; Duda, Y. Properties of films obtained from aqueous polymer dispersions: Study of drying rate and particle polydispersity effects. *Model. Simul. Mater. Sci. Eng.* **2007**, *15*, 355. [CrossRef]
88. Chen, X.; Fischer, S.; Men, Y. Temperature and relative humidity dependency of film formation of polymeric latex dispersions. *Langmuir* **2011**, *27*, 12807–12814. [CrossRef]
89. Limousin, E.; Ballard, N.; Asua, J.M. The influence of particle morphology on the structure and mechanical properties of films cast from hybrid latexes. *Prog. Org. Coat.* **2019**, *129*, 69–76. [CrossRef]
90. Mesic, B.; Cairns, M.; Järnstrom, L.; Joo Le Guen, M.; Parr, R. Film formation and barrier performance of latex based coating: Impact of drying temperature in a flexographic process. *Prog. Org. Coat.* **2019**, *129*, 43–51. [CrossRef]
91. Schulte, M.J.; Wiltgen, J.; Ritter, J.; Mooney, C.B.; Flickinger, M.C. A high gas fraction, reduced power, syngas bioprocessing method demonstrated with a *Clostridium ljungdahlii* OTA1 paper biocomposite. *Biotechnol. Bioeng.* **2016**, *113*, 1913–1923. [CrossRef]
92. Gosse, J.L.; Chinn, M.S.; Grunden, A.M.; Bernal, O.I.; Jenkins, J.S.; Yeager, C.; Kosourov, S.; Seibert, M.; Flickinger, M.C. A versatile method for preparation of hydrated microbial–latex biocatalytic coatings for gas absorption and gas evolution. *J. Ind. Microbiol. Biotechnol.* **2012**, *39*, 1269–1278. [CrossRef] [PubMed]
93. Bernal, O.I.; Bharti, B.; Flickinger, M.C.; Velev, O.D. Fabrication of photoreactive biocomposite coatings via electric field-assisted assembly of cyanobacteria. *Langmuir* **2017**, *33*, 5304–5313. [CrossRef]
94. Chen, Y.; Krings, S.; Booth, J.R.; Bon, S.A.F.; Hingley-Wilson, S.; Keddie, J.L. Introducing porosity in colloidal biocoatings to increase bacterial viability. *Biomacromolecules* **2020**. [CrossRef] [PubMed]
95. Hart, R.; In-na, P.; Kapralov, M.V.; Lee, J.G.M.; Caldwell, G.S. Textile-based cyanobacteria biocomposites for potential environmental remediation applications. *J. Appl. Phycol.* **2021**. [CrossRef]
96. Koszewska, M. Circular economy—Challenges for the textile and clothing industry. *Autex Res. J.* **2018**, *18*, 337–347. [CrossRef]
97. Melki, S.; Biguenet, F.; Dupuis, D. Hydrophobic properties of textile materials: Robustness of hydrophobicity. *J. Text. Inst.* **2019**, *110*, 1221–1228. [CrossRef]
98. Khosravi, A.; King, J.A.; Jamieson, H.L.; Lind, M.L. Latex barrier thin film formation on porous substrates. *Langmuir* **2014**, *30*, 13994–14003. [CrossRef] [PubMed]
99. Sarkar, M.K.; He, F.A.; Fan, J.T. Moisture-responsive fabrics based on the hygro deformation of yarns. *Text. Res. J.* **2009**, *80*, 1172–1179. [CrossRef]
100. Dhiman, R.; Chattopadhyay, R. Absorbency of synthetic urine by cotton nonwoven fabric. *J. Text. Inst.* **2020**, 1–8. [CrossRef]
101. Alassod, A.; Xu, G. Comparative study of polypropylene nonwoven on structure and wetting characteristics. *J. Text. Inst.* **2020**, 1–8. [CrossRef]
102. Ozbolat, I.T.; Peng, W.; Ozbolat, V. Application areas of 3D bioprinting. *Drug Discov. Today* **2016**, *21*, 1257–1271. [CrossRef] [PubMed]
103. Jensen, C.; Teng, Y. Is it time to start transitioning from 2D to 3D cell culture? *Front. Mol. Biosci.* **2020**, *7*, 33. [CrossRef] [PubMed]
104. Zhao, S.; Guo, C.; Kumarasena, A.; Omenetto, F.G.; Kaplan, D.L. 3D printing of functional microalgal silk structures for environmental applications. *ACS Biomater. Sci. Eng.* **2019**, *5*, 4808–4816. [CrossRef] [PubMed]
105. Mehrotra, S.; Kumar, S.; Srivastava, V.; Mishra, T.; Mishra, B.N. 3D bioprinting in plant science: An interdisciplinary approach. *Trends Plant Sci.* **2020**, *25*, 9–13. [CrossRef] [PubMed]
106. Inna, P.; Caldwell, G.S.; Lee, J.G.M. Living textile biocomposites deliver enhanced carbon dioxide capture. *J. Ind. Text.* **2021**. In press.

107. Watkinson, S.C. *The Fungi*; Elsevier Science & Technology: Amsterdam, The Netherlands, 2015.
108. Honegger, R. Lichen-forming fungi and their photobionts. In *The Mycota*; Springer: Berlin/Heidelberg, Germany, 2009; pp. 307–333.
109. Chen, Y.; Su, N.; Zhang, K.; Zhu, S.; Zhu, Z.; Qin, W.; Yang, Y.; Shi, Y.; Fan, S.; Wang, Z.; et al. Effect of fiber surface treatment on structure, moisture absorption and mechanical properties of luffa sponge fiber bundles. *Ind. Crops Prod.* **2018**, *123*, 341–352. [CrossRef]
110. Dautzenberg, F.; Mukherjee, M. Process intensification using multifunctional reactors. *Chem. Eng. Sci.* **2001**, *56*, 251–267. [CrossRef]
111. Joshi, S.; Gogate, P. Process intensification of biofuel production from microalgae. In *Energy from Microalgae*; Springer: Berlin/Heidelberg, Germany, 2018; pp. 59–87.
112. Posten, C. Design principles of photo-bioreactors for cultivation of microalgae. *Eng. Life Sci.* **2009**, *9*, 165–177. [CrossRef]
113. Kupriyanova, E.V.; Sinetova, M.A.; Cho, S.M.; Park, Y.I.; Los, D.A.; Pronina, N.A. CO_2-concentrating mechanism in cyanobacterial photosynthesis: Organization, physiological role, and evolutionary origin. *Photosynth. Res.* **2013**, *117*, 133–146. [CrossRef]
114. Bitog, J.P.; Lee, I.-B.; Lee, C.-G.; Kim, K.-S.; Hwang, H.-S.; Hong, S.-W.; Seo, I.-H.; Kwon, K.-S.; Mostafa, E. Application of computational fluid dynamics for modeling and designing photobioreactors for microalgae production: A review. *Comput. Electron. Agric.* **2011**, *76*, 131–147. [CrossRef]
115. Dangi, A.K.; Sharma, B.; Hill, R.T.; Shukla, P. Bioremediation through microbes: Systems biology and metabolic engineering approach. *Crit. Rev. Biotechnol.* **2019**, *39*, 79–98. [CrossRef] [PubMed]
116. Huang, H.-H.; Camsund, D.; Lindblad, P.; Heidorn, T. Design and characterization of molecular tools for a synthetic biology approach towards developing cyanobacterial biotechnology. *Nucleic Acids Res.* **2010**, *38*, 2577–2593. [CrossRef] [PubMed]
117. Sharma, B.; Dangi, A.K.; Shukla, P. Contemporary enzyme based technologies for bioremediation: A review. *J. Environ. Manag.* **2018**, *210*, 10–22. [CrossRef] [PubMed]
118. Gerotto, C.; Norici, A.; Giordano, M. Toward enhanced fixation of CO_2 in aquatic biomass: Focus on microalgae. *Front. Energy Res.* **2020**, *8*, 213. [CrossRef]
119. Dana, G.V.; Kuiken, T.; Rejeski, D.; Snow, A.A. Four steps to avoid a synthetic-biology disaster. *Nature* **2012**, *483*, 29. [CrossRef]
120. Gressel, J.; van der Vlugt, C.J.; Bergmans, H.E. Cultivated microalgae spills: Hard to predict/easier to mitigate risks. *Trends Biotechnol.* **2014**, *32*, 65–69. [CrossRef] [PubMed]
121. Glass, D.J. Government regulation of the uses of genetically modified algae and other microorganisms in biofuel and bio-based chemical production. In *Algal Biorefineries*; Springer: Berlin/Heidelberg, Germany, 2015; pp. 23–60.

MDPI
St. Alban-Anlage 66
4052 Basel
Switzerland
Tel. +41 61 683 77 34
Fax +41 61 302 89 18
www.mdpi.com

Energies Editorial Office
E-mail: energies@mdpi.com
www.mdpi.com/journal/energies

www.ingramcontent.com/pod-product-compliance
Lightning Source LLC
LaVergne TN
LVHW070926170726
843515LV00005B/882

* 9 7 8 3 0 3 6 5 2 9 0 8 0 *